Nutrient Disequilibria in Agroecosystems: Concepts and Case Studies

Nutrient Disequilibria in Agroecosystems
Concepts and Case Studies

Edited by

E.M.A. Smaling
Wageningen University and Research Centre, Laboratory of Soil Science and Geology, Wageningen, The Netherlands

O. Oenema
Wageningen University and Research Centre, Research Institute for Agrobiology and Soil Fertility, Wageningen, The Netherlands

and

L.O. Fresco
Sustainable Development Department, United Nations Food and Agriculture Organisation (FAO), Rome, Italy

CABI *Publishing*

CABI *Publishing* is a division of CAB *International*

CABI Publishing
CAB International
Wallingford
Oxon OX10 8DE
UK

CABI Publishing
10 E. 40th Street
Suite 3203
New York, NY 10016
USA

Tel: +44 (0)1491 832111
Fax: +44 (0)1491 833508
Email: cabi@cabi.org

Tel: +1 212 481 7018
Fax: +1 212 686 7993
Email: cabi-nao@cabi.org

© CAB *International* 1999. All rights reserved. No part of this publication may be reproduced in any form or by any means, electronically, mechanically, by photocopying, recording or otherwise, without the prior permission of the copyright owners.

A catalogue record for this book is available from the British Library, London, UK.

Library of Congress Cataloging-in-Publication Data
Nutrient disequilibria in agroecosystems : concepts and case studies /
  edited by E.M.A. Smaling, O. Oenema and L.O. Fresco.
    p.   cm.
  Includes bibliographical references.
  ISBN 0-85199-268-4 (alk. paper)
    1. Plant nutrients.   2. Agricultural ecology.   I. Smaling, E. M. A.   II. Oenema, O. (Oene)   III. Fresco, Louise.
  S589.7.N88   1999
  577.5'5--dc21   99-14039
                                                                    CIP
ISBN 0 85199 268 4

Typeset by AMA DataSet Ltd, UK.
Printed and bound in the UK at the University Press, Cambridge.

Contents

Contributors		vii
Preface		xi
1	Nutrient Cycling in Ecosystems versus Nutrient Budgets of Agricultural Systems *M. van Noordwijk*	1
2	Basics of Budgets, Buffers and Balances of Nutrients in Relation to Sustainability of Agroecosystems *B.H. Janssen*	27
3	Upscaling of Nutrient Budgets from Agroecological Niche to Global Scale *K.W. van der Hoek and A.F. Bouwman*	57
4	Uncertainties in Nutrient Budgets due to Biases and Errors *O. Oenema and M. Heinen*	75
5	Technologies to Manage Soil Fertility Dynamics *J.J. Stoorvogel*	99
6	Economic Policy in Support of Soil Fertility: Which Interventions after Structural Adjustment? *A. Kuyvenhoven, N. Heerink and R. Ruben*	119

Contents

7 Integrated Smallholder Agriculture–Aquaculture in Asia:
Optimizing Trophic Flows 141
J.P.T. Dalsgaard and M. Prein

8 Improving Nutrient Management for Sustainable Development
of Agriculture in China 157
Jin Jiyun, Lin Bao and Zhang Weili

9 Nutrient Imbalances following Conversion of Grasslands to
Plantation Forests in South Africa 175
R.J. Scholes and M.C. Scholes

10 Nutrient Balances at Field Level of Mixed Cropping Systems in
Various Agroecological Zones in Mozambique 191
P.M.H. Geurts, L. Fleskens, J. Löwer and E.C.R. Folmer

11 Nutrient and Cash Flow Monitoring in Farming Systems on the
Eastern Slopes of Mount Kenya 211
*J.N. Gitari, F.M. Matiri, I.W. Kariuki, C.W. Muriithi and
S.P. Gachanja*

12 Nitrogen Flows through Fisheries in Relation to the
Anthropogenic N cycle: a Study from Norway 229
M. Azzaroli Bleken

13 Agricultural and Ecological Performance of Cropping Systems
Compared in a Long-term Field Trial 247
P. Mäder, T. Alföldi, A. Fließbach, L. Pfiffner and U. Niggli

14 Comparative Nutrient Budgets of Temperate Grazed Pastoral
Systems 265
K.M. Goh and P.H. Williams

15 Epilogue 295
E.M.A. Smaling, O. Oenema and L.O. Fresco

Index 315

Contributors

T. Alföldi, *Research Institute of Organic Agriculture, Ackerstrasse, CH-5070 Frick, Switzerland*
M. Azzaroli Bleken, *Department of Crop Sciences, Agricultural University of Norway, PO Box 5022, N-1432 Ås, Norway*
A.F. Bouwman, *National Institute of Public Health and the Environment (RIVM), PO Box 1, 3720 BA Bilthoven, The Netherlands*
J.P.T. Dalsgaard, *Danish Institute of Agricultural Sciences (DIAS), PO Box 3950, DK–8830 Tjele, Denmark*
L. Fleskens, *c/o Sponserf 2, 6413 LS Heerlen, The Netherlands*
A. Fließbach, *Research Institute of Organic Agriculture, Ackerstrasse, CH-5070 Frick, Switzerland*
E.C.R. Folmer, *c/o Sponserf 2, 6413 LS Heerlen, The Netherlands*
L.O. Fresco, *Sustainable Development Department, UN Food and Agriculture Organisation (FAO), Viale delle Terme di Caracalla, 00100 Rome, Italy*
S.P. Gachanja, *Kenya Agricultural Research Institute, Regional Research Centre – Embu, PO Box 27, Embu, Kenya*
P.M.H. Geurts, *Sponserf 2, 6413 LS Heerlen, The Netherlands*
J.N. Gitari, *Kenya Agricultural Research Institute, Regional Research Centre – Embu, PO Box 27, Embu, Kenya*
K.M. Goh, *Soil, Plant and Ecological Sciences Division, Lincoln University, PO Box 84, Canterbury, New Zealand*
N. Heerink, *Department of Economics and Management, Wageningen Agricultural University, Wageningen, The Netherlands*

M. Heinen, *Wageningen University and Research Centre, Research Institute for Agrobiology and Soil Fertility (AB-DLO), PO Box 14, 6700 AA Wageningen, The Netherlands*

K.W. van der Hoek, *National Institute of Public Health and the Environment (RIVM), PO Box 1, 3720 BA Bilthoven, The Netherlands*

B.H. Janssen, *Department of Environmental Sciences, Wageningen Agricultural University, PO Box 8005, 6700 EC Wageningen, The Netherlands*

Jin Jiyun, *Soil and Fertilizer Institute, Chinese Academy of Agricultural Sciences, 30 Baishiqiao Road, Beijing 100081, China*

I.W. Kariuki, *Kenya Agricultural Research Institute, Regional Research Centre – Embu, PO Box 27, Embu, Kenya*

A. Kuyvenhoven, *Department of Economics and Management, Wageningen Agricultural University, Wageningen, The Netherlands*

Lin Bao, *Soil and Fertilizer Institute, Chinese Academy of Agricultural Sciences, 30 Baishiqiao Road, Beijing 100081, China*

J. Löwer, *c/o Sponserf 2, 6413 LS Heerlen, The Netherlands*

P. Mäder, *Research Institute of Organic Agriculture, Ackerstrasse, CH-5070 Frick, Switzerland*

F.M. Matiri, *Kenya Agricultural Research Institute, Regional Research Centre – Embu, PO Box 27, Embu, Kenya*

C.W. Muriithi, *Kenya Agricultural Research Institute, Regional Research Centre – Embu, PO Box 27, Embu, Kenya*

U. Niggli, *Research Institute of Organic Agriculture, Ackerstrasse, CH-5070 Frick, Switzerland*

M. van Noordwijk, *International Centre for Research in Agroforestry (ICRAF) – South East Asia, PO Box 161, Bogor 16001, Indonesia*

O. Oenema, *Wageningen University and Research Centre, Research Institute for Agrobiology and Soil Fertility (AB-DLO), PO Box 14, 6700 AA Wageningen, The Netherlands*

L. Pfiffner, *Research Institute of Organic Agriculture, Ackerstrasse, CH-5070 Frick, Switzerland*

M. Prein, *International Center for Living Aquatic Resources Management (ICLARM), MCPO Box 2631, 0718 Makati City, Manila, Philippines*

R. Ruben, *Department of Economics and Management, Wageningen Agricultural University, Wageningen, The Netherlands*

M.C. Scholes, *Department of Botany, University of the Witwatersrand, Private Bag 3, WITS 2050, South Africa*

R.J. Scholes, *Division of Water, Environment and Forest Technology, CSIR, PO Box 395, Pretoria, South Africa*

E.M.A. Smaling, *Wageningen University and Research Centre, Laboratory of Soil Science and Geology, PO Box 37, 6700 AA Wageningen, The Netherlands*

J.J. Stoorvogel, *Wageningen University and Research Centre, Laboratory of Soil Science and Geology, PO Box 37, 6700 AA Wageningen, The Netherlands*

P.H. Williams, *New Zealand Institute of Crop and Food Research, Canterbury Agricultural and Science Centre, Private Bag 4704, Christchurch, New Zealand*

Zhang Weili, *Soil and Fertilizer Institute, Chinese Academy of Agricultural Sciences, 30 Baishiqiao Road, Beijing 100081, China*

Preface

This book is on stocks and flows of carbon and nutrients in the terrestrial and marine agroecosystems of our planet. Although quite an ambitious statement, we feel that this topic will be of major importance over the next decades. In a scientific sense, nutrient stocks, flows and their management bring together different hard and soft sciences, including agronomy, animal production, forestry, fisheries, soil science, economics at macro- as well as micro-level, sociology and communication science. It is not an entirely new subject, but it is clear that its relevance is increasing at different spatial and temporal scales.

Although the Earth can be considered as a more or less closed system, nutrients and products containing them tend to be transported at a strongly increasing rate, from A to B, from plant to animal to man, and from solid to liquid to gaseous phases and back. At a local scale, this has taken place since the early days of agriculture. At present, however, the changes of carbon and nutrient stocks and flows worldwide have become such, that large areas suffer from either severe nutrient depletion or environmental pollution as a result of overuse. This can be attributed to different pressures that have built up during the second half of the 20th century, including population growth, industrialization, intensification and extension of agricultural production, development of high-input livestock systems, and increased use of mineral fertilizers as well as nitrogen-fixing leguminous species. Adverse effects that directly relate to carbon and nutrient imbalances include: poverty, famine and starvation, and erosion on the 'depletion' side; and groundwater pollution, soil acidification, loss of biodiversity, and greenhouse gas emissions on the 'pollution' side. Some examples may help to illustrate the importance of carbon and nutrient stocks and flows:

1. The World Trade Organization (WTO) has market liberalization on top of its agenda. As a consequence, a strong increase can be expected of international trade in agricultural and derived products, and a longer distance between production and consumption sites. This will enhance the differentiation between 'hot spots', where carbon and nutrients accumulate, and 'bleak spots', where they are depleted. The classic example is the massive export of cassava from Thailand to The Netherlands, where it is used to feed dairy animals. At first sight, this is merely a transaction between two countries. At the same time, however, it constitutes a major export of soil fertility from Thailand, and a major contribution to the growing pile of surplus manure in The Netherlands.

2. The slashing and burning of tropical forests is a major flow of carbon and nutrients, the Brazilian Amazon being a case in point. For a soil scientist, this is the more saddening, since most forest ecosystems in the tropics are on very poor and acidic soils, with a very low agricultural potential. The nutrients in these majestic ecosystems are almost all retained in the standing stock and the litter. Encroaching farmers, in the case of Rondonia and Acre, often come from other parts of the country, and, therefore, lack the necessary experiential knowledge to make farming a successful enterprise. In a few years, the former forest land has been turned into grazing grounds of low productivity and low species diversity. Yet, there are developments that give some hope. The Brazilian government recently announced a ban on the felling of mahogany, one of the most wanted species. However, the enforcement of these directives remains a major problem.

3. The nutrient flow 'erosion' has a negative connotation, but it should be realized that entire civilizations have flourished on the sediments that rivers deposited on floodplains in the lower reaches of major river systems. Hence, at field level, erosion may be detrimental, but one should follow the moving nutrients towards their next destination, where they are not necessarily less valuable.

4. A look at farming in The Netherlands over the past decades shows a major increase in the number of rules and regulations, partly stemming from the European Union. Farmers presently have to keep record of their nutrient inputs and outputs, particularly those who have more than 2.5 livestock units ha^{-1}. Also, nitrate levels in groundwater should not exceed 50 mg l^{-1}. Meanwhile, in Africa, the World Bank, the United Nations (UN) Food and Agriculture Organisation (FAO), and the institutes within the Consultative Group of International Agricultural Research (CGIAR) rally behind initiatives for a 'recapitalization' of soil fertility. All are examples of increased legislation and policy development on carbon and nutrient flows.

5. Carbon dioxide (CO_2), methane (CH_4) and nitrous oxide (N_2O) are the most important greenhouse gases. The agricultural sector contributes substantially, amongst others, through irrigated rice culture, high fertilizer use and intensive animal production. During the recent conferences on climate change in Kyoto and Buenos Aires, there was consensus on the fact that an

array of unanswered questions are still on the table, amongst others, on the crucial dual role of soils and ecosystems as sources and sinks of greenhouse gases. These questions cover a wide spectrum, ranging from the functioning of soil biota to the effects of land use change on emissions and sequestration.

6. At microlevel, carbon and nutrient flows are also many. Mineralization turns organic nitrogen into plant-available forms, whereas microbial immobilization works the other way around. Phosphorus follows similar pathways, but can also be strongly retained by kaolinitic clays and sesqui-oxides. Through weathering and dissolution of salts, phosphorus, potassium and an array of other elements turn into labile and plant-available forms. Precipitation and occlusions render them less available. Velocity and direction of these processes are determined by the state and the dynamics of the soil–plant system.

7. Meanwhile, we hope for some breakthroughs on so far fictitious, yet highly desirable nutrient flows. Examples include the biological fixation of nitrogen by other than leguminous species, enhancing knowledge and possible uses of nutrients in marine ecosystems, and the teaching of cattle to keep moving while urinating and defecating. It may seriously reduce leaching and gaseous losses.

We want to point out loud and clear that a natural resource such as 'soil' has so far been treated as a free good worldwide. This situation cannot continue much longer. Soils are not infinite suppliers of nutrients to to-be-harvested plants, nor are they indestructible dumpsites for surpluses of nutrients, including heavy metals and biocide residues. Sister resources such as air, water and biodiversity have been given the same raw deal. Meanwhile, population keeps growing and may well reach 9 billion by the middle of the 21st century, implying that soil, water and biodiversity are more than ever at stake, both in terms of quantity and quality. Agricultural and environmental economists now seem ready for a paradigm shift, in that they include resource depletion or pollution in their sector studies. It will not be an easy job to get this innovation accepted worldwide, but economists, who rule the world after all, have a moral duty to cost natural resources. Incompatibility between something as tangible as 'price' and something much less tangible as 'valuation' cannot remain an excuse forever. Even the US vice-president, Al Gore, held an ardent plea into this direction in his *Earth in the Balance*. Let us hope the early 21st century will really see the coming into being of Gore's environmental 'Marshall Plan', to enhance respect for and safeguard the Earth's natural resources.

This book has two parts. Chapters 1 to 6 are conceptual in nature. They provide the state of the art in the field of nutrient flows and balances from different perspectives. First, a comparison is made between nutrient flows in agricultural and natural systems. Next, the microscale is touched on, i.e. what is the degree of plant–nutrient availability as a function of total nutrients, and which processes influence this ratio. Then, the issue of scale differences is discussed, i.e. nutrient balances at animal or herd level versus nutrient balances for the entire livestock sector in a country. As nutrient balances are made up of different inputs and outputs, the final figure is subject to bias and error (see Chapter 4). Finally, technologies and policies for better nutrient management are discussed, the latter with special attention for the effects of structural adjustment policies, as imposed upon many developing countries by the International Monetary Fund (IMF).

The second part of the book provides case studies of nutrient balance studies. They are cross-continental and multisectoral, i.e. ranging from mixed farming systems in Kenya to marine fisheries in Norway, and from ecological farming in Switzerland to unbalanced fertilizer use in Chinese agriculture. They are of course not meant to give an exhaustive picture of the subject of nutrient flows and balances. They are, however, meant to paint a picture of the wide array of situations where nutrient flows and technologies to change them play a role. The last chapter provides a summary of the highlights of this book, and tries to point at the major issues ahead of us.

Eric Smaling
Oene Oenema
Louise Fresco

Nutrient Cycling in Ecosystems versus Nutrient Budgets of Agricultural Systems

M. van Noordwijk

International Centre for Research in Agroforestry (ICRAF) – South East Asia, PO Box 161, Bogor 16001, Indonesia

INTRODUCTION

Research traditions on 'nutrient cycling' in (semi) natural ecosystems date back to the middle of the 20th century (Odum, 1953), while 'nutrient budgets' of agricultural fields have been made and discussed since the middle of the 19th century (Russell, 1912). This book is thus based on a long research tradition – and on wheels being reinvented. What is new in the present book, apart from some of the rhetoric and way of phrasing classical problems? Global concerns and the relationships between similar phenomena at different scales take a prominent place in current discussions and in this book. Consequences of local nutrient excesses and local nutrient shortages are evaluated from a number of perspectives. The standard recipes of maintaining nutrient flows through agricultural systems by continuously applying new inputs have not worked in practice in major parts of the world and are under environmental scrutiny in the rest of the world where they may have worked too well. An analysis is needed of the social, economic and political context of the various technical solutions. This book should contribute to that debate, on the basis of a conceptual framework and a series of case studies.

In this introductory chapter the evolution of agroecosystems will be considered in relation to nutrient cycles and flows, and agroecosystem nutrient use efficiency will be treated as a scale problem. Scaling in space is linked to 'lateral resource flows' and 'lateral resource capture', while scaling in time gives a new interpretation to 'residual effects'.

RESEARCH QUESTIONS IN AGRONOMIC SOIL FERTILITY RESEARCH

Over the past one-and-a-half centuries, agronomic soil fertility research has gradually evolved from a situation where theories and experiments were largely at odds (mid 19th century), via a century of research where field experiments dominated (the next century), to a situation where models have become integrated with experiments. In the last few decades new questions have emerged, which cannot be directly answered by field experiments.

Von Liebig's theories in the 19th century about nutrient balances ('a farmer must replace by fertilizer all nutrients removed from the field by crop harvest') were soon discredited by experiments of Lawes and Gilbert, which showed that, at least in the initial years of long-term experiments, the best crop growth was obtained with amounts of and nutrient ratios in fertilizer very different from that in the harvested products (Russell, 1912). Was von Liebig's theory, based on nutrient budget calculations, wrong? No, but it was incomplete and did not have the time-scale correct. Yet, the obvious failure of his theoretical predictions and the more tangible results of 'empiricists' such as Lawes and Gilbert, who started the Rothamsted experimental station, have certainly helped to establish a research tradition in soil fertility research based on 'trial and error', rather than models and theory. Research strategy became based on 'the survival of the fitter', by using models which are so flexible that they (in hindsight) can fit any data set and thus survive any validation test. Much of current 'agricultural research design' is still essentially based on the model that the yield of a crop on a given site and in a given year is equal to some intrinsically unpredictable 'control' yield, plus terms for the specific treatment combinations used with coefficients that are unknown beforehand, plus 'error' terms.

When somebody reviewed the list of experiments on nitrogen fertilizer use for lowland rice in Indonesia two decades ago, and noticed that many experiments had been continued for years and years, he was given as an explanation: 'we continue them, because every year we get different results' (Van Keulen, personal communication). He remarked 'then you may as well stop the experiments, because you'll never be able to predict how the response will be in the next year' and then helped establish a method for data analysis focusing on the site differences and between-year variability.

Traditional questions in soil fertility research are (Fig. 1.1):

1. To which extent are nutrients (individually and in combination) limiting crop yield on a given field?
2. How to measure inherent soil fertility (stock of 'available' nutrients)?
3. How effective are recently added nutrients relative to inherent soil fertility, depending on fertilizer type (organic and/or inorganic), timing and

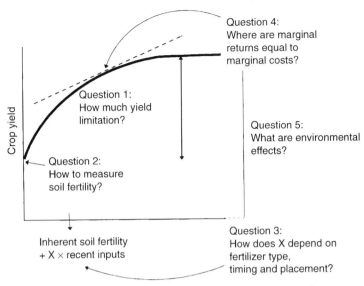

Fig. 1.1. Categories of questions in traditional soil fertility research.

placement? (Questions 2 and 3 together define the x-axis of the yield response curve.)
4. At which input level are additional (marginal) benefits equal to additional costs?

As the answers to these four questions depend on the crop, climate, inherent soil properties, other soil management factors, fertilizer type and application method and prices, it is easily understood that a century of empirical research was not enough to exhaust this field of research.

Later on a fifth question was added:

5. What are the environmental effects of different levels of input use?

The last few decades saw a drastic shift from an overriding importance of field experiments and purely empirical development of soil testing methods, to a balanced approach including theory development and modelling. For question 1, a comparison of real yields and 'potential production' estimates from crop growth models which simply assume the absence of nutrient limitations (Van Keulen and Wolf, 1986) helped in predicting where nutrient problems exist. For questions 1 and 2, more refined spatial extrapolation methods in combination with a functional interpretation of soil survey results helped to establish the broad domains (Sanchez, 1976). For questions 2 and 3, development of a quantitative framework for combining soil chemical, soil physical and plant physiological aspects of the soil–plant system (De Wit, 1953; Nye and Tinker, 1977; Barber, 1984;

De Willigen and Van Noordwijk, 1987; Van Noordwijk and van de Geijn, 1996) helped to support and partly replace the direct empirical approach. Soil biological effects and organic–inorganic interactions remain as the major frontier of data synthesis (Woomer and Swift, 1994). A process-based approach, linked to quantitative models for synthesis of knowledge, was essential in starting to address question number 5.

Question 4 has for a long time been part of agricultural economics and has focused on the financial profitability of fertilizer use. Serious misunderstanding between agricultural economists and agronomists persisted on whether or not different types of fertilizer can substitute for each other (Lanzer et al., 1987). Recently a more truly economical approach is emerging, considering 'costs' and 'benefits' in a broader perspective: both the over-use and the lack of fertilizer use may cause environmental costs.

Overall, system analysis and modelling have allowed us to address broader questions of nutrient cycling and flow in our agroecosystem which the neat rectangles of a classical soil fertility experiment cannot address. The 'sustainability' of our modes of food production is at stake both in excess and in deficit areas (Smaling and Oenema, 1997). In some parts of the globe excessive use of nutrients causes damage to other environmental compartments, in other parts of the globe nutrient stocks are depleted, as the change from a nutrient cycle in a subsistence economy to a nutrient flow after market integration was not accompanied by the use of external nutrient inputs. Economically this is bound to happen where an export economy becomes based on high-volume, low-value products (such as fodder) rather than on high-value commodities.

Except for the negligible amount of nutrients lost in interplanetary traffic, all nutrients of the global ecosystem are recycled at some spatial and temporal scale. Yet, in the current global economy based on mining non-renewable resources many nutrients become tied up in pools from which recycling into agriculture will not be feasible within the human lifetime. A shift will eventually be needed from this approach to one based on recycling all nutrients within urban–rural agroecosystems. Recycling has a substantial energy cost and global energy concerns will probably limit the distance bulk nutrient sources can travel around the globe if recycling becomes mandatory.

Current compartment-flow models applied to a global GIS with discrete colour codes for smoothed polygons provide a first step only in the analysis of a global nutrient economy. As a complement, we will have to include a hierarchy of system boundaries and to scale across fuzzy boundaries in time and space. Also, the traditional phasing of biophysical (questions 1, 2, 3 and 5) and economical (question 4) research should be replaced by a more interdisciplinary analysis of causes and effects of use and misuse of nutrient resources.

NUTRIENT CYCLING IN NATURAL ECOSYSTEMS

The basic concept that natural ecosystems maintain closed nutrient cycles (Chapin, 1980; Jordan, 1985; Brown et al., 1994) depends on spatial and temporal scales. Nutrient cycles will be 'closed' in situations where 'demand' for nutrients by the growing biomass exceeds current supply. In early and late successional stages, however, considerable nutrient losses may occur when a nutrient balance is made at patch scale. In early stages above-ground demand and ability to capture below-ground resources may be too small for making use of all (temporary) nutrient flushes available; in mature systems nutrient demand may be less than current supply. Losses at patch scale can lead to lateral transfers between sites, via groundwater, surface runoff or dust and can support an aggrading phase of vegetation elsewhere. Losses for one patch may provide opportunities for others and a patch-mosaic may therefore be more efficient in nutrient cycling than a summation of supposedly independent units would suggest. On a geological time-scale leaching and erosion on hill slopes dominate soil formation; a transfer occurs from nutrient degrading sites to nutrient aggrading sedimentation zones and colluvial sites, with a net loss of nutrients into the oceans. This differentiation into rich and poor sites increases overall biodiversity.

Nutrient cycling differs between ecological zones and one should be careful with generalizations. Differences are due to climate factors affecting loss rates by erosion and leaching, differences in soil type and in the effective buffering of nutrients in the root zone and differences in the response of vegetation. In mixed vegetation relatively deep-rooted components can provide a 'safety net' for nutrients leaching from topsoil (van Noordwijk et al., 1996). Soil zones with high physico-chemical adsorption constants can, by analogy, be interpreted as 'chemical safety nets'.

In the tropical rain forest zone, water availability may allow roots to focus on the surface layers where most nutrients are available. In fact, on nutrient-poor soils where plants resorb most nutrients from their leaves before litterfall and thus 'litter quality' is low and decomposition slow, a substantial part of the roots may be found in the 'root mats' within the surface layer, on top of the mineral soil. Under these circumstances nutrient cycling can avoid the mineral soil with its strong chemical (aluminium (Al) and iron (Fe)) sinks for phosphorus (P) (Tiessen et al., 1994). Trees with ecto-mycorrhizal associations, such as *Dipterocarpaceae*, tend to dominate in the more mature phases of these systems (Smits, 1994), whereas plants with vesicular-arbuscular (VA) mycorrhiza are more prominent in early successional stages and where the mineral soil is the major nutrient resource (Alexander, 1989; Högberg, 1989). Where individual trees fall and create a gap ('chablis'), the roots of neighbouring vegetation still provide a safety net which can intercept the nutrient flush which follows the inputs to the soil of large amounts of dead organic material.

Nutrient losses from natural ecosystems may occur under the influence of 'extreme events' (fire, cyclone, earthquakes, landslides), when above-ground biomass is disturbed over a large area simultaneously. Fires may mobilize substantial amounts of nutrients locked up in above-ground biomass as well as in resistant organic pools in the soil, and thus bypass the slow decomposition cascade.

The nutrient cycle in natural ecosystems involves uptake by plant roots from (largely) mineral nutrient sources, utilization of these nutrients in plant biomass for a certain period of time, and return to the soil in organic or inorganic form with litterfall, as seeds or as dead plant remains at the end of the plant's life cycle. A considerable part of the nutrients may pass through herbivores before they return to the soil. A fraction of the nutrients may pass through one or more carnivore steps, but eventually most of the nutrients brought above the soil surface by plants return to the soil surface in organic or inorganic form; the remainder may enter riverain and marine ecosystems and return to the soil only at a geological time-scale, or (largely confined to nitrogen and sulphur) become part of atmospheric pools and return as wet and dry deposition or via N-fixation. Most of the nutrients, however, may return to the soil within the direct vicinity of where they were taken up. Decomposition of organic inputs to the soil ecosystem returns the nutrients to the inorganic (mineral) form, but there can be considerable delays involved. Inorganic nutrients are part of physical–chemical equilibria in the soil, governing their current concentration in soil solution. Transfers of nutrients between neighbouring land units can occur, via movement of soil or in water-soluble form, but usually such transfers are small.

NUTRIENT CYCLES BECOME NUTRIENT FLOWS IN AGROECOSYSTEMS

Nutrient losses from natural ecosystems tend to increase under human management. Logging practices lead to serious nutrient losses from tropical forests via stream flow, especially during brief periods after burning; under low-impact management the losses can be largely confined to the nutrients in the harvested wood (Malmer, 1996). Patch size in logging practices influences nutrient losses, as the 'safety net' function of neighbouring tree roots has a limited spatial extent. In small patches the whole nutrient flush can be absorbed, in large patches, only the flush from a border zone (Brouwer, 1995).

Agriculture has its roots in 'slash and burn' techniques for opening forest land, all over the world. 'Fire-clearance husbandry' (Steensberg, 1993) makes use of the nutrient mobilization effect of fire to concentrate resources gradually accumulated during a fallow period into crops grown for a few years (Nye and Greenland, 1960). Again, when this technique is applied on small patches of land, in low population density areas, with a large share of fallow land in the landscape mosaic, overall losses may be small. When slash and burn methods are applied over large areas simultaneously, greater losses are inevitable.

All processes of the nutrient cycle described in the previous section can be recognized in agroecosystems as well, where farmer management is aimed at the nutrient uptake and associated plant productivity of one (or a limited number of) plant and/or herbivore species (Fig. 1.2). A major difference starts at harvest time, however, when products are removed from the field where they grew. In a subsistence economy, most nutrients still stay in the ecosystem from which they were derived in plant uptake and will be returned around the homestead. As long as culture is based on 'shifting homesteads', human systems can stay part of the natural nutrient cycle, and spontaneous vegetation can reclaim the accumulated nutrient stocks from the soil. With increasing integration of agriculture into world markets the distance travelled has increased enormously and in the vast majority of current farms a return flow of nutrients in waste products is no longer feasible. The nutrient cycle has thus become a nutrient flow process based on mining, with substantial on-site losses in each cycle and accumulation in peri-urban areas (Fig. 1.3).

Unless new nutrient inputs are provided and depending on the nutrient buffer capacity of the soil, this nutrient flow results in a contracting spiral with further reductions of plant productivity at each cycle. The more successful the farmer is in manipulating the ecosystem to facilitate nutrient uptake by the desired crops (and/or animals), the faster this depletion proceeds.

Where primary (plant) and secondary (animal) production become spatially segregated on specialized farms, the animal production units will lead to a local surplus of nutrients. In the last decades certain parts of Europe have thus become nutrient-aggrading zones, at the cost of nutrient depletion in the soils from which fodder crops were exported.

The nutrient transfers in the (partial) cycle can be quantified by accounting for all flows or by looking at changes in stocks (pool sizes). In a closed cycle, recording pool sizes may not reveal any activity, and measurements have to focus on the turnover and flows. Where the nutrient cycle has become a nutrient flow, changes in stock sizes may become measurable and may help to predict how long the flow can continue. This raises the question of whether total pool sizes of nutrients in soil are important or only the 'bio-available' fractions.

BIO-AVAILABILITY

Interdisciplinary research on plant–soil relations has probably been hampered by the concept of 'bio-availability'. To many biologists this term suggested that a clear-cut distinction can be made between pools of each substance (water, nutrient, pollutant) which are important, and those which

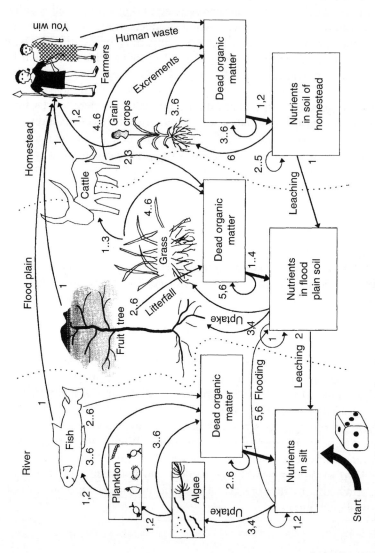

Fig. 1.2. Nutrient cycling in a traditional agricultural system on the floodplain of the Nile in southern Sudan; the diagram can be used as a board game (played with dice, all players simulating nutrients moving over the board; the first players to reach the farmer wins) (van Noordwijk, 1984).

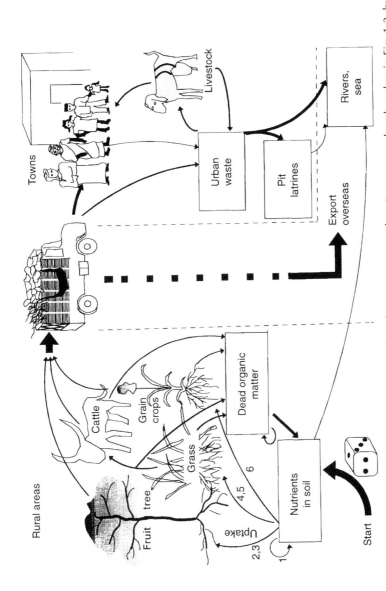

Fig. 1.3. Nutrient flow for an agricultural system producing for urban or export markets (a game can be played as in Fig. 1.2, but it will be of limited duration) (van Noordwijk, 1984).

are not, for uptake processes. To soil chemists and physicists the term suggested that for all biota the same distinction between pools will be relevant. Both interpretations are obviously gross simplifications and may have been misleading. The confusion was increased because the practical methods for measuring 'availability' used actual uptake as the criterion for testing and improving them. These methods are thus useful indicators of the plant–soil interaction under study, but they are not measurements of the pool size implied in the definitions given. In many instances, existing 'availability indices' are treated in models as if they measure pool sizes, and researchers may express surprise when actual plants can take up more of a resource than they thought was present on the basis of an availability index. If an availability index was really a measurement of the available pool, however, it would probably not be a good predictor of actual uptake across a range of plants and conditions. The difference between index and measure can be small for highly mobile nutrients such as nitrate, but will be important for nutrients of lower mobility, such as phosphate.

What is needed to resolve this confusion, is to recognize that both on the plant and on the soil side of the interaction, a large diversity of situations can exist. If bio-availability is really defined as the 'pool from which uptake can occur', it should be accompanied by a concept of 'acquisition strength' of the organisms involved, reflecting the 'part of the available pool which can actually be acquired'. Measurements of the available pool size, especially for nutrients of low mobility, should not be expected to correlate with actual uptake, but should form the basis for comparing the acquisition strength of different biota. A large number of models has been developed for nutrient acquisition which describe transport around single roots, with all its complexities, on a relatively short time span. Scaling up such results to whole plants and complete life cycles is still a major issue. For practical applications of bio-availability indices, a large number of interdependent choices has to be made:

- sampling depth,
- number of samples to be pooled and spatial sampling scheme to be used,
- sampling time,
- method for sample handling and storage,
- method for sample analysis.

Understanding the processes underlying the plant–soil interaction will probably not lead to a better 'universal' method for measuring an index which predicts actual uptake, but it may help to quantify the compromises to be made for any simple index and to predict how existing methods should be modified for new situations. The interpretation of any measurement of a pool size will depend on the types of flow in which we are specifically interested.

AGROECOSYSTEM NUTRIENT USE EFFICIENCY

The term efficiency generally indicates an output/input ratio of transfers (flows) into and out of pools, and thus efficiency depends on the boundaries where inputs and outputs are measured. The term *nutrient use efficiency* is often used without specifying the boundaries of the system in space and time. Efficiency attributes are not necessarily conserved across system scales, however, and thus systems which are efficient from a given perspective may be inefficient when considered with other system boundaries.

It is convenient to use a four-quadrant graphical representation to clarify the various components of nutrient use efficiency on a patch or plot scale (Fig. 1.4). In quadrant III (lower left) of this graph, a relationship is indicated between the inputs of nutrients to the soil (in whatever organic or inorganic form) and the amount of available nutrients in the soil during the growing season. The output/input ratio here can be defined as *'application efficiency'*. This efficiency depends on timing, placement, quantity and quality of the inputs (Cadisch and Giller, 1997). In quadrant IV (lower right) this available amount is related to the amount of nutrients taken up over a given period of time. The output/input ratio here can be defined as *'uptake efficiency'*, and is the domain of acquisition strength and root ecology

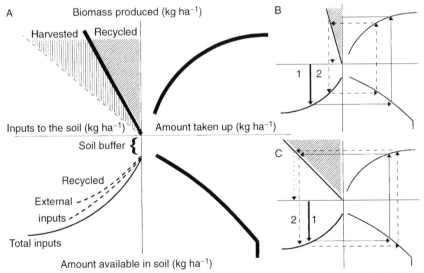

Fig. 1.4. (A) Four-quadrant representation of nutrient flows at a patch/plot level – see explanation in text (modified from van Noordwijk and De Willigen, 1986); (B) and (C) show how sequences of plant growth can follow contracting or expanding spirals, depending on the *replenishment* efficiency of quadrant II (the slope of the line).

(van Noordwijk and De Willigen, 1986; van Noordwijk and Brouwer, 1997). In quadrant I of the graph (upper right) the relation is given between plant biomass production and nutrient uptake; the output/input ratio here can be defined as *'utilization efficiency'*. Finally in quadrant II we can express the conversion of total biomass into harvested products (where the *'harvest efficiency'* or *'harvest index'* gives the conversion efficiency as output/input ratio), and convert the non-harvested part as organic inputs back into the system for the next cycle. For grain crops the term harvest index is normally used to refer to the grain only, but if straw is removed from the field the real harvest index may approach 100% of above-ground biomass, although the harvested products (grain and straw) will differ in value per unit dry weight. The *'agronomic efficiency'* of fertilizer use (increment in yield per unit additional fertilizer input, can now be expressed as a product of four partial (marginal) efficiencies: application, uptake, utilization and harvest efficiency, where the output of each step forms the input to the next one.

The form chosen in Fig. 1.4 allows one to continue into the next cropping cycle, as crop residues from the current crop modify nutrient availability in the next season. The intercept with the availability axis in quadrant III is determined by additions to the available pool from 'external' sources (which may include weathering of minerals, inputs from dry and wet deposition etc.). If the harvested amount of nutrients exceeds these free inputs, one would normally expect a contracting spiral during a sequence of crops – the higher the harvest index, the greater the contraction. During fallow periods a natural or man-made vegetation with a low harvest index may restore fertility by an expanding spiral from year to year, depending on the free inputs from other environmental compartments.

Apart from the organic recycling, based on the non-harvested part, we can also regard a functional link between the harvested part and external organic or inorganic inputs, with a financial linkage of costs and benefits. We can define a *'replenishment efficiency'* as the amount of nutrients acquired by the farmer per unit nutrient in yield products harvested from the field.

Scaling up from patch/plot to field, the input–output relationships will change due to the spatial variability within the field (see below). *Farm-level nutrient use efficiency* can be understood from the nutrient use efficiency of the various components of the farm, but taking due account of which inputs of farm components are based on outputs of other components. Similarly, nutrient use efficiency at the society level depends on the farm level efficiency, but should take transfers among sectors into account. Even if the field-level efficiency of using recycled wastes is lower than that of using 'new' external inputs, the *national-level nutrient use efficiency* can be greatly enhanced by recycling. Agricultural development as exemplified by western Europe is often based on farm specialization, increased distance between production sites and markets and a reduction of recycling. Even if crop-level nutrient use efficiency has been maintained, the global

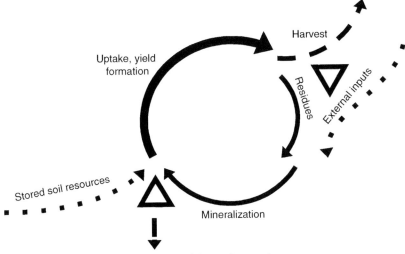

Fig. 1.5. Schematic nutrient flow and/or cycle in agroecosystems, emphasizing four aspects of efficiency: uptake efficiency, utilization efficiency, recycling/replenishment and application efficiency (van Noordwijk and Garrity, 1995).

nutrient use efficiency has decreased and environmental concerns have increased.

Figure 1.5 gives a schematic view of the key processes in nutrient cycling in agricultural systems on a field scale, and focuses on four aspects:

1. (lower left corner) Plant nutrient uptake from stored as well as recently added organic and/or inorganic resources (uptake efficiency).
2. (upper left arc) Internal redistribution in the plant and yield formation (utilization efficiency).
3. (upper right corner) Removal of harvest products, their exchange for external inputs and the recycling of harvest residues in the system (replenishment efficiency).
4. (lower right arc) Increase in available nutrient pool as a result of recycled on-farm or external inputs used to replenish soil fertility (application efficiency).

Aspects 1 and 4 are traditionally studied in soil fertility research (questions 2, 3 and 5 of Fig. 1.1) and aspect 2 in plant physiological research (question 1 of Fig. 1.1). Aspect 3 is normally included in farming systems and agroeconomic studies (question 4 of Fig. 1.1).

Figure 1.6 shows different categories of problems that may reduce the nutrient use efficiency of agroecosystems, depending on climate, soil, topography and cropping system. These problems occur at different spatial scales (van Noordwijk and Garrity, 1995):

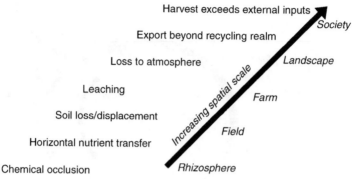

Fig. 1.6. Categories of problems for efficient nutrient use in agroecosystems at different scales (van Noordwijk and Garrity, 1995).

1. Chemical occlusion (and similar soil biological and soil physical phenomena) limits uptake of stored soil resources and/or the utilization of fresh inputs in the root zone at large, or more specifically in the rhizosphere.
2. A number of processes lead to spatial heterogeneity of nutrient supply at the field scale:
  (a) horizontal nutrient transfer by trees, crops or farmer's practices, creating depletion and enrichment zones,
  (b) soil loss and displacement by erosion/deposition cycles, especially on sloping lands, and thus reduce the overall stability (Whitmore and van Noordwijk, 1995).
3. Leaching leading to vertical nutrient transfer to deeper layers, often beyond the reach of shallow-rooted crops.
4. Losses to the atmosphere in gaseous form (especially nitrogen (N) and sulphur (S)), dust (wind erosion) or as particulate ash during fire; the last two lead to deposition elsewhere in the landscape.
5. Export of harvest products beyond the realm where recycling is possible: increasing economic integration of farms and/or hygiene-motivated reductions in waste recycling cause reduction of recycling as part of 'development'.
6. Economic conditions that prevent the use of external inputs to replace the exported nutrients, or that stimulate excessive use of inputs for high-value products.

The relative importance of these six efficiency aspects differs between agricultural systems, for both technical and economic reasons. The farm gate price of agricultural products is not related to their nutrient content, yet the amount of nutrients which have to be used to replenish soil fertility does depend on the exports, as well as on application efficiency. Large differences thus exist in the ratio of financial returns to sale of harvest products and the costs of replenishing the nutrient stock.

Table 1.1. Comparison of nutrient replacement value of various agricultural products and their farm gate price; financial parameters relate to the situation in Lampung (Indonesia) in 1995 (modified from van Noordwijk et al., 1997).

	A Nutrient removal[a] R_i (g kg^{-1} product)			B Nutrient replacement costs $SR_i \times P_i$ (Rp kg^{-1})	C Farm gate value of product (Rp kg^{-1})	D Relative replacement costs, B/C
	N	P	K			
Cassava (fresh tuber)	1.4	0.24	3.6	15	25–100	0.15–0.60
Rice (grain + husk)	12	2.4	2.0	49	400	0.12
Maize (grain)	16	1.8	3.2	49	250	0.20
Cowpea (grain)[b]	34	3.0	15	80	300	0.27
Sugar-cane (cane dry weight)	1.5	0.8	3.5	22	40	0.54
Rubber (kg latex (DRC))	6.3	1.3	4.3	35	2000	0.02
Oil palm (bunches)	2.9	0.4	3.7	19	60	0.31
Replacement costs P_i in Rp g^{-1} nutrient in the exported crop[c]	1.2	12	2.9			

DRC, dry rubber content.
[a]Sources: Ahn (1993) and Höweler (1991).
[b]Assuming that the N harvest index equals the percentage of N derived from the atmosphere, so that no net N is exported from the field.
[c]Replacement costs, P_i, are based on a price of 260, 480 and 400 Rp kg^{-1} for urea, TSP and KCl (based on prices for December 1995, Suyanto, personal communication; for an exchange rate of 2200 Rp/$), a nutrient content of 0.45 (N), 0.20 (P) and 0.46 (K), respectively, and a long-term recovery (kg nutrient in crop kg^{-1} nutrient in fertilizer) of 0.5, 0.2 and 0.3, respectively; combined, these estimates lead to the indicated replacement costs P in Rp g^{-1} nutrient in the exported crop (e.g. 260/(0.45 × 0.5 × 1000) = 1.16).

As a first approximation, the costs of inorganic fertilizer needed for replenishment at reasonable application efficiency can be expressed as a fraction of farm gate price (Table 1.1). Nutrient contents per unit weight of harvested product, R_i, multiplied by the replacement cost per unit nutrient in the crop, p^i, can give the market value of the products which would require all cash obtained to be spent on fertilizer, just to maintain chemical soil fertility. Where this price is more than say 30% of the actual farm gate price (as it is for cassava and sugar-cane in the example of Table 1.1, low-value bulk products with high nutrient contents), fertilizer use is unlikely to be economically justified at any rate. Local prices of both products and inputs may differ from those at the global market, but the range of these 'relative replenishment costs' includes values of more than 50%, such as cassava in Table 1.1, and values close to zero (high-value, low-volume products and/or products of low nutrient content such as rubber).

Existing technical opportunities to increase the agronomic efficiency (output/input ratio) of fertilizer use may not be utilized by farmers if the price ratio of fertilizer and yield products is too low (van Noordwijk and Scholten, 1994). If fertilizer inputs are cheap, no incentive is given to high application efficiency; if fertilizer is expensive, it may not be used at all; in an intermediate range improving fertilizer application efficiency may pay off to the farmer.

Trees can increase nutrient concentrations on small areas of land, at the expense of nutrients elsewhere. If their source of nutrients is deep soil layers (point 3 discussed above), chemically occluded soil nutrient sources (point 1), airborne dust (4) or soil material moving downslope with surface runoff (2b), one may expect that trees increase the nutrient stocks available for other components of the system, such as crops (Buresh, 1995). As long as deep and chemically occluded sources last and as long as wind erosion up-wind and water erosion up-slope continue, these nutrient sources can be sustainable from the agroforest farmers' point of view. None of these processes is easy to prove and quantify, however. If most of the nutrients which the trees absorb come from topsoil layers (point 2a), and this may even extend to 50 m from the tree in some cases, the role of trees is only positive in as far as topsoil nutrients would not be utilized by other components and get lost from the system. The large horizontal spread of tree roots appears to have been neglected in the design of many agroforestry experiments. Positive conclusions about increased nutrient storage and/or crop yields for agroforestry treatments of such experiments, as compared to neighbouring 'control' plots, may in fact be partly due to tree roots mining the soil under the neighbouring plots as well as in their own (Coe, 1994).

The time dimension also causes concern in the definition of nutrient use efficiency of agroecosystems, especially where perennial crops are concerned or for nutrients with strong 'residual effects'.

SPATIAL VARIABILITY

In early stages of fertilizer research, results were interpreted as universal truths, leading to blanket fertilizer recommendation schemes, which differentiate between crops, but not soils. Gradually more detailed recommendations have been developed which are based on soil groups and/or field-level soil analysis. Such schemes can greatly enhance field-level nutrient use efficiency.

Agronomic research has for a long time made the implicit assumption that results of relatively small plots, selected on the basis of their homogeneity for 'proper' experiments, were directly relevant for the field scale.

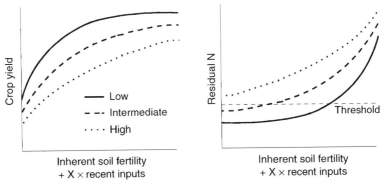

Fig. 1.7. Effects of within-field spatial heterogeneity on the shape of the fertilizer–yield response curve as well as on the amounts of residual mineral N at harvest (based on van Noordwijk and Wadman, 1992).

Van Noordwijk and Wadman (1992) showed that the agronomic efficiency of a crop production system, defined as crop output per unit input, decreased with increasing internal variability of the field, all other parameters being equal (Fig. 1.7). Independently, Cassman and Plant (1992) developed a similar model and applied it to field results. Field heterogeneity does not affect nutrient use efficiency in the linear response range, where it may be most visible in the crop. Heterogeneity affects efficiency especially when the nutrient supply to part of the plants exceeds the requirements for maximum yield.

Field heterogeneity has a direct effect on the production–environment conflict, as the amounts of inputs needed for 'economically optimum' yield increase, while the amounts of inputs which can be tolerated from an environmental point of view decrease. In heterogeneous fields, nutrient use efficiency can be improved by site-specific input decisions. Technical options for such decisions are being developed for large-scale mechanized farming. The small-scale farmer with intimate knowledge of his or her land may be directly inclined to apply nutrients where needed most. This is only possible, however, if the rich and poor strata of the field can be recognized from easily observable patterns.

Spatial variability may under some circumstances help in reducing risks in crop production under uncertain rainfall conditions (Brouwer *et al.*, 1993; van Noordwijk *et al.*, 1994).

SPATIAL EXTRAPOLATION: NOT JUST STRATIFIED SAMPLING

It has become a common phrase that we should study our subject at different spatial and temporal scales. What does this mean in practice? We will discuss two aspects here: 'stratification' and 'scaling rules'.

Stratification

There is a well-established body of theory on 'stratified sampling', in both social and biophysical sciences. Essentially it means dividing the 'world' into different zones (strata) on the basis of a priori knowledge, sampling in each of these zones and deriving weighted averages for the 'world'. If n X_i values are the result for n strata, the overall average value X_t is derived from

$$X_t = \sum_{i=1}^{n} f_i X_i \qquad (1.1)$$

where f_i represents the relative weights of the strata. Often the f_i values are taken to be proportional with the area fractions, a_i, of the various strata:

$$X_t = \sum_{i=1}^{n} a_i X_i \qquad (1.2)$$

Taking area fractions can be a practical way to derive 'time averaged' values for a 'land use system', assuming that it is in 'steady state'. For example, the area fractions under various stages of fallow regrowth and cropping should, under this assumption, represent the relative duration of each phase during the cycle.

Sampling within the strata is normally based on 'minimum representative' sample sizes; these are supposed to be internally heterogeneous, but to be replicated within the stratum. For agronomic experiments there are standard ideas of a 'minimum plot size' required, for soil sampling of a minimum number of replications for pooled samples, in plant ecology the concept of 'minimum area' has a long history.

There is still a lot to be improved on in using this approach in land use research, but we also have to face a more serious challenge: for many, if not all, parameters the use of equation (1.2) can lead to the wrong conclusions. Yet, the vast majority of current GIS applications are based on this area-based extrapolation.

Scaling Up

When we take a more critical look at many phenomena, stocks as well as flows, they are not simply proportional to the area they occupy. Equation (1.2) can be modified by introducing a *scaling factor s*:

$$X_t = \sum_{i=1}^{n} a_i^{0.5 s} X_i \qquad (1.3)$$

The factor 0.5 is introduced here to convert area back to a linear scale; s then represents the appropriate 'dimensionality' of the process (Fig. 1.8). The reasons for s to differ from 2 (and thus for the scaling rule to differ from an area-based one) are manifold, but always appear to involve

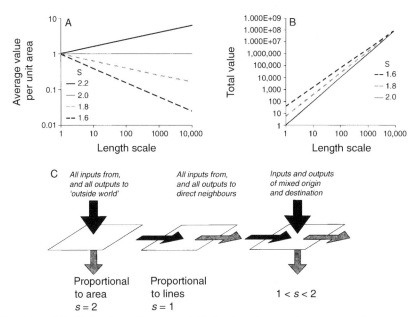

Fig. 1.8. (A), (B) Scaling rules potentially lead to changes in average value per unit area. (C) If all inputs come from, and outputs go to the outside world, an area-based approach with dimensionality of $s = 2$ may apply (left); if all inputs and outputs are laterally transmitted, a linear approach ($s = 1$) is appropriate (middle); if inputs and outputs are of mixed origin and destination, a dimension between 1 and 2 may apply (right).

'interactions'. Spatial interactions are common in 'fluxes' and 'flows', but can also occur in 'stocks', for example in a property such as 'species richness' where only species not yet encountered add to the overall value when scale is increased. Nutrient stocks, however, appear to follow classical ($s = 2$) scaling rules.

In extreme cases we may find that $s = 1$ and lines rather than areas are involved. This may apply, for example, if surface runoff is simply passing through an area with impermeable surface crusts or in 'boundary planting' systems. If the dimensionality is not a whole number (1, 2 or 3) but something in between, we can speak of 'fractal' dimensions. For many phenomena a given fractal dimension may apply across a range of scales, but the scaling rules themselves may change at certain transition points (Lam and De Cola, 1993).

The main reason for the dimensionality to differ from 2, is that there are local *interactions* at a range of scales. If all inputs come from outside the system, and all outputs leave the system, an area-based approach is correct (Fig. 1.8C). If, however, part of the 'inputs' comes from neighbouring areas, and/or some of the outputs go to neighbouring areas, the scaling rule differs from $s = 2$. This makes clear that the 'grain size' of the landscape pattern has an effect on the scaling rule. In fine-grained landscapes, with many boundaries between landscape units, the amount of local interactions will be large and there are many reasons why s differs from 2. In coarse-grained patterns interactions between units will be small. Linking the scaling issue with 'interactions' makes clear that we may have many of the tools needed to tackle the issue.

A simple method exists for generating data with a fractal dimension. If we take a grid of random numbers, some of them negative, others positive, and sample this grid at different scales (1×1, 2×2, 3×3 etc.) with the rule that we record the net value, but take negative values as zero, we will find that the average value per unit area decreases with the scale of sampling. At the 1×1 scale all negative values were perceived as zero, while at larger scales they can increasingly offset positive values. This analogy may help us understand the fractal dimension of net sediment loss in erosion studies as sedimentation sites are recorded as sites of zero (and not negative) erosion.

Conventional agronomic experiments are done in fine-grained patterns, in the form of a patchwork of differently treated plots. It is a continuous source of concern that interactions such as lateral resource capture between the plots are not unduly influencing the results, so that these can be scaled up to the coarser-grained patterns of real world farms. We have to pick up the challenge: we are not only dealing with experimental artefacts, but with important principles and results. In agroforestry techniques such as boundary plantings these interactions are about all there is (scaling up here may be based on s values close to 1.0). We may be able to turn a problem into an opportunity if we can quantify how much lateral resource capture contributes to the productivity of small-scale plots (van Noordwijk and Ong, 1996).

SCALING IN TIME: RESIDUAL EFFECTS AND POLICY IMPLICATIONS

The classical question of fertilizer use of Fig. 1.1 does incorporate a time dimension. As not all the nutrients in the fertilizer recently applied have become available and used during the cropping season 'residual fertility' effects were noted early on. A simple way out of this issue is to define this as part of the 'inherent soil fertility' for the next crop; unfortunately many existing measurement methods for inherent soil fertility do not perform well where partly undissolved fertilizer is present.

Residual effects depend on the mobility, biological and chemical transformation of the nutrient under given soil and climatic conditions. N and P represent the two extremes: the mobility of N in mineral form is so high that residual amounts of mineral N at the end of a cropping season run a risk of leaching out of the profile before the next growing season, especially where a relatively wet period (such as the winter in temperate regions) intervenes.

Residual N effects on the N supply to the next crop thus tend to be small, and there is reason for concern about negative environmental effects after the crop harvest. In contrast, the sorption (adsorption and desorption) equilibria for inorganic P_i in most soils mean that plant roots, even when effective mycorrhizal partners are present, capture only a fraction of what is 'available' in the soil in a physico-chemical sense. Due to the low effective mobility of P, leaching rates will be small. Thus, a substantial part of this 'available' pool of P will remain until the next crop, and the total agronomic value of P fertilization cannot be evaluated after the first cropping season. The low mobility of P also makes P depletion a gradual process, so both the costs of nutrient depletion and the benefits of fertilization tend to be underestimated on the basis of one-season experiments. For other nutrients such as potassium (K) and for N in organic pools a situation intermediate between these two poles can be expected: positive residual effects on a next crop may be expected, but losses to other environmental components cannot be ignored. If P is supplied in such large quantities that the effective marginal sorption capacity becomes small, it may behave like a nutrient such as K.

Calculations of (negative) nutrient budgets of agroecosystems can easily lead to a suggestion that (substantial) fertilizer inputs will be needed to redress past neglect of soil fertility. For P, inputs of inorganic fertilizer of moderate solubility may indeed replenish the capital stock. For N, however, the capital is in organic form and adding inorganic N fertilizer does not directly help to replenish the capital. Maintaining N capital depends on sufficient on-site production of organic residues and/or (labour intensive) inputs from outside the plot. P fertilizer, rather than inorganic N fertilizer, may be needed to stimulate the production of legumes supplying N-rich organic inputs.

In a strictly financial evaluation of profitability of investment in fertilizer use, there is no problem in accounting for future benefits (as net present value), as long as they are sufficiently predictable. Yet, it seems likely that the more immediate benefits farmers perceive on using N fertilizers ('consumables'), rather than P ('capital stock') may lead to a bias in their decisions. As unbalanced fertilizer use may increase the depletion of the soil's capital of other nutrients, the short-term decisions of farmers may contrast with the long-term benefits of sustainable soil management.

The argument that farmers are thus likely to mismanage their soil resources, because of their limited perspective on scaling time, is often taken as a starting point for arguments on the need for government policy interventions. In the early 1970s the UN Food and Agriculture Organization (FAO) and other international agencies recommended government subsidies on inorganic fertilizer as a way to stimulate food production. In these subsidy schemes, however, no distinction was made between N and P fertilizers. Abandoning such subsidy schemes as part of deregulation and structural adjustment programmes has revealed that reliance on subsidies has made farmers dependent and vulnerable. Yet, the problem of past soil fertility depletion remains and may be a major bottleneck for current agricultural development. A new initiative on soil fertility replenishment (Sanchez *et al.*, 1997) rightly focuses on P and organic N as the capital stocks to be replenished. It thus clearly differs from past fertilizer subsidy programmes.

There are obvious arguments for government investment in the 'public good' of understanding general principles of soil management, linked to information on local soils, crops and climate, and in developing efficiency increasing techniques. Arguments for direct government intervention in the supply of inputs, however, depend on how yield response to nutrient additions is scaled in time via residual effects.

Under some circumstances cumulative yield response to total nutrient inputs may have an S-shaped curve. Figure 1.9 shows positive interactions between growth factors only in a certain range upon alleviation of initial toxicities. Only then can other inputs be more fully utilized (De Wit, 1994). In order to get an S-shaped curve of plant growth versus total external nutrient inputs, positive interactions have to go beyond the normal build-up

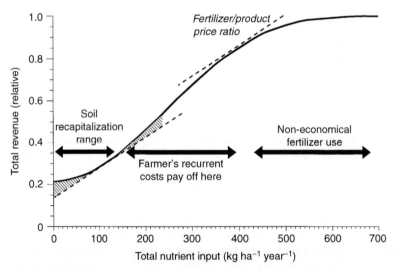

Fig. 1.9. Hypothetical cumulative yield response curve to nutrient addition and domain for government intervention in 'soil fertility replenishment': if the response is S-shaped there are two points where marginal returns equal marginal costs; if one starts from zero, input use is at first uneconomical and in fact only at the point where the two shaded areas are equal in size is input use starting to pay off; a reasonable target for outside help may be to bring the soil up to the first point where marginal returns are non-negative; if the practically relevant part of the curve is not S-shaped the arguments for outside 'soil replenishment' investment will be less convincing.

of residual effects and the expected stimulation of N-fixation by legumes if moderate amounts of P are supplied to the soil.

If the overall response is S-shaped indeed, it may be opportune for governments to help farmers approaching the first point of inflection of the curve (or more specifically, up to the point where marginal returns start to exceed marginal costs). Getting there will improve returns to maintenance fertilization and may start a financially sustainable fertilizer use (up to the second point of the curve where marginal costs equal marginal benefits). 'Economies of scale' in fertilizer use might make future fertilizer use more affordable, once a market has been created, but the empirical evidence that this will happen is not strong. Empirical evidence for S-shaped response curves on farmers' fields is not particularly strong either, and here lies a real challenge for research to address soil fertility from a policy perspective. Perhaps many fields currently used for agriculture have not been depleted to the point that farmers experience an S-shaped response curve to soil nutrient replenishment.

In the absence of S-shaped response curves, if diminishing marginal returns characterize the whole response curve, the debate on soil fertility replenishment is remarkably similar to past fertilizer subsidy schemes and may do little to establish financially as well as ecologically sustainable agricultural systems.

ACKNOWLEDGEMENTS

Thanks are due to Drs Cheryl Palm, Roland Buresh and Thomas Tomich for comments on an earlier draft of the manuscript.

REFERENCES

Ahn, P.M. (1993) *Tropical Soils and Fertiliser Use*. Longman, Harlow, UK, 264 pp.

Alexander, I. (1989) Mycorrhizas in tropical forests. In: Proctor, J. (ed.) *Mineral Nutrients in Tropical Forest and Savanna Ecosystems*. Blackwell Science, Oxford, UK, pp. 169–188.

Barber, S.A. (1984) *Soil Nutrient Bioavailability, A Mechanistic Approach*. Wiley Interscience, New York.

Brouwer, J., Fussell, L.K. and Herman, L. (1993) Soil and crop microvariability in the West African semi-arid tropics: A possible risk reducing factor for subsistence farmers. *Agriculture, Ecosystems and Environment* 45, 229–238.

Brouwer, L.C. (1995) Nutrient dynamics in intact and logged tropical rain forest on an acid sandy soil in Guyana. In: Schulte, A. and Ruhiyat, D. (eds) *Proceedings of International Congress on Soils of Tropical Forest Ecosystems*. Mulawarman University Press, Samarinda, Indonesia, vol. 6, pp. 5–23.

Brown, S., Anderson, J.M., Woomer P.L., Swift, M.J. and Barrios, E. (1994) Soil biological processes in tropical ecosystems. In: Woomer, P.L. and Swift, M.J. (eds)

The Biological Management of Tropical Soil Fertility. John Wiley & Sons, Chichester, UK, pp. 15–46.

Buresh, R.J. (1995) Nutrient cycling and nutrient supply in agroforestry systems. In: Dudal, R. and Roy, R.N. (eds) *Integrated Plant Nutrient Systems.* FAO, Rome, pp. 155–164.

Cadisch, G. and Giller, K.E. (1997) *Driven by Nature, Plant Litter Quality and Decomposition.* CAB International, Wallingford, UK, 409 pp.

Cassman, K.G. and Plant, R.E. (1992) A model to predict crop response to applied fertilizer nutrients in heterogeneous fields. *Fertilizer Research* 31, 151–163.

Chapin, F.S. III (1980) The mineral nutrition of wild plants. *Annual Review of Ecology and Systematics* 11, 233–260.

Coe, R. (1994) Through the looking glass: ten common problems in alley cropping research. *Agroforestry Today* 6(1), 9–11.

De Willigen, P. and van Noordwijk, M. (1987) Roots for plant production and nutrient use efficiency. PhD thesis, Wageningen Agricultural University, Wageningen, The Netherlands.

De Wit, C.T. (1953) A physical theory of fertilizer placement. PhD thesis, University of Wageningen, Wageningen, The Netherlands.

De Wit, C.T. (1994) Resource use analysis in agriculture: a struggle for interdisciplinarity. In: Fresco, L.O., Stroosnijder, L., Bouma, J. and Van Keulen, H. (eds) *The Future of the Land: Mobilising and Integrating Knowledge for Land Use Options.* Wiley, Chichester, UK, pp. 41–55.

Högberg, P. (1989) Root symbioses of trees in savannas. In: Proctor, J. (ed.) *Mineral Nutrients in Tropical Forest and Savanna Ecosystems.* Blackwell Science, Oxford, UK, pp. 121–136.

Höweler, R.H. (1991) Long term effects of cassava cultivation on soil productivity. *Field Crops Research* 26, 1–18.

Jordan, C.F. (1985) *Nutrient Cycling in Tropical Forest Ecosystems.* John Wiley & Sons, Chichester, UK, 190 pp.

Lam, N.S. and De Cola, L. (eds) (1993) *Fractals in Geography.* PTR Prentice Hall, Englewood Cliffs, New Jersey, 308 pp.

Lanzer, E.A., Paris, Q. and Williams, W.A. (1987) A non-substitution dynamic model for optimal fertilizer recommendations. *Giannini Foundation Monograph 41*, University of California, California, 53 pp.

Malmer, A. (1996) Nutrient losses from dipterocarp forests – a case study of forest plantation establishment in Sabah, Malaysia. In: Schulte, A. and Schöne, D. (eds) *Dipterocarp Forest Ecosystems, Towards Sustainable Management.* World Scientific, Singapore, pp. 52–73.

van Noordwijk, M. (1984) *Ecology Textbook for the Sudan.* Khartoum University Press, Khartoum, Sudan.

van Noordwijk, M. and Brouwer, G. (1997) Roots as sinks and sources of carbon and nutrients in agricultural systems. In: Brussaard, L. and Ferrera-Cerrato, R. (eds) *Soil Ecology in Sustainable Agricultural Systems.* CRC Lewis Publ., Boca Raton, Florida, pp. 71–89.

van Noordwijk, M. and De Willigen, P. (1986) Quantitative root ecology as element of soil fertility theory. *Netherlands Journal of Agricultural Science* 34, 273–281.

van Noordwijk, M. and Garrity, D.P. (1995) *Nutrient Use Efficiency in Agroforestry Systems.* International Potash Institute 24th IPI Colloquium, Chiang Mai, 21–24 February 1995, pp. 245–279.

van Noordwijk, M. and Ong, C.K. (1996) Lateral resource flow and capture – the key to scaling up agroforestry results. *Agroforestry Forum* 7(3), 29–31.

van Noordwijk, M. and Scholten, J.H.M. (1994) Effects of fertilizer price on feasibility of efficiency improvement: case study for an urea injector for lowland rice. *Fertilizer Research* 39, 1–9.

van Noordwijk, M. and van de Geijn, S.C. (1996) Root, shoot and soil parameters required for process-oriented models of crop growth limited by water or nutrients. *Plant and Soil* 183, 1–25.

van Noordwijk, M. and Wadman, W. (1992) Effects of spatial variability of nitrogen supply on environmentally acceptable nitrogen fertilizer application rates to arable crops. *Netherlands Journal of Agricultural Science* 40, 51–72.

van Noordwijk, M., Dijksterhuis, G. and Van Keulen, H. (1994) Risk management in crop production and fertilizer use with uncertain rainfall: how many eggs in which baskets. *Netherlands Journal of Agricultural Science* 42, 249–269.

van Noordwijk, M., Lawson, G., Groot, J.J.R. and Hairiah, K. (1996) Root distribution in relation to nutrients and competition. In: Ong, C.K. and Huxley, P.A. (eds) *Tree–Crop Interactions – a Physiological Approach*. CAB International, Wallingford, UK, pp. 319–364.

van Noordwijk, M., Hairiah, K., Partoharjono, S., Labios, R.V. and Garrity, D.P. (1997) Food-crop-based production systems as sustainable alternatives for *Imperata* grasslands? *Agroforestry Systems* 36, 55–82.

Nye, P.H. and Greenland, D.J. (1960) *The Soil under Shifting Cultivation*. Commonwealth Bureau of Soils Tech. Comm. 51, Harpenden, UK, 156 pp.

Nye, P.H. and Tinker, P.B. (1977) *Solute Movement in the Soil–Root System*. Blackwell Science, Oxford, UK.

Odum, E.P. (1953) *Fundamentals of Ecology*. W.B. Saunders Company, Philadelphia.

Russell, E.J. (1912) *Soil Conditions and Plant Growth*. Longman, London, UK.

Sanchez, P.A. (1976) *Properties and Management of Soils in the Tropics*. Wiley, New York, 618 pp.

Sanchez, P.A., Shepherd, K.D., Sopule, M.J., Place, F.M., Buresh, R.J., Izac, A.M.N., Mokwunye, A.U., Kwesiga, F.R., Ndiritu, C.G. and Woomer, P.L. (1997). Soil fertility replenishment in Africa: An investment in natural resource capital. In: Buresh, R.J., Sanchez, P.A. and Calhoun, F. (eds) *Replenishing Soil Fertility in Africa*. Soil Science Society of America (SSSA) Special Publication 51, Madison, Wisconsin, pp. 1–46.

Smaling, E.M.A. and Oenema, O. (1997) Estimating nutrient balances in agroecosystems at different spatial scales. In: Lal, R., Blum, W.H., Valentin, C. and Stewart, B.A. (eds) *Methods for Assessment of Soil Degradation*. Advances in Soil Science, CRC Press, Boca Raton, Florida, pp. 229–252.

Smits, W.T.M. (1994) *Dipterocarpaceae: Mycorrhizae and Regeneration*. Tropenbos Series 9, Tropenbos Foundation, Wageningen, The Netherlands.

Steensberg, A. (1993) *Fire-Clearance Husbandry, Traditional Techniques Throughout the World*. Kristensen, Harning, Denmark, 239 pp.

Tiessen, H., Chacon, P. and Cuevas, E. (1994) Phosphorus and nitrogen status in soils and vegetation along a toposequence of dystrophic rainforest on the upper Rio Negro. *Oecologia* 99, 145–150.

Van Keulen, H. and Wolf, J. (1986) *Modelling of Agricultural Production, Weather, Soils and Crops*. Pudoc, Wageningen, The Netherlands.

Whitmore, A.P.M. and van Noordwijk, M. (1995) Bridging the gap between environmentally acceptable and economically desirable nutrient supply. In: Glen, D.M., Greaves, M.P. and Anderson, H.M. (eds) *Ecology and Integrated Farming Systems*. John Wiley & Sons, Chichester, UK, pp. 271–288.

Woomer, P.L. and Swift, M.J. (eds) (1994) *The Biological Management of Tropical Soil Fertility*. John Wiley & Sons, Chichester, UK, 243 pp.

Basics of Budgets, Buffers and Balances of Nutrients in Relation to Sustainability of Agroecosystems

B.H. Janssen

Soil Fertility and Plant Nutrition Group, Department of Environmental Sciences, Wageningen Agricultural University, PO Box 8005, 6700 EC Wageningen, The Netherlands

ABSTRACT

On the basis of examples of negative, neutral and positive nutrient balances, it is shown that the interpretation of a nutrient budget is less simple than it seems. Several basic theoretical concepts related to nutrient budgets are developed. A distinction is made between 'available' and 'not immediately available (NIA)' nutrients, and the interpretation of their budgets is discussed. The relations between soil nutrient pools of varying stabilities and the budgets of available and NIA nutrients are theoretically analysed. This allows for a quantitative description of the requirements for steady state in general, and for equilibrium and sustainability as special cases of steady state. Changes in crop yields and soil nutrient pools in some non-steady situations are calculated. The results are used to relate chemical soil characteristics to sustainability and to put forward the related concepts of Target Soil Fertility (TSF), Target Yield (TY), and Target Inputs (TI).

It is shown that disproportionate supplies of the different nutrients indirectly but inevitably threatens the sustainability of an agroecosystem. The fate of nutrients is to a great extent determined by physical, chemical and biological processes, which in turn are affected by climate and topography, making nutrient flows and hence nutrient budgets only partly controllable by the farmer. The pros and cons are weighed of 'partial nutrient budgets' comprising only easily measurable inputs and outputs. With the help of the developed criteria, the sustainability of some agroecosystems is valued.

INTRODUCTION

Publications on sustainability often seem implicitly and rather intuitively to assume that agroecosystems are not allowed to change and nutrient balances therefore must be zero. If that is true, both positive as well as negative nutrient balances sound alarming with regard to the future of the system. The notion that steady state is the optimum situation may be naive and asks for a critical examination. For that purpose several basic theoretical aspects related to nutrient budgets will be discussed in this chapter, resulting in a set of quantitative definitions of steady state, sustainability and equilibrium. With the developed tools an attempt is made to evaluate and characterize agroecosystems by sustainability.

By 'nutrient budget' is meant an 'account' of nutrient flows to (inputs) and out of (outputs) a certain, clearly defined, agroecosystem (e.g. field, farm). Nutrient balance is the difference between the sums of nutrient inputs and nutrient outputs. A positive balance (inputs > outputs), sometimes designated as nutrient surplus or excess, points to enrichment, and a negative balance, sometimes designated as 'deficit', to depletion of the system considered.

Nutrient balances alone cannot act as safe and reliable guidelines for farmers, scientists and policy-makers. Depending on soil fertility level, negative and positive balances of a particular nutrient may or may not be desirable. Not only the balance, but also the absolute and relative sizes of individual budget items need attention to appreciate the sustainability of an agroecosystem. This calls for a subdivision in nutrients that are available to plants and nutrients that are not. To avoid the risk of jumping into the classical and never-ending discussions on the concept of 'availability' (van Noordwijk, Chapter 1, this volume), operational definitions are conceived in this chapter and subsequently introduced in the concepts of soil nutrient pools.

Relations between nutrient balances and soil nutrient status are rather seldom involved in sustainability studies, but they will get quite some emphasis in this paper. It is attempted to relate nutrient flows to measurable soil nutrient pools and to describe optimum soil fertility, thus paving the way to the use of chemical soil analysis for the assessment of sustainability. Given that the assessment of nutrient budgets is time consuming, short laboratory procedures would be very welcome.

Another issue that needs attention in an analysis of sustainability is 'balanced nutrition', that is, the situation in which the various nutrients (nitrogen (N), phosphorus (P), potassium (K), secondary and micronutrients) are offered in proportions matching the needs of the crop. It is a prerequisite for optimum nutrient use efficiency and sustainability. 'Mononutrient' studies often overlook the fact that the excess of one nutrient goes together with, and may be the consequence of, the shortage of another nutrient.

INTERPRETATION OF NUTRIENT BUDGETS

Nutrient budgets may relate to different spatial and temporal scales. Spatial scales may vary from continents via countries, districts and watersheds to farms and fields. Within the fields, soil and crop may be distinguished, and in the soil there are compartments like soil solution, soil organic matter, soil minerals and adsorption sites for cations and anions. Usually the appropriate time steps are longer, the larger the spatial dimensions of the system. Many studies on nutrient budgets refer to fields with plot sizes varying from around 0.1 ha to tens of hectares. The corresponding time step is one year, or one growing season in cases where there are two or more crops per year. At each scale, nutrient flows may be classified into inputs, outputs and internal flows (Janssen, 1992; Smaling *et al.*, 1996). A particular flow, for example animal manure, may be an input for arable land, an output for grazing land, and an internal flow for a farm.

Generally, the nutrient budget of a cropped field is composed of five inputs and also five outputs. The nitrogen budgets in Table 2.1 serve as examples for three situations:

A   a negative balance found for Rwanda in the study by Stoorvogel and Smaling (1990);

Table 2.1. Nitrogen budgets in kg ha^{-1} per year. A, arable land in Rwanda (Stoorvogel and Smaling, 1990); B, arable land in The Netherlands (see text); C, grassland in Britain (Ryden, 1986).

		A	B	C
Inputs				
IN 1	Chemical fertilizer	1	160	420
IN 2	Organic manure	2	20	0
IN 3	Atmospheric deposition	5	50	15
IN 4	Biological N-fixation	9	20	0
IN 5	Sedimentation	2	0	0
Total inputs		19	250	435
Outputs				
OUT 1	Harvested products	22	120	30[a]
OUT 2	Removed crop residues	5	40	0
OUT 3	Leaching	4	50	160
OUT 4	Gaseous losses	12	40	160
OUT 5	Erosion	29	0	0
Total outputs		72	250	350
Balance		−53	0	85

[a]Cattle growth.

B   a neutral or zero balance as estimated for the average arable land in The Netherlands; this budget was not measured on a particular field, but it has been compiled from data on average fertilizer and manure application, atmospheric deposition, yields of arable crops, estimated leaching and volatilization losses, and the assumption that leguminous crops are grown once per eight years;

C   a positive balance as estimated for grassland (grazed by oxen) in Britain in the 1980s (Ryden, 1986).

The negative budget of Rwanda is reason for serious concern. The annual decrease of 53 kg N ha^{-1} is equivalent to almost 2% of the N-stock, assuming that the mass of the topsoil was 3×10^6 kg ha^{-1}, and the average N content 1 g kg^{-1}. With a constant relative loss rate of 2% year^{-1}, soil N content will be half the present value after 35 years, and soil productivity probably even less.

The budget for arable farming, reflecting the average situation in The Netherlands, seems satisfying. Some crops, such as vegetables, however, receive far more than the average of 160 kg fertilizer N ha^{-1}, of which a considerable part is prone to leaching. Although also in that case the balance may still be zero, the quantity of leached N in such a cropping system surpasses the limits set by government and the European Union (EU) (Oenema et al., 1997)

The grassland balance shows a surplus of N, which is likely to be converted into soil organic N, resulting in an increasing soil supply of N to future crops. The positive side is that the need for fertilizer will diminish, but the negative side is that leaching and volatilization will increase if the fertilizer application is not accordingly reduced. The mentioned leaching and gaseous losses of 160 kg N each (Table 2.1, column C) are already quite high and larger than presently accepted by the EU.

The negative and positive balances certainly ask for action; also the signal is not necessarily at green for the zero balance for arable land in The Netherlands. To improve soil fertility in Rwanda, inputs must increase or losses decrease. The effects of erosion control differ from those achieved upon reduction of leaching and gaseous losses. Less erosion would imply a less negative balance but probably no immediate yield increase. On the other hand, saving N from leaching and volatilization may result in more absorption by the crop and hence higher yields, but not necessarily in a less negative balance. Similar differences exist among the inputs. More sedimentation would imply a less negative balance but probably no immediate increase in yield, while extra N in fertilizers may be allocated to leaching, volatilization and absorption by the crop, resulting in higher yields but not in a less negative balance. Only in the case where a portion of incoming nutrients is stored in the soil, will the balance become less negative.

The outputs by leaching and gaseous losses of the grassland budget (column C) must be lowered. It has already been experienced in practice that a single measure will not work. By reducing only the leaching losses, gaseous losses may increase. Reduction of both losses theoretically makes more nutrients available for grass production, but actual grass production was probably close to potential production, given the high supplies of N. This means that the effect of reduced N losses is not extra grass production but more accumulation of soil organic N. Diminishing fertilizer application alone will most likely result in lower losses as well as in lower yields. To tackle the problem, fertilizer application and gaseous and leaching losses must be reduced at the same time, which may be achieved, for example, in a system of zero grazing.

The above examples have made clear that it is not justified to use the nutrient balance, alone or in relation to the nutrient stock, as a yardstick for sustainability, or as a guideline to set targets for the future. Not only when the balance is negative or positive, but even when it is neutral, it may represent a non-sustainable situation. As indicated by Smaling *et al.* (1997), nutrient balances should be treated as awareness raisers and are meant to alert policy-makers. At farm level, budgets and flow diagrams are helpful as farm management tools. These are very meritorious functions of nutrient budgets, but do not promote them to absolute criteria for sustainable land use.

The various INs and OUTs of nutrient budgets as presented in the study by Stoorvogel and Smaling (1990) refer to the flows of all nutrients and not of available nutrients only. The budgets were meant to demonstrate soil nutrient depletion, which is caused by the net output of both available and non-available nutrients. A framework for nutrient management aiming at sustainable soil fertility, however, requires a distinction between available and not immediately available nutrients.

NUTRIENT FLOWS AND NUTRIENT AVAILABILITY

In this chapter, available nutrients are conceived as the nutrients that are present in the soil solution at the beginning of the growing season or will enter the soil solution during the season. In general, OUTs 1 to 4 are flows consisting solely of available nutrients. OUT 5 comprises flows of nutrients that are not immediately available, because they are present in solid organic and inorganic particles (erosion), and flows of dissolved and hence available nutrients (runoff). For the inputs the situation is more complex than for the outputs (Table 2.2). The nutrients of IN 4 are available to the crop, as far as it concerns symbiotic biological N-fixation (SBNF), but N fixed by free-living N-fixers (FBNF) is to be considered as soil organic N and is not immediately available to crops. Wet deposition of IN 3 brings in available nutrients, but the nutrients of the dry deposition are not immediately available. The major flow of IN 5 is sedimentation (not immediately available) and a minor flow is runon (available). The nutrients of IN 1 (chemical

Table 2.2. Estimated values of the fraction of available nutrients (f_{ai}) in the various INs and OUTs. See Table 2.1 for explanation of INs 1–5 and OUTs 1–5.

Nutrient	IN 1	IN 2	IN 3	IN 4	IN 5
N	1.0	0.4	1.0	0.9	0.1
P	0.1	0.1	0.5	n.a.[a]	0.0
K	1.0	1.0	0.5	n.a.	0.1
	OUT 1	OUT 2	OUT 3	OUT 4	OUT 5
N	1.0	1.0	1.0	1.0	0.1
P	1.0	1.0	1.0	n.a.	0.0
K	1.0	1.0	1.0	n.a.	0.1

[a]n.a. = not applicable.

fertilizers) and IN 2 (organic manures) are partly or entirely in available form. The N and K in chemical fertilizers and the K in organic fertilizers are usually available. The availability of N and P in organic fertilizers is affected by weather conditions, length of growth season, soil life, and type of manure. The P from water-soluble P fertilizers dissolves easily. Nevertheless, according to our operational definition, only a fraction between 0.05 and 0.2 can be considered as directly available, because most of the dissolved fertilizer P is immediately sorbed on to soil particles.

Each nutrient flow may be split into a fraction f_a of available nutrients and a fraction $(1-f_a)$ of nutrients that are not available within the time step considered. The nutrients that are not immediately available will henceforth be denoted by 'NIA nutrients'. The nutrient budget may also be split into two budgets. The balances of these budgets are indicated by BALAV for available nutrients and BALNIA for NIA nutrients.

$$\text{BALAV} = \sum_{i=1}^{5} f_{ai}\text{IN}_i - \sum_{i=1}^{5} f_{ai}\text{OUT}_i \qquad (2.1)$$

$$\text{BALNIA} = \sum_{i=1}^{5} (1-f_{ai})\text{IN}_i - \sum_{i=1}^{5} (1-f_{ai})\text{OUT}_i \qquad (2.2)$$

The sum of the two is of course:

$$\text{BALNUT} = \sum_{i=1}^{5} \text{IN}_i - \sum_{i=1}^{5} \text{OUT}_i \qquad (2.3)$$

The values of f_{ai} for the various INs and OUTs presented in Table 2.2 must be considered as default values, based on present knowledge. If, in a certain area, other values have experimentally been found, it is better to use such local data.

Table 2.3. Summaries of budgets for available and not immediately available (NIA) nitrogen in kg ha^{-1} year^{-1}. A, arable land in Rwanda; B, arable land in The Netherlands; C, grassland in Britain.

	A	B	C
Available			
Total inputs	15.1	236	449
Total outputs	45.9	250	370
Balance	−30.8	−14	79
NIA			
Total inputs	3.9	14	1
Total outputs	26.1	0	0
Balance	−22.2	14	1
Balance of all nutrients	−53.0	0	80

With the values of f_{ai} for N (Table 2.2), the budgets of Table 2.1 were split into budgets for available N and NIA-N. Table 2.3 shows that practically all N of the grassland budget (C) is available. The surplus of available N is converted into soil organic N, implying an accumulation of NIA-N. A different picture is found in the arable land of The Netherlands: a positive budget, that is accumulation of 14 kg of NIA-N, and an equally large negative budget for available N. Yet the pool of available N will not decrease by 14 kg year^{-1}, and that of NIA-N not increase by 14 kg year^{-1}, because NIA-N is converted into available N by the process of mineralization. Separate budgets of the pools of available N and NIA-N, as discussed below, would show that the negative balance of available N may be compensated for by an extra inflow of mineralized NIA-N, equal to the positive balance of the NIA-N.

In budgets B and C of Table 2.3, available N is far more important than NIA-N for two major reasons: (i) chemical fertilizers, entirely consisting of available N, are by far the largest input; (ii) there is no erosion and hence no output of NIA-N. Quite contrasting is the situation for Rwanda (A), where both balances, for available N as well as for NIA-N, are negative. Repair of the budget of NIA-N may be brought about by erosion control, higher application of organic manure or more sedimentation. This will not immediately give relief to farmers, in contrast to adjustment of the budget of available N, for which application of fertilizer and manure, and introduction of leguminous crops to increase symbiotic biological N-fixation (SBNF) are the most appropriate measures.

So, budgets of available nutrients can be used to predict whether under the prevailing farming management yields can be maintained at the present level, and budgets of NIA nutrients to predict whether the total nutrient store in the soil will change, and yield levels will be maintained in the long run.

RELATION OF NUTRIENT BALANCE TO SOIL NUTRIENT STOCK, A POSSIBLE LAND QUALITY INDICATOR?

As negative nutrient balances are paid by the soil, it is useful to compare their values with the soil nutrient stock. It may be questioned whether different stocks must be used for available, NIA and total nutrients. It is obvious that BALNUT (equation 2.3) is best compared to the total stock of nutrients. There are no principal analytical difficulties in measuring the total stock, but usually it is not a part of the standard package of soil testing laboratories.

It is obvious too that BALNIA (equation 2.2) has to be compared with the stock of NIA nutrients. This stock is practically the same as the total stock. Unlike the situation for available nutrients (see below), the ratio of BALNIA to the nutrient stock cannot be considered as a constant soil property, because the output of NIA nutrients, erosion, is not only a function of soil processes, but depends also on external factors. Nevertheless, the ratios of BALNIA to nutrient stock, or of erosion to nutrient stock, may give interesting information on the seriousness of erosion.

In accord with the operational definition of available nutrients as the nutrients present in the soil solution (SSOL) at the beginning of the growing season or entering SSOL during the season, SSOL must be considered as the stock of available nutrients. For two reasons, however, SSOL cannot act as nutrient stock, at least not at the scale of annual or seasonal nutrient budgets of fields. The first reason is the inappropriate spatial scale. It is true that the INs and OUTs of *available* nutrients of the budget at field scale move into and out of SSOL, but there are also nutrient fluxes into and out of SSOL which should be considered as internal flows at field scale: mineralization of soil organic matter, weathering (dissolving) of minerals, desorption of sorbed ions at the input side; microbial immobilization, chemical precipitation and sorption at the output side. The time steps of some of these processes are expressed in days, hours and seconds rather than in growing seasons or years, and this inappropriate time-scale is the second reason why SSOL cannot act as nutrient stock. The actual nutrient content of SSOL at a certain point in time often is (very) small compared to the fluxes of nutrients into and out of SSOL during a growing season.

It follows that the balance of available nutrients (BALAV) is reflected in changes of the stock of available nutrients (ΔSSOL) as well as in changes in the stock of NIA nutrients. This means that BALAV, just like BALNUT and BALNIA, may be compared to the total stock of nutrients.

The sum of fluxes between the soil stocks of NIA and available nutrients, so the net rate of conversion of NIA nutrients into available nutrients is further denoted by NRM, where M stands for mobilization. The magnitude of NRM depends mainly on the quantities and qualities of soil organic matter and nutrient containing minerals, and can be considered as a constant inherent soil property, apart from seasonal or annual fluctuations.

Not only part of soil NIA nutrients may become available during the time step, also a fraction of the NIA nutrients in the INs may be mobilized. On the other hand, some of the incoming available nutrients ($\Sigma f_{ai} IN_i$) may be immobilized by conversion into NIA nutrients. For the purpose of this chapter, it is assumed that both conversions have already been taken into account in the values of f_{ai}. All together, OUTs of available nutrients may originate from three sources: ΔSSOL, NRM, and $\Sigma f_{ai} IN_i$. Hence the change in soil nutrients (ΔSSOL + NRM) is the opposite of BALAV:

$$\Delta SSOL + NRM = - BALAV \qquad (2.4)$$

In principle, equation (2.4) is generally applicable, but unfortunately the values of f_{ai} are not always known in practice. When the INs of available nutrients are substantial, and the value of BALAV becomes less negative or even positive, it is hardly justified to use equation (2.4). An example of such a case is budget C in Table 2.3.

In many sub-Saharan African countries, the balances of both the available and the NIA nutrients, are negative. Smaling *et al.* (1997) have discussed the possible use of the ratio of the (negative) balance to the nutrient stock as a 'land quality indicator'. They concluded that only the ratio of the balance of available nutrients to total nutrient stock could be meaningful. From the above it follows that the balance must first be corrected for the conversion of incoming available nutrients into NIA nutrients, and that the absolute value of the ratio of the corrected balance to nutrient stock is equal to the sum of the net mineralization and weathering fractions. Some OUTs of the budget are calculated as a function of mineralization and weathering which are not usually measured but estimated from literature data (Stoorvogel and Smaling, 1990). Given these links and at times rather crude estimates, it does not make much sense to use nutrient balances as a possible land quality indicator.

SOIL NUTRIENT POOLS UNDER STEADY STATE AND EQUILIBRIUM

Analyses of the total stock of nutrients in the soil and the nutrients in the soil solution (SSOL) do not belong to the common set of soil tests. The analyses that are usually called tests for 'available' nutrients do not refer to nutrients in the soil solution, but to nutrients in labile pools, that is, to nutrients at sorption or exchange sites for anions and cations. Nutrients continuously move between sorption sites and solution, in equilibrium moving

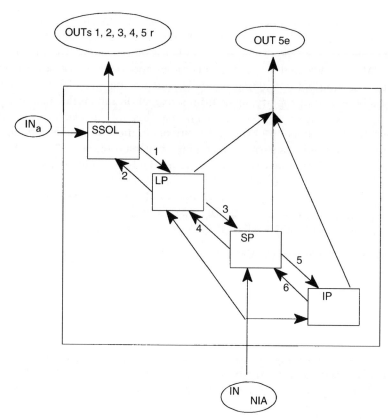

Fig. 2.1. Nutrient INs and OUTs, and nutrient pools and fluxes in the soil system. The numbers alongside the arrows refer to SIFs, i.e. soil internal flows between the various pools in the soil. SSOL stands for soil solution, LP, SP and IP for labile, stable and inert pool, respectively. The subscripts a and NIA refer to available and not-immediately available nutrients. OUT 5e and OUT 5r are erosion and runoff, respectively.

at equal rates in both directions. These and the other flows in the soil are henceforth denoted by soil internal flows (SIFs). Nutrient uptake, changes in soil moisture content, fertilizer application, and so on, very frequently change the magnitudes of the internal flows.

Apart from a labile pool, a stable and an inert pool may be distinguished, as schematically shown in Fig. 2.1. Table 2.4 summarizes the main characteristics of soil nutrient pools. In our concept, soil NIA nutrients change into available nutrients when they move from the labile pool into SSOL. This soil internal flow is SIF 2 in Fig. 2.1. By definition, the nutrient balance is zero in steady state, and the sizes of the various pools remain constant. This means that for each pool the sum of inputs must equal the sum of outputs. Because pools and fluxes remain constant in steady state,

Table 2.4. Description and main characteristics of soil nutrient pools, as conceived in this chapter.

Name	Main soil compartment	Indicative residence time
Available	Soil solution	< 1 year/season
Labile	Adsorption sites 'Free' organic matter	1–10 years
Stable	Clay–humus complex 'Easily' weatherable minerals	10–100 years
Inert	Very old SOM[a] Minerals	> 100 years

[a]SOM = soil organic matter.

the fluxes per time step are fixed fractions of the pools. Pools for NIA nutrients are the labile pool (LP), the stable pool (SP) and the inert pool (IP), and the only pool for available nutrients is the soil solution (SSOL) (Fig 2.1). The fluxes to and from the pools SSOL, LP, SP and IP are given in Equations 2.5–2.8, respectively. $OUT_{5e,LP}$, etc. stands for erosion of the labile pool, and so on. The numbers of SIFs are shown in Fig. 2.1.

$$\sum_{i=1}^{5} f_{ai} \, IN_i + SIF\ 2 = \sum_{i=1}^{5} f_{ai} \, OUT_i + SIF\ 1 \qquad (2.5)$$

$$\sum_{i=1}^{5} (1-f_{ai}) \, IN_{i,LP} + SIF\ 1 + SIF\ 4 = OUT_{5e,LP} + SIF\ 2 + SIF\ 3 \qquad (2.6)$$

$$\sum_{i=1}^{5} (1-f_{ai}) \, IN_{i,SP} + SIF\ 3 + SIF\ 6 = OUT_{5e,SP} + SIF\ 4 + SIF\ 5 \qquad (2.7)$$

$$\sum_{i=1}^{5} (1-f_{ai}) \, IN_{i,IP} + SIF\ 5 = OUT_{5e,IP} + SIF\ 6 \qquad (2.8)$$

In equilibrium, which is a special case of steady state, the two fluxes of a pair of opposite fluxes have the same size. In Fig. 2.1, this means SIF 1 = SIF 2, SIF 3 = SIF 4, and SIF 5 = SIF 6. From Equations 2.5–2.8, it follows that the sum of INs of available nutrients to SSOL then equals the sum of OUTs of available nutrients from SSOL, and that for each of the pools LP, SP and IP, the sum of INs of NIA nutrients equals the erosion of the particular pool.

This makes equilibrium the steady state in which the ratio of the sizes of two pools is the reverse of the ratio of the two fluxes between the pools, expressed as fractions of the pools. For example: SIF 3 = $F_3 \times$ LP and SIF 4 = $F_4 \times$ SP. Because SIF 3 = SIF 4, it follows: SP/LP = F_3/F_4. The value of the leaving flux expressed as a fraction of the pool is approximately reciprocal to the residence time of the particular nutrient in the pool. The flows that leave LP are SIF 2 and SIF 3, and the corresponding fractions are F_2 and F_3.

The residence time of nutrients in LP is 1–10 years (Table 2.4), and hence the value of $F_2 + F_3$ is about 0.2 year^{-1}; for the time being we assume that $F_2 = F_3 = 0.1$ year^{-1}. The residence time of nutrients in SP is 10–100 years, and so $F_4 + F_5$ must be something like 0.02 year^{-1}. Assuming that they are equally important, it follows that $F_4 = F_5 = 0.01$ year^{-1}. So the ratio SP/LP would be 0.1/0.01 = 10. Taking 1000 years as residence time for IP, it follows that IP/SP = 0.01/0.001 = 10. Hence, IP : SP : LP = 100 : 10 : 1 in equilibrium. The total nutrient stock is then: $(1 + 10 + 100) \times$ LP = $111 \times$ LP, or LP/total stock is about 0.01.

These are realistic values. In the case of potassium (K), we may read exchangeable K for LP, fixed K for SP and feldspar K for IP; they are 1–2%, 1–10% and 90–98% of total K, according to Fig. 4-4 (page 166) of Follet et al. (1981). For phosphorus we may consider P-Olsen corresponding to the labile pool. Van der Eijk (1997) found ratios of P-Olsen/total P varying from 0.003 to 0.03 for non-fertilized soils ranging from very poor to extremely rich in phosphorus, and the value of 0.01 for LP/total stock is within this range. For N the labile pool is more than 1% of total N, as follows from annual mineralization rates of 2–8%, but there is no suitable simple chemical method to determine it.

So far, we have not yet considered SSOL. It is fed by SIF 2 and Σf_{ai} IN$_i$. The outgoing nutrients are allocated to OUTs 1–4, if there is no runoff (OUT 5r), and SIF 1. Assuming that the distribution is 60% to OUT 1 + 2, 20% to leaching (OUT 3), and 10% to each of OUT 4 and SIF 1, and that the sizes of LP and SP are 100 and 1000 units, respectively, it follows that in equilibrium SIF 1 = SIF 2 = 10 units year^{-1}, SSOL = 100 units, the values for OUT 1, OUT 2, OUT 3, and OUT 4 are 50, 10, 20 and 10 units year^{-1}, respectively, and Σf_{ai} IN$_i$ is 90 units year^{-1}. The above calculations give scope for renewed interpretation of soil analytical data.

TOWARDS 'TARGET' VALUES FOR INPUTS, OUTPUTS, YIELDS AND SOIL FERTILITY

If any emission of nutrients is considered an undesired burden to the environment, the 'perfect agroecosystem' would be a steady-state system where no nutrient losses do occur, and the quantity of nutrients added (Σ IN) is sufficient to maintain soil fertility at the particular level at which crop growth is not limited by nutrient deficiency and crops can just absorb all available nutrients. This implies that: (i) the yield is the maximum obtainable with the given cultivar, and under the prevailing climate and soil physical conditions; (ii) the nutrient balance is zero; and (iii) there are no other outputs than OUTs 1 and 2:

$$\sum_{i=1}^{5} \text{IN}_i = \text{OUT 1} + \text{OUT 2} \qquad (2.9)$$

Soil fertility level is then optimum. The quantity of input nutrients (Σ IN), the yield level, the sum of OUTs 1 and 2, and the soil fertility level of this 'perfect agroecosystem' are henceforth denoted by target input (TI), target yield (TY), target output (TO) and target soil fertility (TSF), specified as TSF_N, TSF_P and TSF_K, for N, P and K, respectively. Because TY changes with soil physical conditions, climate and cultivar, also TO, TSF and TI vary; they may be higher for clayey than for sandy soils, in tropical highlands than in tropical lowlands, and for modern than for traditional cultivars. If soil fertility level (SFL) deviates from TSF, it is still possible to get the target yield (TY) without any nutrient emission to the environment, but that requires another Σ IN than TI. The Σ IN that is needed to produce the target yield at any SF we call the 'Model Σ IN'. It must result in a nutrient output equal to TO, consisting of OUTs 1 and 2 only.

As said before, outgoing nutrients partly originate from the soil nutrient stock and partly from INs. In graph A of Fig. 2.2, the nutrient output from the soil nutrient stock is assumed to be linearly related (sloping bold line) to soil fertility level (SFL). The horizontal dotted line represents Σ OUT equal to TO (solely consisting of OUT 1 + OUT 2). Where this horizontal line intersects the sloping bold line, SFL is equal to SSF (standing for 'saturated soil fertility'), and all nutrients needed for TO can be derived from the soil alone. Where SFL is above SSF, the nutrient supply from the soil exceeds TO, and a part of the nutrients is emitted to the environment as OUTs 3, 4, 5. Where SFL is below SSF, the nutrient supply from the soil is not sufficient for TO and a part of the nutrients for these OUTs 1 and 2 has to be derived from Σ IN.

In graph B of Fig. 2.2, the 'Model Σ IN' is represented by the sloping bold line that intersects the x-axis at SSF. It is inversely proportional to SFL. Theoretically, it must be negative where SFL is above SSF, to avoid emission of nutrients (see graph A). Also in graph B, the horizontal dotted line represents the quantity of nutrients equal to TO. Where the horizontal line intersects the sloping line, Model Σ IN equals TI (= TO), and SFL is equal to TSF; it is the situation of the 'perfect agroecosystem'. Where SFL is less than TSF, Model Σ IN is higher than TI.

The INs are allocated to crop and soil. The INs allocated to soil may be used entirely for 'soil fertility maintenance', that is, to keep SFL at TSF; this is the case when SFL is between TSF and SSF. When SFL is less than TSF, a part of the INs allocated to soil is spent on 'soil fertility repair'; this repair implies a heightening of SFL from its actual level to TSF; it is a process of net accumulation of nutrients in the soil. It may take several years before SFL has reached TSF. Soil fertility repair is a form of 'gradual recapitalization'; compare Fig. 1.9 in van Noordwijk (Chapter 1, this volume). Where SFL is higher than TSF, Model Σ IN is less than TI; then the INs are smaller than the OUTs and the soil is mined.

The consequence of soil repair and soil mining is that, irrespective of the original value of SFL, future values gradually approach TSF, as schematically indicated by the lines for years 1, 2, and 3 in graph C. The line for year 0 is of course the 1 : 1 line.

INs may be distinguished into controllable and uncontrollable INs (see below). The uncontrollable INs are indicated by the interrupted horizontal line in graph B; between this line and the line for Model Σ IN are the controllable INs which have to be added by the farmer.

The Model Σ IN is such a quantity that all INs are spent in a useful way (crop, soil fertility repair and maintenance), and none are lost. Not shown in graph B is what happens if the actual Σ IN is lower or higher than the Model Σ IN: if it is lower, yields will be less than the target yield and maintenance and repair of soil fertility will be less than needed, and if it is higher the superfluous nutrients are lost from the soil or accumulate in the soil.

In the 'perfect agroecosystem', OUT_{5e} of equations (2.6) to (2.8) is zero. When the INs of NIA nutrients ($\Sigma (1 - f_{ai}) IN_i$) are negligible, which is practically true in many situations, only the SIF terms are remaining. Thus SIF 5 becomes equal to SIF 6 (equation 2.8), and hence SIF 3 = SIF 4 (equation 2.7), and SIF 1 = SIF 2 (equation 2.6). Such a system is the equilibrium, as explained above, with IP/SP = F5/F6 and SP/LP = F3/F4.

Sustainability can only be evaluated if higher hierarchical levels than the system under consideration are also taken into account (Fresco and Kroonenberg, 1992). For an agroecosystem, the 'environment' is (part of) a higher hierarchical level. The situations described for 'Model Σ IN' (graph B of Fig. 2.2) seem sustainable because no nutrients are emitted to the environment. It is not always possible to manage nutrients in such a model way. Moreover, it is questionable whether agroecosystems are not allowed to have any emission to be sustainable. Emissions are not always undesired, but may even be necessary inputs to keep the 'environment' sustainable. Theoretically the emissions from the agroecosystem to the environment should equal the inputs in the reverse direction (IN 3 + IN 4 + IN 5 = OUT 3 + OUT 4 + OUT 5). This may justify the use of 'acceptable levels of losses' in practical policy (Oenema et al., 1997).

SOIL NUTRIENT POOLS UNDER NON-STEADY STATE

If the balance of INs and OUTs of available nutrients (BALAV, equation 2.1) is negative, SSOL and, as a consequence SIF 1, will go down. This leads to a reduced size of LP. The decrease in LP results in lower values of SIFs 2 and 3, and by that in a further decrease of SSOL and in a decrease of SP. If LP : SP : IP = 1 : 10 : 100 (see above), the relative reduction of SP is approximately one-tenth the relative reduction of LP. As a result of the smaller SP, SIF 5 becomes smaller and hence also IP. The relative reduction of IP is approximately one-tenth the relative reduction of SP and 0.01 times that of LP. The reduction of IP lags behind that of SP, which lags behind that of LP,

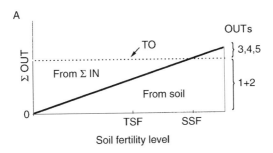
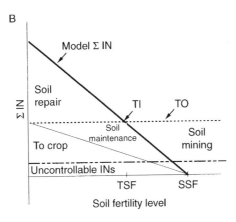
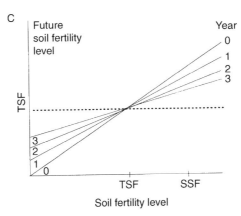

Fig. 2.2. Relations between soil fertility level and (A) soil-derived nutrient outputs (bold line), and required nutrient inputs to reach the target output (TO); (B) Model Σ IN needed to obtain the target yield, and allocation of inputs to crop, soil maintenance and soil repair; (C) soil fertility level after 1, 2 or 3 years of Model Σ IN of graph B.

which in turn lags behind that of SSOL. These lags and the ever smaller relative reductions (SSOL > LP > SP > IP) give the soil its buffering capacity against nutrient depletion and nutrient accumulation. Because crop yields are directly related to SSOL, crop response to a change in the INs and OUTs of available nutrients is much more pronounced than crop response to a change in the INs and OUTs of NIA nutrients.

In non-steady situations, SIFs cannot be calculated as simply as in steady state, where the outgoing SIFs per year are constant fractions of the pools to which they relate. In reality some SIFs are mainly driven by microbial activity, others by electrochemical gradients. Because of the complicated feedback mechanisms, the relations between fluxes and pools and among pools do not remain constant and are more difficult to describe quantitatively. Nevertheless we made some simple calculations to show the effects changing inputs have on crop yields and on the ratio of the labile pool (LP) to the total stock of nutrients, in other words to show the effects of disturbances of the equilibrium situation. The pool sizes in the original equilibrium situation were set at 100, 100, 1000 and 10,000 for SSOL, LP, SP and IP, respectively. As above, $\Sigma f_{ai} IN_i$ was 90 units year^{-1}, SSOL was 100 units, and the values for OUT 1, OUT 2, OUT 3 and OUT 4 were 50, 10, 20 and 10 units year^{-1}, respectively. Calculations were made for the following cases assuming a change of the inputs (one at a time) of 50 units year^{-1} by:

A   decrease of INs of available nutrients to SSOL, e.g. by reducing fertilizer N application;
B   addition to IP of dust or volcanic ash;
C   addition to SP of sparingly soluble fertilizer, e.g. rock phosphate;
D   addition to LP of easily soluble (P) fertilizer, e.g. superphosphate, assumed to be sorbed;
E   addition to SSOL of easily soluble fertilizer, e.g. N fertilizers.

Table 2.5 shows the relative values (rounded) of crop yields during the first year and of the ratio of labile pool to total stock at the end of the first year of the changes A to E. Relative means in relation to the equilibrium situation which was set at 100%. For the relative values of crop yields, the relative values of OUT 1 were calculated; they are equal assuming that OUT 1 and yield are proportional.

Although the calculations are over-simplified, we may draw some evident, important conclusions. Of course, yields increase from A to E, and are not noticeably affected by extra input to SP or IP. Compared to their values in equilibrium, the values of LP/total nutrient stock are for the cases A to E:

A   lowered when the balance of available nutrients is negative;
B   unchanged when the balance of NIA nutrients is positive as a result of addition of dust;

Table 2.5. Relative values (rounded) of the yield during the first year, and of the ratio of labile pool to total stock at the end of the first year after the indicated nutrient flows have been changed by 50 units year^{-1}. Values for equilibrium situation are set at 100. For an extensive description of the codes A–E see text.

Code	Changed nutrient flow Short description	Yield	LP/total stock
A	Less fertilizer N	50	95
B	Dust, volcanic ash	100	100
C	Rock phosphate	100	100
D	Superphosphate	105	139
E	More fertilizer N	150	105

C practically unchanged when the balance of NIA nutrients is positive as a result of application of sparingly soluble fertilizers;

D strongly raised when the balance of available nutrients is positive as a result of application of soluble nutrients that are sorbed;

E raised when the balance of available nutrients is positive as a result of application of soluble nutrients.

Only in case D does the ratio LP/total nutrient stock change considerably. This case represents the usual situation after application of soluble phosphorus fertilizers. The ratio LP/total nutrient stock may therefore be used to indicate depletion or enrichment of the agroecosystem, in other words negative or positive nutrient balances. It would be very valuable if chemical soil analysis could be used for that purpose.

CHEMICAL SOIL ANALYSIS FOR SUSTAINABILITY

An essential condition for the use of chemical soil analysis in nutrient balance studies is that the various pools can be identified. For that purpose, several quite successful soil P fractionation procedures have been developed in the past (Hedley *et al.*, 1982; Tiessen *et al.*, 1984; Beck and Sanchez, 1994), but they are too laborious for routine analyses. For labile phosphorus, two frequently used methods are those of Olsen and Bray-I. Usually their values are of the same order of magnitude, but the relation is affected by soil properties such as pH and the presence of oxides and carbonates. The method of Dabin is roughly a combination of the two and is quite frequently applied in West Africa. It extracts approximately twice as much P as either Olsen or Bray-I. Each of the methods is used in the literature to denote 'available P' (with a different meaning than in our concept), and requires calibration for different agroecosystems. Table 2.6 lists some values of their relation to total P, as derived from literature.

In Table 2.7, it is attempted to relate the nutrient balance to the ratios P-Olsen/total P and exchangeable K/CEC. It may be regarded as a

Table 2.6. Labile P as a percentage of total P, according to three analytical methods, and some details on agroecosystem and location.

Method	LP/total P	Details	Location	Reference
Olsen[a]	0.3–0.6	P-limited	Kenya, Kisii	Van der Eijk, 1997
	0.6–3.3	Weathered sandy	Côte d'Ivoire	Van Reuler, 1996
	0.8–4	N-limited, weathered soils	Kenya Coast	Smaling and Janssen, 1987
	1.8–3.5	P-rich	Kenya, Kisii	Van der Eijk, 1997
	4–9	600 DAFA[d]		
	6–13	20–60 DAFA		
	13	Long-term trials	Rothamsted	Johnston, 1996
Bray-I[b]	1.0–5.9	Weathered sandy	Côte d'Ivoire	Van Reuler, 1996
	2.5–5	Loamy sand, forest	Suriname	Boxman and Janssen, 1990
	5–9	Loamy sand, 5–10 years of fertilizer P		
Dabin[c]	2.1–23.2	Weathered sandy	Côte d'Ivoire	Van Reuler, 1996
	5.5–11.0	Weathered sandy secondary vegetation	Côte d'Ivoire	Fritsch, 1982 cited in Van Reuler, 1996

[a] 0.5 M $NaHCO_3$
[b] 0.03 M NH_4F + 0.025 M HCl
[c] 0.5 M $NaHCO_3$ + 0.5 M NH_4F
[d] DAFA = days after fertilizer application; soils samples were taken just below the spots where fertilizer granules had been placed.

Table 2.7. Tentative interpretation of labile P (P-Olsen) as a percentage of total P, and exchangeable K as a percentage of CEC.

LP/total P (%)	Exchangeable K/CEC (%)	Nutrient balance	Comments
< 0.6	< 1	Strongly negative	Strongly depleted
0.7–1.5	1–2	Negative	Depleted
1.5–3	3–4	Neutral	Natural steady-state
3–6[a]	5–6[a]	Positive	Fertilized or other nutrient limiting
6–10	> 6	Strongly positive	Heavily fertilized
> 10		Long-term positive	Very heavily fertilized; risk of pollution

[a] Target soil fertility (TSF_P, TSF_K) is within these ranges.

preliminary diagnostic table for P and K. Nutrient balance refers to the balance in the recent history; the chemical data are a result of that past balance. Exchangeable K is often considered as the pool of available or labile K. It has been shown by Hemmingway (1963) among others, that crops may take up more K than is present as exchangeable K in some situations, and that in other situations the quantity of exchangeable K may be larger than can be absorbed by the crop. The availability of exchangeable K depends on its percentage of the total of exchangeable cations and on the type and quantities of the other cations, hence on CEC and pH. The interpretation of exchangeable K as percentage of CEC (relative K saturation) in Table 2.7 is derived from values mentioned in textbooks and the author's personal experience. Also the relation of exchangeable K to fixed K or to total K could give meaningful information, but fixed K and total K are seldom measured.

In Table 2.7, the values of LP/total P, and exchangeable K/CEC corresponding to target soil fertility (TSF_P and TSF_K) are found at (slightly) positive nutrient balances. This indicates that some soil enrichment was needed to bring soil fertility from 'natural steady state' to TSF. Once the soil is at TSF, a neutral balance must suffice to maintain TSF. It is the level needed for sustainability as discussed above. Levels of P and K below or above TSF are results of negative or strongly positive balances, both evidences of non-sustainable systems.

For N, the situation is more complicated. Many attempts have been made to measure 'mineralizable N', which could act as a measure of LP, but until recently these were in vain. New hope has come from Hassink (1995), who developed a number of indices based on the observation that soils in equilibrium have 'saturation' contents of organic N (org. N) and microbial N (MB-N) that are related to soil texture, i.e. to the clay + silt ($< 50 \mu m$) content. The N supply of (grassland) soils is related to the difference between that equilibrium content and the actual content (Δ org. N and Δ MB-N). The larger that difference is, the less saturated or the more depleted is the soil. Hassink (1995) gives data on saturation contents of organic N and microbial N for grasslands in The Netherlands, which could be considered as TSF_N values. Data for arable land are almost absent, certainly in the tropics. That is why it is not yet possible to relate them to nutrient balances and, hence, why no values of N-pools are provided in Tables 2.6 or 2.7.

The above relates to pools and flows of available nutrients. The output of NIA nutrients is by erosion. Because erosion is not directly related to soil nutrient pools, chemical soil analysis is not a suitable tool to indicate soil nutrient depletion by erosion. A discussion of procedures for the assessment of soil erodibility is beyond the scope of this chapter. We refer to, for example, Kilasara *et al.* (1995), who used topsoil depth as a criterion for the extent of past erosion.

BALANCED NUTRITION

Nutrients move through the soil into the soil solution through a chain of reactions (Fig. 2.1), which are mainly triggered off by a lowering of the nutrient concentration in the soil solution, for example, as a result of nutrient uptake by crops. When crop demand for a particular nutrient is low, for example because another nutrient is limiting crop growth, the soil solution is not 'emptied' with regard to the particular nutrient; the successive reactions slow down, and part of that nutrient remains in the labile and stable pools and does not move in the direction of the soil solution. That may be the reason for the relatively high ratios of LP/total P, found in Kenya Coast in soils that were N limited (Table 2.6).

After fertilizer application, crops are not always able to absorb the entire quantity of nutrients that is available during the growing season. Sometimes weather conditions make it impossible for the crop to use the available nutrients. Under optimum weather conditions, however, quite often only some of the nutrients are almost completely taken up and others are not. In such cases the supplies of the various nutrients are not well balanced. An example is given in Table 2.8, which summarizes data from relatively high-P and low-P soils in Kenya. When no P was applied, the uptake of N in the low P-Olsen soils was only one-quarter of the uptake found when P was applied, while in the relatively high P-Olsen soils the application of P did not promote the uptake of N. It is possible that less N was mineralized in the low P soils when no P was applied, but it is possible as well

Table 2.8. Nitrogen uptake by maize (kg ha^{-1}) as affected by phosphorus application on soils with different organic carbon and P-Olsen values (Janssen et al., 1990).

Field code	Organic C (g kg^{-1})	P-Olsen[b] (mg kg^{-1})	N-uptake (kg N ha^{-1})		Ratio −P/+P
			No P applied	Fertilizer P applied	
RG[a]	23	1.6	24	94	0.26
MK	11	2.6	30	80	0.38
IB[b]	35	2.4	87	153	0.57
SH	17	3.5	34	52	0.65
CS	5	4.4	27	41	0.66
MS	9	4.6	36	54	0.67
LS	22	4.5	42	42	1.00
MZ	5	5.1	34	30	1.13

[a]These soils received 80 kg fertilizer N ha^{-1}.
[b]P-Olsen values refer to soils that did not receive fertilizer P.

that the N that was not taken up has been leached, denitrified, volatilized and microbially immobilized, or has simply stayed in the soil solution.

If the various nutrients are not in balance with the crop's demand, and the superfluous amounts of a particular nutrient leave the solution by leaching or volatilization, the total output of that nutrient may not change compared to a balanced situation, but there is a shift from 'useful outputs' to 'non-useful outputs'. It is practically impossible to maximize the uptake of all nutrients at the same time. It is more likely that the total available quantity of only the most limiting nutrient is taken up, and less than the total quantity of others. It has been calculated that the nutrition of maize is perfectly balanced when the proportions of N : P : K in the crop are 7.8 : 1 : 5.6 (expressed in mass units). The uptake of each nutrient is then 96% of maximum uptake (Janssen *et al.*, 1994; Janssen, 1998). This implies that for the plant as well as for the environment balanced nutrition is optimum. Disproportions in nutrient supplies aggravate depletion of one nutrient and emissions to the environment of others.

CONTROLLABLE NUTRIENT FLOWS AND PARTIAL NUTRIENT BUDGETS

Only some nutrient flows can easily be controlled by the individual farmer. SIFs are mainly regulated by chemical and microbiological laws and can hardly and only indirectly be affected by humans; an example is the enhancement of the rate of mineralization by soil tillage. Among the INs, atmospheric deposition (IN 3) is beyond human control; farmers may play a role in capturing the incoming deposition, for example by planting trees, but in the nutrient budget the effect of such an activity is reflected in less runoff and erosion (OUT 5). Farmers have direct and complete command of IN 1 and IN 2 (mineral and organic fertilizers), and indirect and partial command of IN 4 (biological N-fixation) by planting leguminous crops. Checking of IN 5 (sedimentation and runon) and OUT 5 (erosion and runoff) usually requires joint action of the farmers of a village or watershed. OUTs 1 to 4 refer to nutrients that are made available by both autonomous processes in the soil (SIFs) and controllable and uncontrollable INs. Therefore, the farmer has not much influence on the total magnitude of OUTs 1–4, but he or she may at least partly and indirectly modify the allocation of nutrients among the various OUTs, via the choice of crops, the time of planting, the method and time of fertilizer application, the balance of nutrient supplies, the cultivation of catch and cover crops and so on.

Climate and relief are other determinants for the distribution of OUTs. Their influence is different for the different nutrients. A flat relief may give rise to waterlogging and hence reduced conditions. This enhances denitrification, and thus OUT 4 of N (Table 2.9). This process is not completely

Table 2.9. Relative importance of OUTs 4, 3 and 5 for N, P and K, per relief form. Climate is supposed to be humid.

OUT	Flat[a]	Subnormal	Normal	Excessive
Nitrogen				
4 Gaseous	***[b]	*	*	0
3 Leaching	*	***	**	0[b]
5 Erosion, runoff	0	0	**	****
Phosphorus[c]				
3 Leaching	**[b]	*	0	0
5 Erosion, runoff	0	0	*	**
Potassium				
3 Leaching	***	***	**	0[b]
5 Erosion, runoff	0	*	**	***

[a]Relief forms according to Soil Survey Staff (1962).
[b]See text.
[c]Losses of P are always far less important than those of N and K.

beyond human control, as was recently demonstrated by Sigunga (1997) who found that drains of 40 cm depth considerably curtail denitrification in flat Vertisols. Another consequence of anaerobic conditions is the change of ferric (hydr)oxides into the better soluble ferrous (hydro)oxides, whereby occluded P may be 'freed' and partly lost by leaching (OUT 3). On the other hand, leaching losses are small in hilly and mountainous terrain (in Table 2.9, relief form: excessive), as runoff dominates as the major loss pathway of available nutrients.

Leaching is of course more important in (sub-)humid climates than in (semi-)arid climates, but in the latter there may be leaching during the short, but often pronounced wet season. In (semi-)arid climates, wind erosion and runoff at the onset of the rains constitute the most important loss mechanisms (Table 2.10). It should be noticed that this table has an indicative meaning only and is restricted to normal relief.

It is not easy to assess nutrient flows, certainly so in low-data environments. Relatively easy to determine are the controllable INs 1 and 2 (mineral and organic fertilizers), and OUTs 1 and 2 (removed biomass in harvest and crop residues). All four flows are strongly human-influenced. Disadvantages of 'partial' nutrient budgets comprising only these four flows are that potentially important flows are ignored (Smaling et al., 1997). INs 1 and 2 form by far the major part of the INs in high external input agriculture (HEIA) without sedimentation (IN 5) and erosion (OUT 5). In such situations, BAL (1 + 2), i.e. the balance of (IN 1 + IN 2) − (OUT 1 + OUT 2) reflects the leaching and gaseous losses to the environment plus the accumulation of nutrients in the soil. In sustainable agriculture, soil fertility

Table 2.10. Relative importance of OUTs 4, 3 and 5 for N, P and K, per climate type. Relief is supposed to be normal.

OUT	Perhumid	Humid	Subhumid	Semi-arid
Nitrogen				
4 Gaseous	**	*	*	0
3 Leaching	***	**	*	0
5 Erosion	*	**	***	****
Phosphorus[a]				
3 Leaching	*	*	*	0
5 Erosion	*	**	**	***
Potassium				
3 Leaching	***	**	*	0
5 Erosion	*	**	***	****

[a]Losses of P are always far less important than those of N and K.

is at the optimum level (TSF) and yields are the maximum possible (TY) with the used cultivar, and hence there is no need to further increase the nutrient stock in the soil. If no emission to the environment is allowed, the value of BAL (1 + 2) must be zero, or even less to account for the nutrients added via atmospheric deposition and biological N-fixation. Higher fertilizer applications can only be permitted if the extra added nutrients are entirely utilized, and that implies the introduction of new, higher yielding cultivars with efficient root systems (Cassman *et al.*, 1995).

In low external input agriculture (LEIA), the partial nutrient balance is usually negative, and its significance as a system quality indicator is limited. Because fertilizer input is small, the 'natural' INs 3, 4 and 5 are relatively important and cannot stay out of sight. Moreover, under LEIA, soil fertility is often below TSF. For maximum nutrient use efficiency, soil fertility must be at TSF. If it is lower, part of nutrient inputs are used for soil fertility repair, requiring a positive value of the total nutrient budget, as explained above. Information on just the partial budget could be misleading, as a positive partial budget can still go together with a negative total budget.

In LEIA systems the reuse and recycling of biomass receive much emphasis. Their effects may be different for the different fields of a farm. They may show up as a decrease of OUT 2, for example when straw is worked into the soil instead of being removed from the field. In the case of prunings of improved-fallow trees, OUT 2 of (the parts of) the fields where the trees are growing and IN2 of the fields receiving the prunings increase at the same time. When such measures result in less leaching and gaseous nutrient losses from decaying organic material, the overall effect may be that the farm nutrient budget becomes less negative or more positive. The effect of recycling for the national nutrient budget and the national-level nutrient use efficiency may be considerable (van Noordwijk, Chapter 1, this volume).

EVALUATION OF THE SUSTAINABILITY OF SOME AGROECOSYSTEMS

Sustainable agriculture implies the succesful management of resources for agriculture to satisfy changing human needs while maintaining or enhancing the quality of the environment and conserving natural resources (TAC/CGIAR, 1989).

It is not possible to rank agroecosystems by sustainability or by nutrient balance. Apart from climate and relief, the economic environment of the farmers and their families, and above all their skill and zeal, have strong effects on the nutrient budgets. Therefore, similar agroecosystems may differ widely in nutrient balance values. Nevertheless, this chapter attempts to value a number of agroecosystems on the basis of criteria discussed above: nutrient balances, nutrient emissions, ratio of labile pool to total soil nutrient stock, balanced nutrition and nutrients expected to be in shortage or excess. The overall judgement is in terms of high, medium and low (H, M and L) separately for sustainability at system and 'environmental' level. In Table 2.11, the agroecosystems are arranged from left to right approximately in order of increasing intensity.

Shifting cultivation, mostly seen as a sustainable form of land use, may not be as efficient in nutrient use as suggested. The practice of burning causes volatilization of N and S. As long as the field cleared from forest or fallow vegetation is not yet covered by crops, a substantial part of the nutrients in the ashes from the burnt vegetation may get lost by leaching or erosion. Such losses are not very high when averaged over the whole fallow–cultivation cycle and expressed ha^{-1} $year^{-1}$, but that picture drastically changes when nutrient use efficiency is considered per unit of produce. Usually, only part of the vegetation is burnt, and only part of the nutrients in ash is taken up by the crops. The remaining nutrients stay in the ecosystem, provided the unburnt woody parts of the fallow vegetation are not removed from the field, and may gradually become available to next fallow vegetations and crops. Nevertheless, a major part of the nutrients never ends up in crops, but moves from fallow vegetation to fallow vegetation. That makes shifting cultivation not very efficient, if the nutrient use per unit of produce is taken as a criterion. Van Reuler and Janssen (1993) found that the effect of ashes on yield was noticeable during not more than three successive crops (upland rice – maize – upland rice). Between burnt and non-burnt plots, differences in N contents in the above-ground plant parts of the three consecutive crops together amounted to 7–9% of the N originally present in the fallow vegetation. The portions were 28–30% for P, 15–20% for K, 1–2% for Ca, and 8–9% for Mg. The unrecovered nutrients were still present in the unburnt parts

Table 2.11. Valuation of agroecosystems by various 'sustainability criteria'.

Criterion	Shifting	Perennials	Agroforestry	Annuals			
				LEIA	HEIA	Greenhouse	Bio-industry
BALAV[a]	–	–	–+	--0	++	++	++
BALNIA[a]	+	+	0	--0	0	0	++
BALNUT[a]	-0	-0	-0	-0	+	++	+++
LP/Total	LM	M	LM	L	MH	H	H
Emissions[b]							
Leaching	*	0	*	*	**	**	**
Gaseous	**	0	0	0	*	*	*
Erosion	*	0	*	0	**	0	0*
Bal.nutr.[c]	No	No	No	Var	Yes	Yes	No
Rel.deficit[d]	P	P	P	P	n.a.	n.a.	N
Rel.excess[d]	K	?	N	N	n.a.	n.a.	P
Sustainability[e]							
System	M	H	M	L	H	H	MH
Environment	M	H	MH	MH	LM	LMH	LM

[a] Nutrient balances for available, NIA and all nutrients. They may be negative, zero or (strongly) positive.
[b] Emissions may be absent (0), moderate (*) or severe (**).
[c] The agroecosystems may have (Yes) or may not have (No) balanced supplies of N, P and K, or it may vary within the system (Var).
[d] In cases where the nutrient supply is not balanced, the likely nutrient in relative short or excessive supply is mentioned; for balanced supply this is not applicable (n.a.).
[e] The overall evaluation of sustainability is made at the system level and at the higher hierarchical level of the environment. L, M and H stand for low, medium and high.

of the vegetation (stems and thick branches) and in poorly soluble (P) compounds in the ash, or had been lost during burning and by leaching from the ash. The nutrients Ca and Mg had a low recovery because they were not in short supply in the soil and therefore there was no need to add them via the ash. This shows that there is no balanced nutrition in shifting cultivation.

At first glance, systems with *perennials* seem to have the best chance of being sustainable, because the permanent ground cover reduces erosion and leaching. Tree crops usually have only modest outputs of nutrients in the harvested products, and hence do not need large inputs. Nutrients stored in the wood are saved from removal in harvested products or from leaching and volatilization losses. Moreover, leguminous trees add extra N to the system. This picture, however, is too bright because it overlooks the fact that trees are harvested at the end of their life cycle, for timber or fuel, and the stored nutrients leave the field. Another point is that some perennial crops have large seasonal outputs of nutrients; an example is the output via the harvested leaves of sisal. In Tanzania, soil fertility sharply declined under sisal, and fallowing could not adequately reverse the decline (Hartemink, 1995; Hartemink *et al.*, 1996).

Agroforestry in the form of planting of leguminous trees in combination with food crops, sometimes advocated as a cheap and environmentally sound substitute for fertilizers, works in that way only when N is the limiting nutrient. If P or another nutrient is limiting, leguminous trees may jeopardize the nutrition of food crops, because they compete for the limiting nutrient (besides the 'normal' competition for water and light). When P is limiting, legumes usually do not fix much N but take up N from the soil. Where they do fix N, the disproportion among the nutrients becomes more severe and so does the risk of N emission to the environment.

It is obvious that *continuous arable cropping*, especially of highly demanding vegetables and root and tuber crops, will give very negative nutrient balances and strong soil fertility depletion when no external inputs are added. Demanding, marketable crops are often the first to receive fertilizers. The sum of INs to such crops is often much larger than the sum of OUTs, and hence nutrients accumulate in the soil. This is especially the case for crops that are harvested when green, like leaf vegetables, tea, grass and silage maize, because the green, assimilating plants require high N concentrations in the soil solution up to the time of harvest. After harvest, there is a severe risk of N losses by leaching, and catch crops are needed to prevent or mitigate these losses. Far less risky is the cultivation of cereal crops because they practically stop the uptake of nutrients after flowering. It is feasible to organize fertilizer applications to these crops in a way that the soil solution is almost 'empty' after harvest.

Industrial agriculture does not necessarily create an environmental problem. Almost 100% of the added nutrients are recovered in the crops in greenhouse horticulture where nutrient solutions are recycled. If the nutrient circuit is not closed, however, nutrient emissions can be extremely high, making such agroecosystems unsustainable.

Bio-industry is known for wasting nutrients. The problem lies in the large quantities of manure, the nutrients of which have been derived from a large area outside the farm or even outside the country. The nutrients are concentrated in a much smaller area, and because the soil is unable to hold the total quantity, part is leached or escapes as gas. The same occurs in extensive grazing systems, when cattle stay overnight in corrals. Here the emitted nutrients are above all an unnecessary waste for the farmer, and of minor environmental concern because the density of corrals is low. In Western Europe, the bio-industry is locally so dense that the burden to the environment makes this agroecosystem non-sustainable.

Summarizing, ideally sustainable land use requires, as far as it concerns nutrients, an optimum soil fertility status (TSF), a permanent soil cover of growing plants, and an application rate of fertilizers that is exactly tailored to the crop's need and the soil's buffering capacity. When changing human needs require higher yields, the best option would be to introduce new cultivars utilizing nutrients more efficiently. Trying to increase yields of the current cultivars by extra fertilizer application inevitably creates pollution problems. There are circumstances where climate, topography, or economic conditions make it impossible to practice agriculture without emissions to the environment. For instance, farmers cannot avoid leaching losses of recently mineralized N in humid winters. In other cases, population growth compels farmers to cultivate land on steep slopes, making erosion unpreventable.

There are also situations where, despite favourable climatic, topographic and economic conditions, agriculture is practised in a non-sustainable way, because of lack of knowledge, interest and keenness of farmers or authorities. In such cases, farmers and authorities should at first be made aware of their malpractices. Usually they voluntarily improve their management, but if not other measures must be considered, as discussed by Kuyvenhoven *et al.* in Chapter 6.

FINAL REMARKS

The intuitive opinion that sustainable land use is secured when nutrient balances are zero is too simple. If the balance of a particular field is zero, its fertility does not change and at the field level the system seems sustainable. At a higher hierarchical level, the system is not sustainable if food production is too low to feed the population, or if a substantial part of the OUTs consists of losses (OUTs 3, 4 and 5). In that respect, it seems more justified, certainly in high external input agriculture, to demand a zero balance of the 'partial budget' consisting of only IN 1, IN 2, OUT 1 and OUT 2. There are farmers and politicians who consider this unfair, arguing that one has to accept some unavoidable nutrient losses. On the other hand, some INs (atmospheric deposition, N-fixation by free-living microorganisms) are unavoidable as well. Compared to the huge INs 1 and 2 and OUTs 1 and 2 found in Western Europe, the unavoidable INs and OUTs are small. Certainly when the balance of the partial budget is neutral and soil fertility is at the optimum level, the difference between unavoidable INs and OUTs cannot be large. Hence, the requirement of 'equilibrium fertilization' (IN 1 + IN 2 = OUT 1 + OUT 2), in other words of a zero partial balance, often is not unjustly severe. If some emissions of available nutrients are to be accepted, their levels must be in the order of size of the inputs by atmospheric deposition and biological nitrogen fixation.

The recovery of added fertilizer nutrients is frequently disappointingly low. Sometimes this can be ascribed to disproportional supplies of the various nutrients, sometimes to waterlogging and subsequent denitrification and leaching, sometimes to erosion and runoff, sometimes to low fertility level. In low fertility soils, a part of the added nutrients is spent on soil restoration instead of plant nutrition. Other possible causes of low recovery in crops are the use of low-potential cultivars, too wide spacing of plants, or inappropriate methods of fertilizer application. Balanced nutrition is a prerequisite for efficient nutrient use, and this implies that farmers must be able to buy the right cocktail of fertilizers. Poor combination of fertilizer stocks, as is often seen in the tropics, certainly contributes to nutrient wastes.

This chapter advocates the use of the concepts of labile, stable and inert nutrient pools, and optimum or target soil fertility (TSF), if wished distinguished into TSF_N, TSF_P, TSF_K, etc., and related target yield (TY) and target input (TI). Interpretation of chemical soil analysis should be guided in this direction. As is already done in, for example, The Netherlands, fertilizer recommendations should strive at TSF, either by adding extra fertilizer if the soil is depleted, or by reducing the fertilizer rate if actual soil fertility is higher than TSF. Once TSF has been reached, the practice of 'equilibrium fertilization' and the demand of zero partial nutrient balances make sense.

REFERENCES

Beck, M.A. and Sanchez, P.A. (1994) Soil phosphorus fraction dynamics during 18 years of cultivation on a Typic Paleudult. *Soil Science Society of America Journal* 58, 1424–1431.

Boxman, O. and Janssen, B.H. (1990) Availability of nutrients and fertilizer use. In: Janssen, B.H. and Wienk, J.F. (eds) *Mechanized Annual Cropping on Low Fertility Acid Soils in the Humid Tropics: a Case Study of the Zanderij Soils in Suriname*. Wageningen Agricultural University Papers 90–5, Wageningen, pp. 73–99.

Cassman, K.G., Steiner, R. and Johnston, A.E. (1995) Long-term experiments and productivity indexes to evaluate the sustainability of cropping systems.

In: Barnett, V., Payne, R. and Steiner, R. (eds) *Agricultural Sustainability in Economic, Environmental and Statistical Considerations*. John Wiley & Sons, London, pp. 241–244.

Follet, R.H., Murphy, L.S. and Donahue, R.L. (1981) *Fertilizers and Soil Amendments*. Prentice-Hall, New Jersey, pp. 163–170.

Fresco, L.O. and Kroonenberg, S.B. (1992) Time and spatial scales in ecological sustainability. *Land Use Policy*, July, 155–168.

Hartemink, A.E. (1995) *Soil Fertility Decline under Sisal Cultivation in Tanzania*. ISRIC Technical Paper No 28, Wageningen, 67 pp.

Hartemink, A.E., Osborne, J.F. and Kips, Ph.A. (1996) Soil fertility decline and fallow effects in Ferralsols and Acrisols of sisal plantations in Tanzania. *Experimental Agriculture* 32, 173–184.

Hassink, J. (1995) Organic matter dynamics and N mineralization in grassland soils. PhD thesis, Wageningen Agricultural University, Wageningen, 250 pp.

Hedley, M.J., White, R.E. and Nye, P.H. (1982) Plant induced changes in the rhizosphere of rape (*Brassica napus* var. Emerald) seedlings. III. Changes in L value, soil phosphate fractions and phosphatase activity. *New Phytologist* 91, 45–56.

Hemmingway, R.G. (1963) Soil and herbage potassium levels in relation to yield. *Journal of Science Food and Agriculture* 14, 188–196.

Janssen, B.H. (1992) *De Betekenis van Nutriëntenbalansen en Gebalanceerde Plantevoeding voor Duurzaam Landgebruik in de Tropen*. Verslagen en Mededelingen 1992–3. Vakgroep Bodemkunde en Plantevoeding. Landbouwuniversiteit, Wageningen, 60 pp.

Janssen, B.H. (1998) Efficient use of nutrients: an art of balancing. *Field Crops Research* 56, 197–201.

Janssen, B.H., Guiking, F.C.T., Van der Eijk, D., Smaling, E.M.A., Wolf, J. and Van Reuler, H. (1990) A system for quantitative evaluation of the fertility of tropical soils (QUEFTS). *Geoderma* 46, 299–318.

Janssen, B.H., Braakhekke, W.G. and Catalan, R.L. (1994) Balanced plant nutrition: simultaneous optimization of environmental and financial goals. In: Etchevers B., J.D. (ed.) *Transactions 15th World Congress of Soil Science, Acapulco, Mexico, 10–16 July, 1994*. International Society of Soil Science, Mexico, Volume 5b, pp. 446–447.

Johnston, A.E. (1996) Phosphorus: essential plant nutrient, possible pollutant. *Meststoffen* 1996, 53–63.

Kilasara, M., Kullaya, I.K., Kaihura, F.B.S., Aune, J.B., Singh, B.R. and Lal, R. (1995) Impact of past soil erosion on land productivity in selected ecological regions in Tanzania. *Norwegian Journal of Agricultural Sciences (Suppl.* 21), 71–79.

Oenema, O., Boers, P.C.M., Van Eerdt, M.M., Fraters, B., Van der Meer, H.G., Roest, C.W.J., Schröder, J.J. and Willems, W.J. (1997) The nitrate problem and nitrate policy in The Netherlands. Nota 88. Research Institute for Agrobiology and Soil Fertility (AB-DLO), Haren, The Netherlands, 21 pp.

Ryden, J.C. (1986) Gaseous losses of nitrogen from grassland. In: Van der Meer, H.G., Ryden, J.C. and Ennik, G.C. (eds) *Nitrogen Fluxes in Intensive Grassland Systems*. Martinus Nijhoff Publishers, Dordrecht, pp. 59–73.

Sigunga, D.O. (1997) Fertilizer nitrogen use efficiency and nutrient uptake by maize (*Zea mays* L.) in Vertisols. PhD thesis, Wageningen Agricultural University, The Netherlands, 207 pp.

Smaling, E.M.A. and Janssen, B.H. (1987) Soil fertility. In: Boxem, H.W., De Meester, T. and Smaling, E.M.A. (eds) *Soils of the Kilifi Area, Kenya.* Agricultural Resource Report 929, Pudoc, Wageningen, pp. 109–117.

Smaling, E.M.A., Fresco, L.O. and De Jager, A. (1996) Classifying, monitoring and improving soil nutrient stocks and flows in African agriculture. *Ambio* 25, 492–496.

Smaling, E.M.A., Nandwa, S.M. and Janssen, B.H. (1997) Soil fertility in Africa is at stake! *Special Publication American Society of Agronomy* 51, 47–61.

Soil Survey Staff (1962) *Soil Survey Manual.* Handbook 18, USDA, Washington, DC, pp. 155–172.

Stoorvogel, J.J. and Smaling, E.M.A. (1990) *Assessment of Soil Nutrient Depletion in Sub-Saharan Africa: 1983–2000.* Report 28. The Winand Staring Centre for Integrated Land, Soil and Water Research, Wageningen, Vol. I: Main report, 137 pp. Vol. II: Nutrient balances per crop and per Land Use System, 228 tables.

TAC/CGIAR (1989) *Sustainable Agricultural Production: Implications for International Agricultural Research.* FAO Research and Technology Paper No 4, FAO, Rome.

Tiessen, H., Stewart, J.W.B. and Cole, C.V. (1984) Pathway transformations in soils of differing pedogenesis. *Soil Science Society of America Journal* 48, 853–858.

Van der Eijk, D. (1997) Phosphate fixation and the response of maize to fertilizer phosphate in Kenyan soils. PhD thesis, Wageningen Agricultural University, The Netherlands, 187 pp.

Van Reuler, H. (1996) Nutrient management over extended cropping periods in the shifting cultivation system of south-west Côte d'Ivoire. PhD thesis, Wageningen Agricultural University, The Netherlands, 189 pp.

Van Reuler, H. and Janssen, B.H. (1993) Nutrient fluxes in the shifting cultivation system of south-west Côte d'Ivoire. I. Dry matter production, nutrient contents and nutrient release after slash and burn for two fallow vegetations. II. Short-term and long-term effects of burning on yield and nutrient uptake of food crops. *Plant and Soil* 154, 169–188.

Upscaling of Nutrient Budgets from Agroecological Niche to Global Scale

K.W. van der Hoek and A.F. Bouwman

National Institute of Public Health and the Environment (RIVM), PO Box 1, 3720 BA Bilthoven, The Netherlands

ABSTRACT

Agricultural systems can be stratified into a range of space and time scales. When studying nutrient cycles in the agricultural production system, problems in selecting the appropriate scale and delineation of the system boundaries are obvious. Apart from selection of scale, there is a very general problem of data availability. To illustrate the process of up-scaling nutrient budgets, nutrient balance sheets for nitrogen were constructed for three different scales, ranging from a cultivated field or animal category to the farm scale, and finally regional scale (country and global). At the regional scale balance sheets were constructed for the arable land production system including grass and forage production, livestock production and the total agricultural system. Balance sheets of an individual field reveal which production processes and management factors influence losses. Up-scaling from individual to farm scale shows how surpluses and efficiencies can be varied by changing management. A considerable loss of information is occurring during up-scaling. In general, the results for certain scales are based on the individual items in the balance sheets derived for the lower scales, including spatial and temporal variability. Overall system efficiencies can be studied on the basis of regional balances. The paradox is, however, that (improvements in) the efficiency can only be studied using knowledge on individual animal categories and farm scale. In the examples discussed, the results obtained from independent bottom-up scaling of different balance sheets can be validated by comparing the aggregated balance sheets.

INTRODUCTION

Nutrient budgets are important tools in agriculture for understanding the nutrient cycle, for minimizing nutrient losses and for optimizing the uptake and conversion within the production process.

Agricultural systems can be stratified into a range of space and time scales. Problems of selecting the appropriate scale, delineating the system boundaries, and data availability emerge when studying nutrient cycles in an agricultural production system. In particular, when data for the targeted scale have to be derived from a scale in the hierarchy, the method of aggregation or generalization may seriously affect the results.

This chapter focuses on problems associated with up-scaling information between nutrient budgets. A wealth of literature is available on the theoretical backgrounds of scaling (e.g. O'Neill, 1988 and references therein; Risser, 1989; Ehleringer and Field, 1993; Bouwman and Asman, 1997). We will summarize briefly the various scaling problems and backgrounds, and illustrate the up-scaling of nutrient budgets on three scales, including individual crop and animal, farm, and regional scales. The information obtained from nutrient budgets depends on the scale considered.

The nutrient budgets are presented as nutrient balance sheets. Nitrogen is selected as the central nutrient, but the same methodology applies for other elements like phosphate and potassium.

SOME FUNDAMENTAL ASPECTS OF NUTRIENT BUDGETS

A number of important issues concerning the construction of nutrient budgets and their up-scaling include the definition of system boundaries, aggregation and generalization in relation to data availability and the scale considered.

The first step in the construction of nutrient budgets is the determination of the system boundaries, which define which terms are considered as inputs and outputs. In the case of livestock production systems, animal manure is an output, while it can be an input for crop production systems. The delineation of the system boundary should be done in such a way that differences in structure, composition and properties coincide with the system or process studied.

Information obtained for individuals can be aggregated to describe a whole population. For example, information on individual farms can be used to aggregate data on a higher scale, for example, a landscape. This involves grouping farms on the basis of one or more common properties, such as farm size and management. The frequency distributions of certain properties can be used to describe the variation within the group.

During aggregation, the temporal scale should also be taken into account. For example, to understand the nutrient intake and excretion in livestock production, knowledge on individual animals, the lowest level in the scale hierarchy, is required to describe properties of a population of animals (herd), which would be the next level. At any given moment, the result on a larger scale is the integral of the property studied on the smaller scale taken over the frequency distributions of the different individuals. The size of the population is important, because small systems (e.g. individual farms) change more rapidly and drastically over time than large systems (e.g. a country animal population). However, if the temporal variability is also taken into account, the above premise may not be correct, as frequency distributions of a property may change over time. Hence, the size of the population or the level in the hierarchy, should be appropriate for the temporal and spatial variability.

Any scaling exercise is embedded in the data. Quantitative data with adequate spatial and temporal resolution on the use and management of fertilizers, animal excreta and environmental data (including soil, climate and weather) are often very sparse. At the national and global scales agricultural data from the UN Food and Agriculture Organisation (FAO) (1996, 1997) are appropriate for constructing nutrient budgets. On a lower scale there is a wealth of information available. However, in many cases these subnational data cannot be used to extrapolate to other regions, because they do not cover the full range of the environmental and management conditions met in the field. In many cases such statistical information is not available, and 'representative' or generalized types are used to describe group properties. It is obvious that part of the information available for the lower scale is lost during this type of generalization.

It is very important to select the proper scale for the basic data best suited to the scale level for which the aggregation or generalization is performed. A different scale level of the basic data may lead to different results of aggregation. For example, the average nitrogen surplus in arable cropping systems calculated from the data for the individual fields within a landscape, or from the farm averages within that same landscape, may lead to different results. This problem occurs when information to calculate the property weighted on the basis of farm area included in the population is not available. This may be a trivial problem, but such errors often occur, in many cases due to paucity or an absolute lack of data.

NUTRIENT BALANCE SHEETS FOR THREE SCALES

Scale 1: Crops and Animals

Balance sheet for arable crops
Photosynthesis and the supply of water and nutrients enable plants to build up biomass. For their nitrogen supply, plants utilize nitrogen released from mineralization processes in the soil as well as nitrogen from external

sources such as fertilizer. For the balance sheet approach only the external nitrogen supplies are taken into account. The balance sheet for individual crops contains the following entries on the input side: fertilizer, animal manure, atmospheric precipitation and biological nitrogen fixation. Biological nitrogen fixation is considered as an input because it introduces nitrogen into the agricultural system (Peoples and Craswell, 1992; Thomas, 1992, 1995; Peoples et al., 1995). On the output side there is one entry for the part of the harvested crop effectively removed from the field. Crop residues which remain in the field contribute to the soil organic matter nitrogen pool and are therefore not considered. The nitrogen surplus is the difference between total nitrogen supplied and nitrogen removed in harvested crops. This surplus of nitrogen is subject to processes such as volatilization, leaching, wind or water erosion and incorporation into the soil organic matter.

On the global average, about 50% of the applied nitrogen fertilizer is taken up by crop plants during the growing season. Of the remaining 50% of the applied nitrogen, half stays in the soil nitrogen pool and half is lost from the soil–plant system (Hauck, 1984; Newbould, 1989; Powlson, 1993).

Balance sheet for grasslands

The balance sheet for grasslands contains the same entries as described for arable crops. While arable crops are fertilized in most parts of the world, grasslands receive nitrogen fertilizers mainly in western Europe and North America (IFA/IFDC/FAO, 1994).

Balance sheet for animals

Animals require energy, water and nutrients for maintenance, growth, reproduction, production of milk, eggs and wool and for labour (draft animals). In The Netherlands the excretion of nutrients via dung and urine has been calculated according to the balance sheet method since 1990. For every animal category a balance sheet is created using statistical data for consumption of concentrates, grass and roughage, and production of animal products like meat, milk and eggs (Van Eerdt, 1995). The ratio between nitrogen output and nitrogen input broadly ranges from 10 to 40%, and can be considered as nitrogen efficiency at animal level (Table 3.1).

The nitrogen excretion levels for dairy cattle depend on the level of milk production and the animal live weights. The nitrogen excretion rates for cattle on European and global scales have been calculated using standard feeding tables for minimum protein requirement for maintenance and production, adjusted to the mean European and global situation (Table 3.2). For pigs and poultry the Dutch nitrogen excretion data are used as it is assumed that the biological efficiency for producing meat and eggs does not differ much on a global scale (van der Hoek and Couling, 1996; Bouwman et al., 1997; van der Hoek, 1997).

Table 3.1. Nitrogen intake and nitrogen withdrawal in kg N per animal year^{-1}[a] for animal products in some important animal categories in The Netherlands. Source: Van Eerdt, 1995.

Animal category	N in feedstuff	N in animal products	N in excretion	N efficiency[b] (%)
Dairy cattle	177.0	36.5	140.5	20.6
Young cattle 0–2 years	74.1	6.5	67.7	8.7
Young cattle 0–1 years	49.7	6.7	43.0	13.5
Young cattle 1–2 years	98.6	6.2	92.4	6.3
Suckling cows	123.0	12.5	110.5	10.2
Beef cattle 0–16 months[c]	51.3	10.7	40.7	20.8
Beef cattle 0–12 months	41.6	11.2	30.4	26.9
Beef cattle 12–16 months	80.5	9.1	71.5	11.3
Sheep (ewe and 1.6 adherent lamb)	27.6	2.6	25.0	9.4
Dairy goats (doe and 1.8 adherent kid)	23.7	3.7	19.9	15.6
Fattening pigs (25–108 kg)	21.0	6.1	14.9	29.0
Sows (including adherent piglets)[d]	44.3	14.2	30.1	32.0
Laying hens	1.17	0.37	0.81	31.2
Broilers	1.03	0.46	0.57	44.7

[a]All figures are for the year 1994, except for young cattle, suckling cows, sheep and goats (1990). For all the animals mentioned the difference between 1990 and 1994 is very small.
[b]N efficiency is calculated as (N in animal products)/(N in feedstuff).
[c]Beef cattle are fed with maize grown for fodder and concentrates and have therefore a higher nitrogen efficiency than young cattle fed with mainly grass and silage.
[d]The adherent piglets are included in the sows. The figures for young sows (25 kg till first mating) are about the same as for fattening pigs: the N excretion is 13.6 kg per animal per year.

Scale 2: Farm-scale Operations

The balance sheet for farm-scale operations includes input entries such as purchased fertilizers, concentrates, roughage and animals, and output entries such as sold crops, milk, animals, roughage and animal manure. Items produced and used on the farm, such as animal manure and fodder crops, do not appear on the balance sheet. The system boundary for this balance sheet is the farm gate, and it is therefore called the farm gate balance approach.

The up-scaling of information from the lowest scale of individual fields and animal categories consists of data aggregation. The nitrogen balances for fields are summed to obtain the farm balance; results for different farms

Table 3.2. Nitrogen excretion in kg N per animal year^{-1} for the main animal categories for Europe and for two world regions. Source: van der Hoek and Couling, 1996; Bouwman et al., 1997; van der Hoek, 1997.

Animal category	Europe	Region I[a]	Region II[b]
Dairy cattle	100	80	60
Non-dairy cattle	50	45	40
Young cattle	46	30	25
Suckling cows	80	70	60
Beef cattle	40	25	
Draft cattle			40
Buffalo		45	45
Camels		55	55
Horses	50	45	45
Sheep	10	10	10
Goats	9	9	9
Pigs	11	11	11
Poultry	0.5	0.5	0.5

[a]Region I includes Europe, the former USSR, North America, Australia and New Zealand, Israel and Japan.
[b]Region II includes Latin America, Oceania excluding Australia and New Zealand, Africa and Asia excluding former USSR.

are subsequently aggregated to balances for 'typical' and 'generalized' farms. This latter generalization step can be included because results for individual farms may vary considerably due to differences in farm management.

Arable farms

The nitrogen surplus of an individual crop contributes partly to the soil organic matter pool; the mineralized nitrogen can be used by the succeeding crop. The official Dutch fertilizer recommendations take into account the amount of plant-available nitrogen in the soil (Neeteson, 1995). The variation in the nitrogen surplus within a region for a group of farms with different crops is much smaller than that between individual crops. A disadvantage of farm-scale nutrient budgets is that they do not indicate where (and when) surpluses exactly occur. However, budgets at the level of arable farms can be used to describe the differences between different management systems, as illustrated in Table 3.3. Farms that use only mineral fertilizers have much lower nitrogen surplus than farms that use both mineral fertilizers and animal manure obtained from livestock production systems. Nitrogen surpluses appear to be much lower on arable farms in the United Kingdom than on arable farms in The Netherlands.

Table 3.3. Nitrogen input and nitrogen withdrawal in kg N ha^{-1} year^{-1} in crops for arable farms in The Netherlands and the United Kingdom.

	N input	N fertilizer	N animal manure	N products	N surplus	Reference
NL arable farms, 1989/1990	304	158	83	119	185	Baltussen et al., 1992
Without use of animal manure	215	153	—	109	106	Baltussen et al., 1992
With use of animal manure	340	160	119	121	219	Baltussen et al., 1992
UK cereal farms, 1990/1991	169			110	59	Brouwer et al., 1995
UK general cropping farms, 1990/1991	192			115	77	Brouwer et al., 1995

Dairy farms

The crop and animal production processes are combined on dairy farms. During the grazing period of the cattle, nitrogen excretion is unevenly distributed in spots over the meadow. In these spots the nitrogen supply per unit area is very high, resulting in larger losses than losses occurring after homogeneous spreading of the same amount of nitrogen over the field.

Table 3.4 presents nitrogen surpluses on specialized Dutch dairy farms. The intensity of animal production determines the nitrogen efficiency as well as the nitrogen surplus. Increasing milk production from 8000 to 17,700 kg ha^{-1} improves the nitrogen efficiency from 11.7 to 16.0%. At the same time the nitrogen surplus increases by 190 kg ha^{-1}. This increase in milk production per hectare has been mainly achieved by purchasing feedstuffs grown on other farms. If the nitrogen losses generated during the production of these imported feedstuffs were to be accounted for, the nitrogen surplus would be even higher.

Scale 3: Regional and Global Approach

In theory, it is possible to collect data from every farm within the region. However, generally it is more convenient to study a region having sufficient statistical data. Three nutrient balance sheets are constructed, each from a different point of view, so that information from one can be used as input for the next, and so that the results of the three nutrient balances can be validated. These balance sheets are based on animal production, crop

Table 3.4. Nitrogen input and nitrogen withdrawal in kg N ha^{-1} year^{-1} in animal products for dairy farms in The Netherlands and the United Kingdom.

	N input	N fertilizer	N feedstuff	N products	N surplus	Reference
NL dairy farms, 1983–1986, sandy soil	568	331	181	82	486	Aarts et al., 1992
Extensive (8000 kg milk ha^{-1})	454	311	89	53	401	Van Keulen et al., 1996
Intensive (17,700 kg milk ha^{-1})	706	359	289	113	592	Van Keulen et al., 1996
NL dairy farms, 1989/1990, all soil types	514	320	120	90	425	Baltussen et al., 1992
NL dairy farms, 1992/1993, all soil types	467	267	126	85	382	Poppe et al., 1994
UK model dairy farm, calculations	337	250	50	67	270	Jarvis, 1993

Notes: A fixed national milk quota and increasing milk yield per dairy cow result in decreasing numbers of dairy cattle. Environmental concern and the establishment of a national milk quota have resulted in decreasing nitrogen surpluses on Dutch dairy farms (Aarts et al., 1992; Van Keulen et al., 1996).

production and the total agricultural system. To illustrate this approach, nitrogen balance sheets were constructed for the agricultural sector in The Netherlands, United Kingdom, China and finally the global agricultural system. Primary statistical data from the FAO (1997) were used, complemented with local information where necessary.

The up-scaling of information between the generalized farm and regional scales consists of aggregating the arable and livestock production systems within a region or country.

Balance sheet for animal production
The balance sheet for animal production systems lists output terms in the form of animal products and animal manure production. The statistical data from the FAO (1996, 1997) provide information on animal production; the nitrogen content of the products was taken from the literature. Manure production was calculated by using nitrogen excretion factors per animal category as presented in Table 3.2. Input terms include feedstuffs (concentrates) and grass/silage. The first was taken from the FAO food balance sheets (1997), which give country estimates of the amount of available feed from crop and animal origin. We have added to this term the difference between import and export in the major feedstuffs (brans, cakes and meal, meat meal), and the oilmeals produced during processing of the available

oilcrops. The processing of 1000 kg of oilcrops yields on average about 500 kg oilmeals (OECD, 1994).

For The Netherlands and the United Kingdom the remaining demand for feed is supplied by grass and fodder maize. For China we assumed 33% of the feed for pigs and poultry to come from feedstuffs, complemented by household wastes and other residues (Simpson *et al.*, 1994). In Table 3.5 these wastes and residues are included in the term 'unaccounted for', that is, not originating from the agricultural sector. For the global scale, 50% of the demand of pigs and poultry is covered by feedstuffs and 50% is not accounted for at all. For China and the world animal production, the remaining feedstuffs not fed to pigs and poultry are fed to cattle. The remaining demand for cattle is supplied by grass. The share of feedstuffs (concentrates) in the ration for cattle and corresponding grassland productivity are presented in Table 3.6.

Balance sheet for crop production
The balance sheet for crop production systems focuses on arable crops and grass consumed by animals. In both cases only the parts of the plant that are removed from the field are considered. Crop residues and grass that are not consumed remain part of the soil system. The output terms, crop products, were derived from the FAO database, while the grass consumption was taken from the balance sheet for animal production. The input terms, fertilizer and animal manure, were taken from the FAO database and the balance sheet for animal production, respectively. The nitrogen input from biological fixation by leguminous crops was taken from local data (The Netherlands: Van Eerdt *et al.*, 1996; United Kingdom: Lord and Anthony, 1997). Data from Galloway *et al.* (1995) were used for the global scale. The biological fixation in Chinese soils was calculated from the global estimate proportional to the agricultural area.

Balance sheet for the agricultural system
The balance sheet for the agricultural system combines the animal and crop production. This implies that certain items on the previous balance sheets are now considered as internal terms and therefore do not appear on the balance sheet for the agricultural system. These internal terms are grass and animal manure. The items, animal products, fertilizer, nitrogen fixation and 'unaccounted for', which either leave or enter the agricultural system, are taken from the two foregoing balance sheets. For crop products and feedstuffs the situation is more complex because some of these terms are used or produced internally. For crop products a deduction was made for internal consumption by animals. The term feedstuffs consists of the net import of brans, cakes and meal, meat meal and the oilmeals produced in the region (see also the balance sheet for animal production).

Table 3.5. Nitrogen balance sheets for the animal production, crop production and the agricultural sector, for The Netherlands, United Kingdom, China and the world, 1994. Figures are in millions kg of nitrogen.

The Netherlands animal production

Feedstuffs	445	Animal products	177
Grass and fodder maize	380	Animal manure	648
Unaccounted for	0		
Total	825	Total	825

The Netherlands crop production

Fertilizer	380	Crop products	67
Animal manure	648	Grass and fodder maize	380
N-fixation	15		
Unaccounted for	0	Surplus	596
Total	1,043	Total	1,043

The Netherlands agricultural sector

Fertilizer	380	Animal products	177
N-fixation	15	Crop products	50
Feedstuffs	418		
Unaccounted for	0	Surplus	586
Total	813	Total	813

United Kingdom animal production

Feedstuffs	517	Animal products	241
Grass	904	Animal manure	1,180
Unaccounted for	0		
Total	1,421	Total	1,421

United Kingdom crop production

Fertilizer	1,412	Crop products	456
Animal manure	1,180	Grass	904
N-fixation	105		
Unaccounted for	0	Surplus	1,337
Total	2,697	Total	2,697

United Kingdom agricultural sector

Fertilizer	1,412	Animal products	241
N-fixation	105	Crop products	295
Feedstuffs	301		
Unaccounted for	0	Surplus	1,282
Total	1,818	Total	1,818

China animal production

Feedstuffs	2,924	Animal products	2,049	
Grass	7,131	Animal manure	13,032	
Unaccounted for	5,026			
Total	15,081	Total	15,081	

World animal production

Feedstuffs	20,533	Animal products	12,013
Grass	83,040	Animal manure	102,385
Unaccounted for	10,825		
Total	114,398	Total	114,398

China crop production

Fertilizer	19,791	Crop products	8,900
Animal manure	13,032	Grass	7,131
N-fixation	4,500		
Unaccounted for	0	Surplus	21,292
Total	37,323	Total	37,323

World crop production

Fertilizer	73,599	Crop products	49,122
Animal manure	102,385	Grass	83,040
N-fixation	45,000		
Unaccounted for	0	Surplus	88,822
Total	220,984	Total	220,984

China agricultural sector

Fertilizer	19,791	Animal products	2,049
N-fixation	4,500	Crop products	8,900
Feedstuffs	1,294		
Unaccounted for	5,026	Surplus	21,148
Total	30,611	Total	30,611

World agricultural sector

Fertilizer	73,599	Animal products	12,013
N-fixation	45,000	Crop products	38,977
Feedstuffs	9,600		
Unaccounted for	10,825	Surplus	88,034
Total	139,024	Total	139,024

Notes:
1. In China traditionally cattle and horses are used as work animals with an estimated share of 25% in the total agricultural energy consumption (Mengjie and Yi, 1996). This means that feed (energy and protein) is not converted into animal products, but instead animal power is produced.
2. Chinese figures for total N + P_2O_5 + K_2O consumption with fertilizer and organic manure are 31.5 and 15.15 million tons respectively in 1993 (Portch and Jin, 1995).

Table 3.6. Feed demand of animals and grassland use, 1994.

	The Netherlands	United Kingdom	China	World
Feed demand all animals (million kg N)	825	1,421	15,081	114,398
Pigs and poultry				
Total feed demand (million kg N)	309	244	7,517	21,635
Of which feedstuffs (million kg N)	309	244	2,481	10,818
Share feedstuffs in ration (%)	100	100	33	50
Cattle, sheep etc.				
Total feed demand (million kg N)	516	1,177	7,574	92,755
Of which feedstuffs (million kg N)	136	273	443	9,715
Of which consumed grass (million kg N)	380	904	7,131	83,040
Share feedstuffs in ration (%)	26	23	6	10
Total area grasslands (million hectare)	1.25	11.1	400	3,400
Consumed grass[a] (kg N ha^{-1})	300	81	18	25

[a]Total nitrogen intake with consumed grass divided by the total area of grasslands results in kg consumed grass-nitrogen ha^{-1} grassland. In reality not all grasslands are utilized and not all grass is taken up by animals. For China and the world this figures corresponds with about 20% of the nitrogen production in grass. The figures given for China are in line with data from the literature (Jin et al., 1990; Cao et al., 1995; Simpson and Li, 1996).

Validation of the balance sheets

The balance sheets for crop production and for the agricultural system have in common that the calculated surplus has to be identical. For the crop production system the equation is:

fertilizer + animal manure + N-fixation = crop products + consumed grass + surplus$_1$ (3.1)

For the agricultural system the equation is:

fertilizer + N-fixation + imported feedstuffs (including 'unaccounted for') = animal products + crop products (minus internal use for feeding) + surplus$_2$ (3.2)

For the animal production system the equation reads:

feedstuffs + consumed grass = animal products + animal manure   (3.3)

Combining these three equations shows that surplus$_1$ = surplus$_2$. As the balances for the crop production system and the agricultural system were derived independently, validating the overall result of the two balance sheets is possible.

UP-SCALING AND LOSS OF INFORMATION

Crop Production

Nutrient balances for arable and fodder crops provide information about the nutrient use efficiency of the crop itself. On the farm scale the surplus reflects the net result of surpluses for different crops. Hence, information from the lower scale is required to interpret the result at the level above. A similar up-scaling problem occurs for animal manure in arable crops. Table 3.3 shows a larger surplus when animal manure is used. Nutrient balances for a region show a nutrient surplus, but generally it is not clear whether the resulting surplus is the effect of applying animal manure or a specific mix of crops, or both. The conclusion is that going from the lower to the next lowest scale, information is lost and so the interpretation is only possible on the basis of information from the lower scales.

Animal Production

For animal production a similar phenomenon is observed. This can be demonstrated especially for cattle. Dairy cattle produce milk and meat with an efficiency of nitrogen use of about 20% (Table 3.1). However, the herd of dairy cattle is partly replaced yearly with young dairy cattle. Young cattle have a lower nitrogen efficiency (Table 3.1). This means that when looking at the farm scale, where the dairy herd consists of all age categories of cattle, the nitrogen efficiency ranges between 10 and 20%.

The nutrient balance constructed for animal production in a region gives an overall surplus. This surplus is dependent on the composition of the dairy herd and on the share of the other animal categories, like pigs and poultry, within that region. Without detailed information on this topic, it is difficult to interpret the regional surplus. The impact of measures to increase the nitrogen efficiency must also be studied on the lowest scale, otherwise it is difficult to draw the right conclusions.

Regional and Global Scale

Nutrient balances for regions having animal production as well as plant production show the net surplus of both production systems. Measures to improve the nutrient efficiency for a region can only be studied on the individual crop or animal category scale. Only when the agricultural situation is more or less comparable in two regions, can the nutrient surplus of both regions be compared.

Paradox

The paradox is that up-scaling and loss of information are very closely connected. Constructing the nutrient balance for the higher scale is facilitated by information for the lower scale. But when the nutrient surplus of the higher scale is determined, information is lost. Research and recommendations for improving the nutrient efficiency needs information obtained on the lower scale. Then the effect of the improved efficiency can be assessed by drawing up a nutrient balance for the higher level. When presenting nutrient balances it is therefore recommended to provide the information from the lower scales used for constructing the nutrient balance of the scale considered.

DISCUSSION

Nutrient balance sheets have been constructed for three different scales, ranging from the cultivated field or animal category to farm, region or country consisting of a variety of different farms, and the world. For the regional and global scales, three different balance sheets have been constructed, including both the livestock and arable production systems (including grass and forage) and the total agricultural system.

The balance sheets on the individual field or animal scale tell us where nutrient losses occur, and by which processes the losses are influenced. For example, efficiencies for crops are seemingly higher than 100% calculated as nitrogen in harvested crop : fertilizer application. However, in this calculation the nitrogen supplied from mineralization of nitrogen already present in the soil is not accounted for. Output : input for individual animals is a better expression of efficiency, clearly reflecting the characteristics of each animal and the conditions within the production system for each animal category.

The data collected on the individual scale can be aggregated to the farm scale, where we can see again, but now for a large group of different farms, how surpluses and efficiencies can be varied by changing management. For example, in arable farms an additional use of animal manure is apparently not found in the product, and is partly lost from the system or accumulates in the soil. For dairy production farms, the influence of intensification causes both an increase in efficiency and in the surpluses.

By aggregation of the farm-scale data to regional or national scales, as shown by combining farm-scale data with national data, new balance sheets can be made. This has been shown for three different countries and for the world. These balances can be used to assess overall efficiencies of production systems and some major components in the nitrogen budget of a country. The data confirm the comparison for The Netherlands and United Kingdom at farm level that surpluses per unit area in The Netherlands are far higher than in the United Kingdom. In livestock production on the global scale the efficiency is roughly 10%, while in China it is 13%. In The Netherlands and the United Kingdom efficiencies are 21% and 17%, respectively. However, the surpluses are higher in these countries than on the global scale. The efficiencies for crop production and total agricultural system are far higher than in livestock production. On the global scale these are 37% for the total agricultural system and 60% for crop production.

What can we learn from the up-scaling of nutrient balance sheets as discussed in this chapter?

In the first place, a considerable loss of information occurs during up-scaling. Two methods have been discussed: generalization and aggregation. The purpose of the scaling exercise is to determine how far this information loss is a disadvantage. In general, interpreting the results on a certain scale requires information on the individual items in the balance sheet obtained on the lower scales, including spatial and temporal variability underlying the results. For example, when the purpose is to assess the overall efficiency of nutrient use of an agricultural production system, the regional nutrient balance can be used. The paradox is, however, that (improvements on) the efficiency can only be studied of the basis of knowledge of the individual animal category and the farm scales.

Secondly, information obtained for individual animal categories or arable fields can be aggregated to represent the population, or a landscape. This data can be aggregated by grouping individuals according to one or more criteria. Frequency distributions can be used to describe the variability on a group scale. When a generalization is performed the characteristics of a group or population are described by a representative individual. The scale associated with the basic data is very important when data are aggregated or generalized since different scales may lead to different results on the subsequent scale used, in particular when data needed to calculate weighted group averages in time or space are lacking.

Thirdly, top-down scaling approaches are useful as a check on the overall result. We have shown that results obtained from independent bottom-up scaling of different balance sheets can be validated by comparing the aggregated balance sheets.

REFERENCES

Aarts, H.F.M., Biewinga, E.E. and Van Keulen, H. (1992) Dairy farming systems based on efficient nutrient management. *Netherlands Journal of Agricultural Science* 40, 285–299.

Baltussen, W.H.M., Hoste, R., Daatselaar, C.H.G. and Janssens, S.R.M. (1992) *Differences in Mineral Surpluses between Dairy and Arable Farms in The Netherlands.* Onderzoekverslag 101, Agricultural Economics Research Institute, The Hague, 64 pp. (in Dutch).

Bouwman, A.F. and Asman, W.A.H. (1997) Scaling of nitrogen gas fluxes from grasslands. In: Jarvis, S.C. and Pain, B.F. (eds) *Gaseous Nitrogen Emissions from Grassland.* CAB International, Wallingford, UK, pp. 311–330.

Bouwman, A.F., Lee, D.S., Asman, W.A.H., Dentener, F.J., van Der Hoek, K.W. and Olivier, J.G.J. (1997) A global high-resolution emission inventory for ammonia. *Global Biogeochemical Cycles* 11, 561–587.

Brouwer, F.M., Godeschalk, F.E., Hellegers, P.J.G.J. and Kelholt, H.J. (1995) *Mineral Balances at Farm Level in the European Union.* Onderzoekverslag 137, Agricultural Economics Research Institute, The Hague, 141 pp.

Cao, M., Ma, S. and Han, C. (1995) Potential productivity and human carrying capacity of an agroecosystem: an analysis of food production potential of China. *Agricultural Systems* 47, 387–414.

Ehleringer, J.R. and Field, C.B. (eds) (1993) *Scaling Physiological Processes. Leaf to Globe.* Academic, London, 388 pp.

FAO (1996) *FAOSTAT.* FAO Publications Division, Food and Agriculture Organisation of the United Nations, Rome.

FAO (1997) *FAOSTAT* database collections, Food and Agriculture Organisation of the United Nations, Rome. Available on the Internet (http://aps.fao.org/default.html).

Galloway, J.N., Schlesinger, W.H., Levy II, H., Michaels, A. and Schnoor, J.L. (1995) Nitrogen fixation: anthropogenic enhancement–environmental response. *Global Biogeochemical Cycles* 9, 235–252.

Hauck, R.D. (ed.) (1984) *Nitrogen in Crop Production.* American Society of Agronomy, Crop Science of America, Soil Science of America, Madison, Wisconsin, 804 pp.

van der Hoek, K.W. (1997) *Calculation of N-excretion and NH_3 Emission by Animal Production on European and Global Scale.* RIVM report 773004005, RIVM, Bilthoven, The Netherlands.

van der Hoek, K.W. and Couling, S. (1996) Manure management, SNAP code 100500. In: McInnes, G. (ed.) *Joint EMEP/CORINAIR Atmospheric Emission Inventory Guidebook.* European Environment Agency, Copenhagen, pp. B1050-1–B1050-16.

IFA/IFDC/FAO (1994) *Fertilizer Use by Crop.* Report ESS/MISC/1994/4, Food and Agriculture Organisation of the United Nations, Rome, 44 pp.

Jarvis, S.C. (1993) Nitrogen cycling and losses from dairy farms. *Soil Use and Management* 9, 99–105.

Jin, Y., Xiong, Y. and Ervin, R.T. (1990) Energy efficiency of grassland animal production in Northwest China. *Agriculture, Ecosystems and Environment* 31, 63–76.

Lord, E. and Anthony, S. (1997) *Nutrient Balances as Environmental Indicators for the UK.* Table document for Working Group 'Statistics of the environment', Agricultural Development Advisory Services, Wolverhampton, UK, 33 pp.

Mengjie, W. and Yi, D. (1996) The importance of work animals in rural China. *World Animal Review* 86(1), 65–67.

Neeteson, J.J. (1995) Nitrogen management for intensively grown arable crops and field vegetables. In: Bacon, P.E. (ed.) *Nitrogen Fertilization of the Environment.* Marcel Dekker Inc., New York, pp. 295–325.

Newbould, P. (1989) The use of nitrogen fertilizer in agriculture. Where do we go practically and ecologically? In: Clarholm, M. and Bergstroem, L. (eds) *Ecology of Arable Land Perspectives and Challenges.* Kluwer Academic Publishers, Dordrecht, pp. 281–295.

OECD (1994) *The World Oilseed Market: Policy Impacts and Market Outlook.* OECD, Paris, 69 pp.

O'Neill, R.V. (1988) Hierarchy theory and global change. In: Rosswall, T., Woodmansee, R.G. and Risser, P.G. (eds) *Scales and Global Change.* SCOPE, John Wiley & Sons Ltd, Chichester, UK, pp. 29–45.

Peoples, M.B. and Craswell, E.T. (1992) Biological nitrogen fixation: investments, expectations and actual contributions to agriculture. *Plant and Soil* 141, 13–39.

Peoples, M.B., Herridge, D.F. and Ladha, J.K. (1995) Biological nitrogen fixation: an efficient source of nitrogen for sustainable agricultural production? *Plant and Soil* 174, 3–28.

Poppe, K.J., Brouwer, F.M., Welten, J.P.P.J. and Wijnands, J.H.M. (eds) (1994) *Agriculture, Environment and Economics.* Periodieke Rapportage 68–92, Agricultural Economics Research Institute, The Hague, 180 pp. (in Dutch).

Portch, S. and Jin, J. (1995) Organic manure and biofertilizer use in China. *Fertiliser News* 40(12), 79–83.

Powlson, D.S. (1993) Understanding the soil nitrogen cycle. *Soil Use and Management* 9, 86–94.

Risser, P.G. (1989) The movement of nutrients across heterogeneous landscapes. In: Clarholm, M. and Bergstroem, L. (eds) *Ecology of Arable Land Perspectives and Challenges.* Kluwer Academic Publishers, Dordrecht, pp. 247–251.

Simpson, J.R. and Li, O. (1996) Feasibility analysis for development of Northern China's beef industry and grazing lands. *Journal Range Management* 49, 560–564.

Simpson, J.R., Cheng, X. and Miyazaki, A. (1994) *China's Livestock and Related Agriculture. Projections to 2025.* CAB International, Wallingford, UK, 474 pp.

Thomas, R.J. (1992) The role of the legume in the nitrogen cycle of productive and sustainable pastures. *Grass and Forage Science* 47, 133–142.

Thomas, R.J. (1995) Role of legumes in providing N for sustainable tropical pasture systems. *Plant and Soil* 174, 103–118.

Van Eerdt, M.M. (1995) Animal manure production, excretion of nutrients and nutrients in animal manure, base year 1994. *Kwartaalbericht Milieustatistieken* 4, 11–21 (in Dutch).

Van Eerdt, M.M., Olsthoorn, C.S.M. and Fong, P.K.N. (1996) Nutrient balance sheets in agriculture. In: *Environment-related Agricultural Statistics in the Nordic Countries.* Tekniske rapporter 62, Nordisk statistik sekretariat, Koebenhavn, pp. 112–126.

Van Keulen, H., Van der Meer, H.G. and De Boer, I.J.M. (1996) Nutrient balances of livestock production systems in the Netherlands. In: Groen, A.F. and Van Bruchem, J. (eds) *Utilization of Local Feed Resources by Dairy Cattle.* EAAP Publication No. 84, Wageningen Pers, Wageningen, pp. 3–18.

Uncertainties in Nutrient Budgets due to Biases and Errors

O. Oenema and M. Heinen

Wageningen University and Research Centre, Research Institute for Agrobiology and Soil Fertility (AB-DLO), PO Box 14, 6700 AA Wageningen, The Netherlands

ABSTRACT

Nutrient budgeting involves the estimation of nutrient pools and of the flows between these pools. The accuracy and precision of these estimates depend on many factors, including the agroecosystem under consideration, budgeting approach and data acquisition strategy. There is often a considerable amount of uncertainty in budgets, due to biases and errors in the estimates of inputs and especially outputs. This chapter firstly reviews the various possible sources of biases and errors in nutrient budgets, and secondly illustrates that uncertainty is tractable by analysing three case studies.

Bias is defined as systematic deviation, error as random variation. We have distinguished five sources of biases and two sources of errors. The five sources of biases are personal bias, sampling bias, measurement bias, data manipulation bias and fraud. The two sources of errors are sampling and measurement errors. Both biases and errors in nutrient budget estimates may lead to wrong conclusions. Bias can be avoided by system analyses, testing of assumptions and by proper planning and application of well-adopted techniques and procedures. Errors can be minimized via appropriate sampling and analytical procedures.

The first case illustrates that both biases and errors occurred in the nutrient budget study of a simple one-dimensional culture solution system, that is, a tomato plant growing on a nutrient solution. The results clearly indicate that the initial assumptions did not hold. The second case shows that uncertainties in the nutrient budget at farm level are dominated by errors in the estimation of outputs via nutrient losses and changes in the nutrient stocks. Uncertainties in the farm budget appear to be smaller than uncertainties in the budgets at compartment and field levels. However, the nutrient budgets at compartment level do illustrate more clearly the sites and pathways of nutrient losses and also illustrate more clearly where nutrient use efficiency can be increased. The third case discusses the use of the nutrient accounting system MINAS as a policy instrument to regulate nitrogen and phosphorus surplus for agricultural land in The Netherlands, from 1998 onwards. The economic and environmental consequences of biases and errors in MINAS can be large, because it is a regulatory policy instrument of nutrient surpluses at farm level. Proper guidelines, registration and accredited bureaux are required to minimize bias, particularly fraud. Thus far, experience indicates that there are a number of potential sources of biases and errors in MINAS. Through information and cooperation with farmers, the negative effects of these errors and bias can be minimized.

INTRODUCTION

For decades, nutrient budgets have been a powerful tool for scientists to facilitate the understanding of nutrient flows in ecosystems at various scales. These studies have made clear that nutrient inputs and outputs seldom balance (cf. van Noordwijk, Chapter 1, this volume; Janssen, Chapter 2, this volume). This is in particular the case for agricultural systems. In intensively managed agricultural systems, nutrient inputs often exceed nutrient outputs via useful products, while the opposite is true for extensively managed and low-input agricultural systems. Systems with severe imbalances are not sustainable in the long term; a nutrient surplus may contribute to pollution and eutrophication of neighbouring systems, while a nutrient deficit leads to impoverishment of the system and to decreased crop production. Human interferences with the nutrient cycle increased tremendously, in scale and intensity, during the second half of the 20th century.

There is now an increasing amount of evidence to suggest that imbalances in nutrient budgets, especially for nitrogen and phosphorus, have shifted in scale from what was once a local problem to what is now one of regional and continental dimensions (e.g. Heathwaite *et al.*, 1993). It is the shift in scale that has recently given the impetus to use nutrient budgets as indicators and policy instruments for nutrient management planning. Indeed, nutrient budgets are increasingly used by policy-makers, extension workers and farmers as an instrument for planning and control of nutrient management plans directed towards sustainable agricultural production (e.g. Romstad *et al.*, 1997). But how accurate are nutrient budgets? How susceptible are nutrient budgets to biases and errors? When nutrient budgets are being used as a control instrument, how can it be enforced, and is budgeting susceptible to fraud?

In this chapter, we will consider these questions and provide background information about uncertainties in nutrient budgets, though without embarking on exercises of statistics. Nutrient budgets often summarize the results of a large number of measurements, estimates and assumptions, often from various sources, in one table. All these measurements, estimations and assumptions are possibly cursed with biases and errors, due to insufficient information, poor data and because of natural variability. Unfortunately, there is often little quantitative information available about biases and errors and, consequently, about uncertainties in the overall budget. Moreover, an uncertainty analysis to examine the possible effect of uncertain inputs and outputs on the surplus or deficit appears to be the exception rather than the rule. This is surprising, because it can be a perilous undertaking to make nutrient management plans without considering the uncertainties and assumptions in the nutrient budget.

Uncertainties arise from biases and errors. *Bias* is defined here as misrepresentation, leading to a systematic deviation of the measurement mean from the true (scientific) mean. *Error* is random variation around the true mean, leading to a confidence interval around the measurement mean. Bias in nutrient budgets may lead to wrong conclusions, whilst errors complicate the making of conclusions, and in turn may lead to wrong conclusions as well. The term *'accuracy'* is used here as a measure for bias; a high accuracy means that the deviation of the measurement mean from the true mean is small. The term *'precision'* is used here as a measure for error; a high precision means that the variance of repeated measurements is small.

In this chapter, we firstly summarize the possible sources of biases and errors in nutrient budget studies. Secondly, we analyse three case studies to illustrate how uncertainties in inputs and outputs have their effect on the calculated surplus and deficit of the budget. The first case deals with the nutrient budget of a solution culture with tomatoes in a greenhouse, which in fact is a one-dimensional study involving nutrient transport from solution to the plant only. The second case deals with 'De Marke', a 55 ha large experimental dairy farm in The Netherlands. This is a three-dimensional study, involving vertical and lateral transfers of nutrients between different compartments and fields, including vertical and lateral movement of nutrients in the soil profile. The third case deals with the nutrient accounting system MINAS, which was introduced by The Netherlands' government in 1998 as a regulatory control instrument for nutrient management planning and for decreasing the nitrogen and phosphorus surpluses of agricultural land. Our intent is not to present definite numbers for specific systems but, rather, to demonstrate that uncertainty is tractable and that careful analyses and experimental design can improve both the accuracy and precision of nutrient budget estimates.

BIASES AND ERRORS IN NUTRIENT BUDGETS

Differences in Aims, Systems and Budgets

The *required* accuracy and precision of a nutrient budget are dependent on the objectives and the originators of the study. They define the necessary accuracy and precision of the estimates of nutrient pools and flows, and thereby the required methodology and expertise, equipment, facilities and labour. The *achievable* accuracy and precision, in turn, depend to a large extent on the complexity of the ecosystem and on the understanding of nutrient cycling and nutrient transformation processes. The complexity of nutrient budgeting is often large when grazing animals are involved, and when insight is required into the *fate* of the nutrients entering the system. Evidently, understanding the nutrient cycle is a prerequisite for accurate budgets. Understanding of nutrient cycling processes facilitates the selection of suitable methodologies, selection and timing of appropriate sampling sites, and ultimately the establishment of accurate and precise budgets.

Budget studies may differ in purpose, budgeting approach, scale and data acquisition strategy (Smaling and Fresco, 1993; Smaling and Oenema, 1997). The purpose of the study defines the approach, scale and data acquisition strategy. A *farm gate balance* is a black-box approach; it records the amount of nutrients in all kinds of products that enter and leave the farm via the farm gate. A farm gate approach is often preferred in policy analyses, because of its simplicity and ease of data acquisition. A *surface balance* records all nutrients that enter the soil and that leave the soil via crop removal, both expressed in kg ha^{-1} year^{-1}. The surplus is often further partitioned over various nutrient loss pathways (e.g. Janssen, Chapter 2, this volume). The difference in surplus between the farm gate balance and surface balance can be considerable, particularly for nitrogen, as the surface balance includes inputs via atmospheric deposition and biological fixation, whilst the farm gate balance does not. A surface balance is required when the total loading of the soil with nutrients needs to be known. Note that the demarcation of the system is different for a farm gate balance and a soil surface balance (Fig. 4.1). The *system balance approach* also allows partitioning between the various nutrient loss pathways and the storage and/or depletion of nutrients in nutrient compartments within the system (Fig. 4.2). The overall surplus of the system balance is an estimate for the net enrichment or depletion of the system. A system balance is required when nutrient inputs have to be partitioned over storage in the systems, outputs in useful product and losses. Confusion and, hence, bias may occur when data from, for example, a farm gate balance are transferred into a surface balance approach and vice versa. Evidently, proper guidelines and conventions are needed for nutrient budgeting (e.g. Smaling and Fresco, 1993; PARCOM, 1995; Smaling and Oenema, 1997).

Uncertainties in Nutrient Budgets

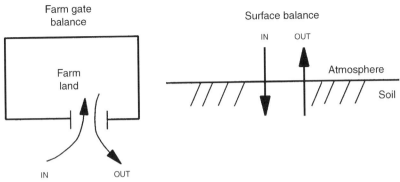

Fig. 4.1. Boundaries and direction of nutrient flows for a farm gate balance and a surface balance. Nutrient flows are generally expressed in kg ha^{-1} year^{-1}.

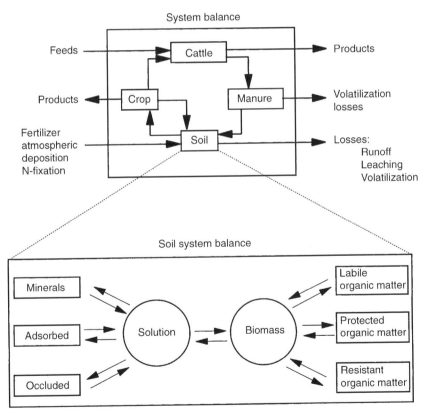

Fig. 4.2. Boundaries and direction of nutrient flows for a systems balance. Conveniently, the soil compartment is subdivided into eight pools, i.e. four inorganic pools and four organic pools. Nutrient flows are generally expressed in kg ha^{-1} year^{-1}.

Nutrient budgeting can be undertaken at any scale, the choice of which depends on the problem of interest. From a reductionist point of view, the system should be studied at a small scale where the basic rules of physics, chemistry and biology govern the rate and direction of processes and flows. Information on large scales should then be derived subsequently from aggregation of small-scale studies (e.g. van der Hoek and Bouwman, Chapter 3, this volume). This can be cumbersome and is not always practical. However, at each scale, appropriate generalizations can be made so that the macroscopic behaviour of the nutrient cycling processes is explained in sufficient detail and at the required accuracy, without bothering about the microscopic rules at underlying scales. The consequence is that a surface balance budget at national scale, based on data statistics, can be made for a particular year as easily and accurately as a surface balance of a single field, based on measurements.

How to quantify the sources of uncertainty? This depends on the nature of the uncertainty and on the available information. Usually, it is convenient to start with a system analysis to check whether all pools, inputs and outputs have been identified properly. Classification of uncertainties in important and less important sources, based on available information, is a second step. The third step is specification of distributions of probabilities of the various sources, using mean, standard deviation, percentile values, and specification of correlations between sources. The effect of uncertainties in individual sources on the overall budget must be calculated next and can be specified in means and variances, frequency distributions and confidence intervals. Various procedures may be used for uncertainty analyses; the reader is referred to the appropriate literature (e.g. Janssen *et al.*, 1990; Heuvelink, 1998; Jansen, 1998).

Awareness of possible sources of biases and errors in nutrient budgeting is the first step to minimize uncertainty.

Sources of Biases

There are five possible sources of biases, which lead to inaccuracies in nutrient budgets: (i) personal biases; (ii) sampling biases; (iii) measurement biases; (iv) data manipulation biases, including guestimation; and (v) fraud, which is the conscious introduction of biases.

Personal biases

Personal biases are a most complex source of biases. The conceptual interpretation and simplification of an agroecosystem and of the nutrient flows into and out of the system can be different by different scientists. Our knowledge of nutrient cycling in ecosystems is limited. This limited knowledge is the cradle of personal biases. When measurement data and data statistics are lacking, guestimations have to be made. These estimates can be different for different scientists. Examples of different conceptual notions of farming systems can be found in Frissel and Kolenbrander (1977), Clark (1981) and Karlovsky (1981). Moolenaar *et al.* (1997) compared three balance approaches for calculating the accumulation of heavy metals in agroecosystems as a results of different fertilizer strategies. Differences between the static balance, the dynamic balance and the dynamic soil composition balance in long-term predictions of cadmium contents in soil were as large as a factor of three for a particular fertilizer strategy. Similar conclusions may hold for phosphorus and other immobile elements.

A well-known example of personal biases is the 'enigma of the nitrogen system balance' (Allison, 1955), that is the measured nitrogen input is larger than the measured total nitrogen output, including nitrogen losses. This suggests that in the system balance nitrogen sinks have been overlooked, outputs have been underestimated, inputs have been overestimated, or both. The nitrogen surplus can be as large as 40 to 60% of the total input, especially on intensively managed grazed grassland (e.g. Garrett *et al.*, 1992). There is now evidence to suggest that nitrogen losses via volatilization are often underestimated, in part due to neglect of trace gas emissions, and due to artefacts in denitrification measurements (e.g. Bollman and Conrad, 1997; McKenny *et al.*, 1997). Storage in the soil via immobilization of nitrogen in soil organic matter is sometimes also neglected, in part because of the poor methodology to quantify accurately the immobilization of the nitrogen from roots and stubble in soil organic matter.

Sampling biases
Quantification of nutrient pools and nutrient flows requires sampling and subsequent analysis of these pools and flows. It requires determination of both volume and nutrient content. A sample usually consists of a small collection from an aggregate or population. The sample is examined but it is the aggregate or population that has our interest. The sample can be soil, water, crop, animals or a farm and landscape. Unless samples are representative, facts about the sample cannot be taken as the facts about the aggregate or population. There is an endless variety among environments, sites and farms so that successive samples are usually different. Hence, the key problem is where and when to sample to obtain representative samples.

Sampling is a large potential source of bias in studies that aim at quantifying all nutrient losses, including those via leaching, volatilization, erosion and runoff. This is especially the case in sloping areas and in grazed areas where transfer of nutrients occurs via runoff and grazing animals, and where nutrient losses often occur via 'hot spots' and via preferential flow patterns in soils. Stratified and random sampling designs are often required, at the appropriate scale of investigation in combination with key variables that facilitate the up-scaling to the aggregate level of interest (e.g. Dumanski *et al.*, 1998; Wagenet, 1998).

Measurement biases

Measurements of the volume and nutrient content of samples can be carried out quite accurately, generally, when use is made of well-calibrated equipment. Measuring the nutrient content often requires wet chemical analyses. A number of factors may contribute to measurement biases, such as incomplete dissolution and extraction, precipitation, volatilization, matrix effects, and insufficient sensitivity of the methods at low concentration. All these potential artefacts can be minimized by good laboratory practice and by careful calibration of the methods. Inclusion of appropriate standard samples in routine analyses and participation in sample exchange programmes are very useful for identifying and preventing systematic deviations in the chemical analyses. Results of the International Soil and Plant Analytical Exchange (Houba and Uittenbogaard, 1994) indicate that median and mean values for major nutrients in plant materials, as determined by 250 laboratories from 64 countries, are about equal, and that coefficients of variations are less than 10%, suggesting little bias. This was, however, not the case for major elements at low concentrations, and for sulphur and for trace elements at all concentration levels. Similar results are usually obtained for soil samples.

Data manipulation biases

A potential source of bias is data manipulation and estimating the mean of replicate analyses. Generalization, averaging and up-scaling may lead to loss of information and to bias (e.g. van der Hoek and Bouwman, Chapter 3, this volume). Frequency distributions of replicate measurements of nutrient losses via leaching, denitrification, runoff and volatilization tend to be skewed. In studies on grazed grassland, Velthof and Oenema (1995) compared four commonly used estimators for calculating the mean N_2O flux from replicated measurements. They observed that differences between the calculated means were up to one order of magnitude. These large differences complicate comparisons among studies in which different estimators have been applied. Unfortunately, no proper guidelines are available yet for dealing with (comparison of) skewed distributions.

Biases due to fraud

When nutrient budgets are being used as policy instrument to enforce a nutrient management strategy with possible economic consequences for farmers and other stakeholders, it will be obvious that some of these stakeholders may try to manipulate the budget so as to minimize the economic consequences. Implementation of such a nutrient budget system in practice, with repressive measures, requires a careful registration and also certification of all nutrient flows. The greater the repressive economic consequences, the greater the need for manipulation of inputs and outputs. As will be illustrated in more detail later, there are a number of possible routes to manipulate budgets. The only message presented here is that biases may be larger in cases where nutrient budgets are being used as a regulatory control instrument instead of a management instrument.

Sources of Error

Errors occur as a result of random variations, and show up as variance in repeated determinations. Two types of errors can be distinguished: sampling and measurement errors. When the precision is low, and determinations are carried out only once, errors and biases merge into one another.

Sampling error
Sampling errors originate from 'within-plot' heterogeneity, from spatial variations or temporal variations or both. Soils, crops and animal wastes are notoriously variable in space and time, even within well-defined plots and areas, and require well-designed sampling strategies. The flow of nutrients in water and air leaving the system via leaching, runoff and volatilization commonly has a very large spatial and temporal variability. In many cases, 90% of the total annual loss occurs during peak events via hot spots and preferential flows (e.g. Sharpley and Rekolainen, 1997). As a consequence, frequency distributions of the losses in space and time tend to be highly skewed, and the variance extremely large. In these cases, errors are large. Quantification of the sampling error requires repeated sampling following appropriate sampling designs, and repeated analyses.

Measurement error
Measurement errors originate from variations introduced during the determinations of the volume and composition of the sample, that is, through sub-sampling, pre-treatment and chemical analysis. The measurement error shows up as the variance of repeated measurements of one sample. Usually, the measurement error is much smaller than the sampling error. Qualified laboratories present the results of the chemical analyses with a relative error of less than 5% for nitrogen and phosphorus.

Total error in a nutrient budget
Because of the interactive nature and the covariance between various nutrient flows, the total variance in the final budget is less than the sum of the variances of the individual inputs and outputs. The total variance is equal to the sum of the variances of the various flows plus twice the covariance of all possible two-way combinations of these flows. Because there is often a negative correlation between the sizes of the various outputs, the covariance is negative. In statistical terms, the total variance of the budget on the balance is equal to (Snedecor and Cochran, 1967):

$$\begin{aligned}\text{Var (budget)} = &\text{ var } (I_1) + \ldots + \text{var } (I_x) + \text{var } (O_1) + \ldots + \text{var } (O_y) + \\ &2 \text{ cov } (I_1, I_2) + \ldots + 2 \text{ cov } (I_1, O_1) + \ldots + 2 \text{ cov } (O_{y-1}, O_y)\end{aligned} \quad (4.1)$$

where, I_1, I_x = inputs of nutrients, a total of x items; O_1, O_y = outputs of nutrients, a total of y items; var = estimated variance; and cov = estimated covariance.

The total variance of the budget depends to a large extent on the size and variance of major items and on the covariance between these items. If precise information on the variance and covariance is lacking, and one is interested in the possible variance of the budget, it is recommended to perform an uncertainty analysis. This may be done via a Monte Carlo type of simulation, using frequency distributions for the various items based on literature data and best guesses. Again, proper assessment of the interactions between the various inputs and outputs is crucial for obtaining unbiased and precise estimates of the frequency distribution of the budget, and hence of errors.

Table 4.1 summarizes the relative errors of the various items of the phosphorus (P) budget for animal farms in The Netherlands. Three categories may be distinguished: (i) items with a relative error of less than 5%, such as mineral fertilizers; (ii) items with a relative error of 5–20%, such as harvested crops, slurry, forage, and atmospheric deposition; and (iii) items with a relative error of more than 20%, i.e. losses via leaching, runoff or volatilization.

Summarizing, there are a number of possible sources of biases and errors. Biases are generally a more serious problem than errors, because biases in the various inputs and outputs accumulate to a much larger extent in the surplus or deficit of the budget than do errors. Biases must be prevented, to preclude unwise conclusions. It requires proper systems analysis, testing of assumptions and proper data acquisition and handling. Errors are

Table 4.1. Approximate values for the relative errors of phosphorus and nitrogen budgets of farms in The Netherlands (Oenema, unpublished data). Estimates for the relative error of the total input and output are based on average farms, with little exchange of manure and slurry.

Input	Error (%)	Output	Error (%)
Fertilizers	1–3	Milk	2–8
Manure	10–20	Meat	2–10
Plant material	5–20	Manure	10–20
Atmospheric deposition	10–30	Crops	5–10
Concentrates	5–10	Leaching	50–200
Forages	5–10	Runoff	50–200
		Volatilization	50–200
Total	5–15		
		Total	10–20

inherent in the variability of nature. Prevention of errors requires proper sampling and standardized measurement procedures.

CASE 1: TOMATO PLANTS GROWN ON A NUTRIENT SOLUTION

Glasshouse horticulture in The Netherlands covers more than 10,000 ha. Nutrient input and yields in the intensively managed horticultural systems are large, but nutrient losses are also large. To decrease nutrient losses and to further improve yields, soil-less cultures with recirculating nutrient solutions have been succesfully introduced in the 1980s. Nowadays, 60% of glasshouse horticulture has soil-less cultures with recirculating nutrient solution systems. However, nutrient budget studies indicate that even for soil-less cultures, nutrient inputs via nutrient solutions exceed nutrient output via crop uptake. This is puzzling because nutrient losses via leakages, leaching and volatilization are, with good reasons, assumed to be negligible.

The purpose of the study presented here was to obtain a balanced nutrient budget of a simple, one-dimensional soil-less culture, and to identify possible biases and errors in the measurements. At four institutes an identical solution culture experiment was carried out according to one 'working plan' using the same initial plant material and nutrient solution (Heinen et al., 1996). Four different institutes were involved to gain insight into the possible biases, and four identical experiments were performed to gain insight into possible errors. In the experiment in each of the four participating institutes, four tomato plants were grown on a nutrient solution in greenhouses for 38 days. The nutrient solution was aerated (Fig. 4.3). It was assumed that nutrients only disappear from the nutrient solution through uptake by the crop. Hence, it was assumed that the nutrient depletion of the solution is equal to the nutrient accumulation in the tomato plants. Deviations from a perfect mass balance were expressed as the dimensionless quantity L (Loss) given by:

$$L = \frac{U_d - U_a}{U_a} \qquad (4.2)$$

where U_a (mmol) and U_d (mmol) represent the accumulation in the crop and depletion from the nutrient solution, respectively. U_a was defined as:

$$U_a = A_f - A_i \qquad (4.3)$$

where A_f and A_i are the final and initial amounts (mmol) present in the crop, respectively. It was assumed that A_i is negligible with respect to A_f so that U_a is equal to the final concentration in the crop: $U_a = A_f$. U_d was defined as:

$$U_d = C_i V_i - C_f V_f + D \qquad (4.4)$$

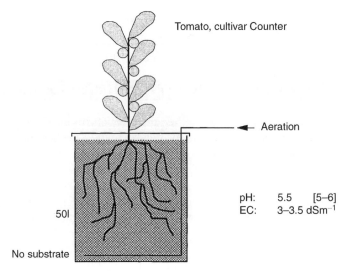

Fig. 4.3. Case 1. Schematic representation of the nutrient solution cropping system.

where C_i and C_f are the initial and final concentrations (mmol l^{-1}), respectively, V_i and V_f are the initial and final volumes (l) of the nutrient solution, respectively, and D is the amount (mmol) of nutrient added during the experiment. At the beginning of the experiment samples of the nutrient solution were sent to each of the four participating institutes for chemical analysis. At the end of the experiment samples of the nutrient solution and samples of oven-dried crop material (shoot, root, intermediately removed/fallen-off parts of the crop) were analysed by all four institutes. The computations were done by one of the participants.

Table 4.2 shows the results of the four identical experiments carried out at the four institutes separately. In most cases nutrient depletion from the solution exceeded by far nutrient accumulation in the plant ($U_d > U_a$). Deviation from a perfect mass balance was relatively small for chloride, potassium and sodium and relatively large for nitrogen, phosphorus, sulphur, calcium and magnesium. Differences between institutes were large, as illustrated by the high coefficients of variation belonging to the averaged values of L (Table 4.2). For chemically reactive elements, L was large suggesting that other processes occurred in addition to nutrient uptake. Hence, the initial assumption, that nutrient uptake by the crop was the only process that contributed to depletion of the nutrient solution, was incorrect. Possible processes involved at crop level are volatilization from leaves, for example as fragrance, or excretion of greasy substances at leaves which are removed by touching animals and/or humans. Possible processes involved at solution level are volatilization, denitrification, precipitation

Table 4.2. Results of case 1; deviation from perfect mass balance, 100% × L (equation 4.3 (%)), for all elements per institute, and the averages plus coefficient of variation (cv (%)).

Nutrient	Institute				Average	CV
	A	B	C	D		
N-tot	16.4	15.2	6.9	13.6	13.0	32.8
S	61.5	39.7	−17.9	28.6	28.0	119.8
P	48.5	0.4	50.5	21.3	30.2	79.2
Cl	−3.2	7.7	10.0	3.0	4.4	133.5
K	−4.6	19.4	−2.5	12.8	6.3	186.7
Ca	14.0	27.9	−12.2	41.1	17.7	128.7
Mg	3.1	3.5	68.9	33.0	27.1	115.0
Na	−29.8	115.8	−107.3	30.2	2.2	4213.0

and/or immobilization. None of these possibilities had been considered before as a significant factor.

Results of the analyses by the four institutes per experiment are shown in Fig. 4.4. Differences within an experiment appear often as large as the differences between experiments. Contrasting results were obtained for the experiment at Institute B on the one hand and experiments at Institutes A, C and D on the other hand. These results indicate that both measurement biases and measurement errors contributed to uncertainties in the overall nutrient balances. Additional data manipulation and discarding extreme values gave no improvement and a better balance between nutrient depletion from the solution and nutrient accumulation in the crop (Heinen *et al.*, 1996).

Summarizing, the results confirm the experiences of the soil-less glasshouse horticulturists that nutrient depletion of the recirculating solution is larger than the nutrient accumulation in the growing crop. These results challenge statements like 'glasshouse horticulture with recirculating nutrient solution can adopt balanced fertilization', that is, a nutrient input via fertilization that does not exceed nutrient removal in harvested crops.

CASE 2: EXPERIMENTAL DAIRY FARM 'DE MARKE'

The 55 ha large experimental farm for husbandry and environment, 'De Marke', has been designed to develop economically and environmentally sound dairy farming on dry sandy soils in The Netherlands. The farming system should meet all strict environmental targets for nitrate and phosphorus in groundwater and surface waters, and irrigation and ammonia

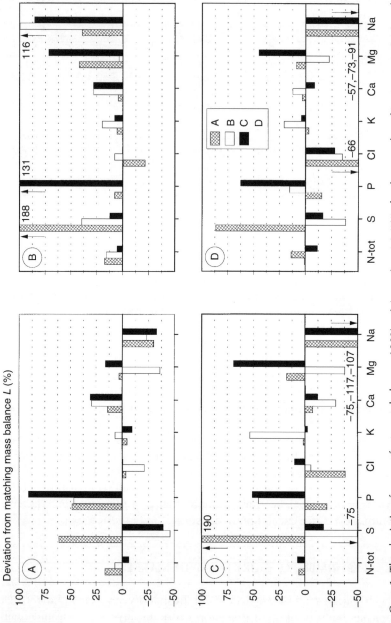

Fig. 4.4. Case 1. The deviation from perfect mass balance, $100\% \times L$ (equation 4.2 (%)), for the samples of institutes A, B, C and D as analysed by the four institutes.

volatilization targets (Aarts et al., 1992). Milk production (12,000 kg ha^{-1} year^{-1}) and the surpluses of nitrogen and phosphorus have been chosen as strategic instruments for the control of long-term productivity and environmental sustainability. Target surpluses of nitrogen and phosphorus are 128 and 0.5 kg ha^{-1} year^{-1}, respectively. Model calculations indicate that nitrogen losses via runoff, leaching, ammonia volatilization and denitrification are acceptably low when the nitrogen surplus is less than 128 kg ha^{-1} year^{-1}. An intensive research programme, involving the sampling and analyses of all significant nutrient flows and pools, has been set up to verify and test the model calculations and to identify possible constraints.

Nutrient budgets have been established at farm level, at field level and at compartment level, for the compartments cattle, manure, soil and crop. Budgets at compartment and field scales are being used to analyse the nutrient use efficiency, and to test the assumptions. Budgets at farm level have been derived in two ways: via aggregation of the field and compartment budgets, and directly using the farm gate approach (Aarts, 1996). These two approaches allow the consistency of the budgets to be checked. The budgets at farm level are being used to verify the performance of the farm at farm level.

In 1993, the second year, nitrogen surplus at farm level was 140 kg ha^{-1}. This surplus must be equivalent to the sum of the net nitrogen storage in the soil plus nitrogen losses via ammonia volatilization, denitrification, runoff and leaching (e.g. Goh and Williams, Chapter 14, this volume). The sum of the estimates for ammonia volatilization, denitrification, runoff and leaching amounted to 99 kg N ha^{-1}. The net storage of organic nitrogen in the soil was estimated at 17 kg ha^{-1}. The difference (24 kg ha^{-1}) between the surplus at farm level and the sum of nitrogen losses and net storage was attributed to measuring errors and to the neglect of possible losses from ensiled feed. To explore the possible influence of variance on the gap of 24 kg ha^{-1}, we conducted a Monte Carlo simulation. The variances were estimates from the results of the measurements. When estimates of the variances were lacking (e.g. biological nitrogen fixation, nitrogen immobilization mineralization turnover) we made best guesses (guesstimates).

Results of the simulations are shown in Fig. 4.5 as cumulative frequency distributions for the total input and total output of the system balance at farm level. It is clear that the variance is much larger for the total output than for the total input. Means and standard deviation of the total input were 222 ± 16 and for the total output 224 ± 44. Inputs with a relatively large variance were biological nitrogen fixation through clover, and outputs with a relatively large variance were ammonia volatilization, nitrate leaching, denitrification and net storage of nitrogen in the soil. The results suggest that a gap of 24 kg ha^{-1} between the mean surplus at the farm level and the mean sum of the nitrogen losses can easily be the result of the variance in the sum of the total losses.

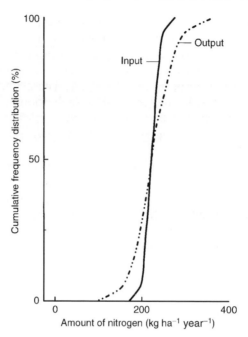

Fig. 4.5. Case 2. Cumulative frequency distributions of the sum of all measured nitrogen inputs and outputs at farm level of the dairy farming system De Marke in 1993.

Similar simulations were carried out to explore the variance in the estimates of the four compartments cattle, slurry, soil and crop (Fig. 4.6). The cumulative frequency distributions of total inputs minus total outputs indicate that the variance is large for the soil compartment. Cumulative frequency distributions for the compartments cattle, slurry and crop were rather similar; the total variance in these cases was dominated by the variance associated with nitrogen losses through ammonia volatilization.

Summarizing, estimates of the net storage of nitrogen in the soil, and of nitrogen losses through leaching, ammonia volatilization, denitrification and runoff at farm level are cursed with relatively large variances. The relatively large variance is probably the reason for the gap between the surplus calculated at farm level and the sum of measured nitrogen losses. An increase in the number of samples and analyses per year will not decrease this variance. Results of the monitoring programme on inputs, outputs and losses in subsequent years (Aarts, personal communication) indicate that the variance of the estimated net storage of nitrogen in the soil was even larger than assumed (in the simulation runs) on the basis of the results of the first year. Clearly, monitoring of nutrient budgets at field, compartment and farm levels over a number of years is required to further analyse uncertainties in nutrient budgets and to discriminate between biases and errors.

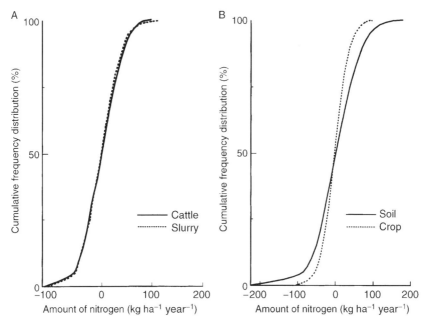

Fig. 4.6. Case 2. Cumulative frequency distribution of the measured nitrogen inputs and outputs at compartment level of the dairy farming system De Marke in 1993. Data for cattle and manure are shown on the left, data for soil and crop on the right.

CASE 3: NUTRIENT ACCOUNTING AS A POLICY INSTRUMENT

Agricultural land in The Netherlands, covering 60% of the total surface area, is intensively managed. Inputs of nitrogen and phosphorus into the agricultural system via purchased fertilizers and feeds have boosted agricultural production to a high level, but have also contributed to very large nutrient surpluses and nutrient losses. A series of policies and measures have been implemented by the government from 1987 onwards, to decrease nutrient losses to environmentally acceptable levels. The last step in this series is the implementation of the nutrient accounting system MINAS (Minerals Accounting System). From 1998 onwards, livestock farms with a livestock density of more than 2.5 livestock units ha^{-1} will have to register and report at farm level all the incoming and outgoing nitrogen and phosphorus with imported and exported products on an annual basis. Other livestock farms will follow in 2000 and arable farms and vegetable growers in 2002.

The MINAS accounting system is a farm gate approach. All nitrogen and phosphorus entering the farm gate and leaving the farm gate in products is quantified. Inputs like atmospheric deposition, net mineralization of soil organic nitrogen and free-living nitrogen fixing soil organisms, which do not pass the farm gate and which are unmanageable by the farmer, are excluded from the budget calculations. Biological nitrogen fixation via leguminous crops has also been excluded so far, because this nitrogen input is very difficult to quantify and certify accurately.

There are levy-free surpluses for nitrogen and phosphorus, which steadily diminish between 1998 and 2008 until surpluses have been achieved that are considered acceptable by the government. These levy-surpluses will be uniformly applied on all agricultural soils, except for sandy soils sensitive to nitrate leaching. Levies will encourage farmers not to exceed levy-free surpluses. It is the height of the levies and the height of the levy-free surpluses together which regulate the total nitrogen and phosphorus surpluses, and hence the total nitrogen and phosphorus losses. Clearly, the government intends to manage the surpluses in practice by simply managing the height of the levy-free surpluses and the height of the levies. Initially, levy-free surpluses are a compromise between 'what is environmentally acceptable and what is agronomically possible'. MINAS will have severe economic consequences for intensive livestock production systems. At present, nitrogen and phosphorus surpluses of intensive livestock production systems are a factor of two and more higher than the target levy-surpluses. To adjust to the levy-free surpluses, intensive livestock farms have to improve their nutrient management or properly dispose of the excess manure to other farms or cut the number of livestock, or both.

Averaged over the whole agricultural sector, nitrogen surpluses have to decrease from a mean of 299 in 1995 to a mean of 147 in 2008. Phosphorus surpluses have to decrease from 31 to 8.7 kg P ha^{-1} year^{-1}. As a result, the surface area of agricultural land with more than 50 mg of nitrate in the shallow groundwater will drop from 35 to 12% (Oenema et al., 1998). Ultimately, losses of nitrogen from agricultural land to surface waters will decrease by more than 50% (Oenema and Roest, 1998). Losses of phosphorus from agricultural land to surface waters will not decrease so much. These studies indicate that MINAS is an effective regulatory instrument for nutrient losses from agricultural land to the wider environment. The studies also indicate that more tight policies and measures are required region-wide.

Although it was initially the farmers' organization that convinced the government to choose a nutrient accounting system at farm level as a policy instrument for the regulation of nitrogen and phosphorus losses from agriculture, it is clear now that many farmers are reluctant. All farmers face an increasing amount of administration and registration. Intensive livestock production farms face large economic consequences such as increasing levies, increased costs for disposal (distribution) of animal manure and/or lower production because of a decrease in livestock number. Because of these large economic consequences, they may find alternative routes for lowering the nitrogen and phosphorus surpluses. To circumvent illegal routes, the nutrient accounting system needs to be accurate and controllable. An accurate and controllable accounting system requires proper guidelines and standardized registration and administration.

Thus far, the following lessons have been learned from the introduction of the nutrient accounting system MINAS as a regulatory policy instrument in practice:

- All agricultural land and all farms have to be included, also farms which have already acceptably low surpluses of nitrogen and phosphorus, to prevent unwanted problem shifting.
- Ownership and leasehold of all land needs to be documented officially.
- Only accredited manufacturers and organizations are allowed to purchase, sell and analyse nutrient-containing products, including animal manures, from and to farms; the volume and nutrient content of all products need to be registered by both purchaser and seller.
- Certified procedures are needed for the sampling and analysis of nutrient stocks and nutrient flows.
- A special Bureau has to check the budgets and charge farmers when surpluses exceed levy-free surpluses.
- The height of levy-free surpluses needs to be a compromise initially between what is environmentally acceptable and agronomically possible; farmers need time to adopt more efficient nutrient management planning.
- The height of levies needs to be a compromise between justice and pragmatism.

There are a number of possible flaws that have not been solved yet:

- Nitrogen input via biological nitrogen fixation has not been included in the accounting system because this input is difficult to quantify accurately. However, the present mono-species grasslands will probably change into clover-grassland when levy-free nitrogen surpluses steadily decrease. This may introduce a large bias in the registered budget.
- Yields and nutrient outputs via harvested crops are higher on fertile soils than on soils low in fertility, provided that fertilizer input is low. Hence, farmers benefit from the nutrients accumulated in the soil during previous manuring practices. Moreover, nutrient losses will be higher from the heavily manured soils than from the properly manured soils. Unless MINAS budgets are coupled with the soil nutrient store (see Janssen, Chapter 2, this volume), the environment still pays the penalty and the farmer receives the benefits from manure disposal practices in the past.
- Farmers will only purchase high quality fertilizers with high efficiency, and will be reluctant to use waste products. However, these waste products, for example from the processing industry, have to be recycled somewhere in order to prevent problem shifting.

- To minimize costs for sampling and chemical analysis, and to facilitate the registration of inputs and outputs, default values for a large number of products have been made. Farmers may choose to use either default values or data from the sampling and analysis by accredited organizations. However, annual variations and deviations between default values and actual values can be large, and there exists the possibility for systematic deviations (bias).
- The nutrient accounting system MINAS corrects for changes in the stock of nutrients in cattle, ensiled feed and animal manure stored in basins, but not for changes in the soil nutrient stock, which in fact is the largest store and buffer of nutrients. Systematic deviations (bias) in the overall budget will show up in the soil nutrient stock, in the long term, and analyses of these stocks may provide a check of MINAS.

Summarizing, the nutrient accounting system MINAS is the first experience with nutrient budgets as a repressive policy instrument in practice. When used as such, unwanted and wanted errors and bias in the budgets have direct economic and environmental consequences. Careful procedures and documentation are required to minimize these consequences. Thus far, experience indicates that there are many potential sources of errors and bias in MINAS in practice, but through tact and close cooperation with farmers, the effects of these can be minimized.

CONCLUSIONS

Different approaches exist for making nutrient budgets of agroecosystems. These approaches can be grouped into: (i) farm gate or black-box balance; (ii) soil surface balance; and (iii) system balance. Application of these approaches to an agroecosystem may yield different results. Evidently, the approach used for nutrient budgeting must be clearly indicated when presenting results.

Uncertainty in nutrient budgets traces back to a number of possible biases and errors, which depend largely on the complexity of the ecosystem under study and on the understanding of nutrient cycling and nutrient transformation processes. Proper guidelines and uniform procedures for nutrient budgeting are required to minimize biases and errors and to facilitate comparison of budgets between ecosystems.

The flow of nutrients into and out of an agroecosystem is generally a small fraction of the total stock of nutrients within that system. As a result, changes in the nutrient stock can be inferred more accurately from the net inflow of nutrients than from direct measurements of changes in the nutrient stock. Further, the flow of nutrients associated with agricultural produce, fertilizers and manures can be determined much more accurately than nutrient losses via leaching, volatilization, erosion and runoff. Systems analysis and classification of items into different categories of uncertainty is essential for cost-effective nutrient budgeting.

Uncertainty in nutrient budgets is tractable and biases and errors may occur simultaneously. The surprisingly large uncertainty due to both errors and biases in the nutrient budget of the very simple soil-less culture in case 1 is partly due to the absence of a nutrient buffer in the system, partly also due to the summing up of a number of individual analyses into one item. The study also shows that uncertainties in the budget differ greatly between different nutrients, being largest for reactive ones. Evidently, this case needs further study. Analysing the nutrient budgets of different compartments, and comparing these budgets with the overall budget of the system provides a check of the consistency in the budgets and is very helpful to trace back possible biases and errors, as shown in case 2.

Nutrient budgets are manipulatable. Fraud may occur when budgets are used as a regulatory policy instrument and when economic consequences for the manager of the agroecosystem are significant. Education and careful procedures and documentation are needed to prevent juggling with budgets, as illustrated in case 3. On the other hand, the fact that juggling with nutrient budgets occurs is in itself an indication that nutrient budgeting is a powerful instrument.

The soil is the largest buffer and stock of nutrients, but is highly variable. Nutrient losses from soil to the wider environment occur often in hot spots during peak events. Quantifying these losses requires flow proportional sampling and analysis for prolonged periods. Whilst the sampling and analysis of individual samples can be done reasonably accurately and precisely, the uncertainty in the sum of all individual samples remains large compared with, for example, the uncertainty in the estimate of nutrient output via agricultural produce. Monitoring nutrient budgets of the various compartments of an agroecosystem, including the monitoring of nutrient losses from and nutrient stocks in the compartments, over a number of years, provides a very powerful check of the consistency in the measured nutrient flows, as illustrated by 'De Marke' in case 2. Evidently, long-term studies are needed, both for testing the sustainability of improved nutrient management strategies in practice, and for testing and improving the accuracy of nutrient budgets.

REFERENCES

Aarts, H.F.M. (1996) *Efficient Nutrient Management in Dairy Farming on Sandy Soils*. Technical results of the experimental farm 'De Marke' for the years 1993/94 and 1994/95. Rapport no. 18, AB-DLO rapport 67, Wageningen, 24 pp.

Aarts, H.F.M., Biewinga, E.E. and van Keulen, H. (1992) Dairy farming systems based on efficient nutrient management. *Netherlands Journal of Agricultural Science* 40, 285–299.

Allison, F.E. (1955) The enigma of soil nitrogen balance sheets. *Advances in Agronomy* 7, 213–250.

Bollman, A. and Conrad R. (1997) Acetylene blockage techniques leads to underestimation of denitrification rates in oxic soils due to scavenging of intermediate nitric oxide. *Soil Biology and Biochemistry* 29, 1067–1077.

Clark, F.E. (1981) The nitrogen cycle, viewed with poetic license. In: Clark, F.E. and Rosswall, T. (eds) *Terrestrial Nitrogen Cycles. Processes, Ecosystem Strategies and Management Impacts*. Ecological Bulletins No 33, Swedish Natural Science Research Council (NFR), Stockholm, pp. 13–24.

Dumanski, J,. Pettapiece, W.W. and McGregor, R.J. (1998) Relevance of scale dependent approaches for integrating biophysical and socioeconomic information and development of agroecological indicators. *Nutrient Cycling in Agroecosystems* 50, 13–22.

Frissel, M.J. and Kolenbrander, G.J. (1977) The nutrient balances. *Agro-Ecosystems* 4, 277–299.

Garrett, M.K., Watson, C.J., Jordan, C., Steen, R.W.J. and Smith, R.V. (1992) *The Nitrogen Economy of Grazed Grassland*. The Fertilizer Society Proceedings No. 326, pp. 1–32.

Heathwaite A.L., Burt, T.P. and Trudgill Burt, S.T. (1993) Overview – the nitrate issue. In: Burt, T.P., Heathwaite, A.L. and Trudgill, S.T. (eds) *Nitrate: Processes, Patterns and Management*. John Wiley & Sons Ltd, Chichester, UK, pp. 3–12.

Heinen, M., Sonneveld, C., Voogt, W., Baas, R., Keltjens, W.G. and Veen, B.W. (1996) *Mineral Balance of Young Tomato Plants Grown on Nutrient Solutions*. AB-DLO report 66, Haren, The Netherlands, 47 pp.

Heuvelink, G.B.M. (1998) Uncertainty analysis in environmental modeling under a change of spatial scale. *Nutrient Cycling in Agroecosystems* 50, 255–264.

Houba, V.J.G. and Uittenbogaard, J. (1994) *Chemical Composition of Various Plant Species*. International Plant-Analytical Exchange (IPE). Wageningen Agricultural University, Wageningen, The Netherlands, 226 pp.

Jansen, M.J.W. (1998) Prediction of error through modelling concepts and uncertainty from basic data. *Nutrient Cycling in Agroecosystems* 50, 247–253.

Janssen, P.H.M., Slob, W. and Rotmans, J. (1990) *Sensitivity and Uncertainty Analyses; an Inventory of Ideas, Methods and Techniques* (in Dutch). RIVM report 958805001, Bilthoven, The Netherlands, 119 pp.

Karlovsky, J. (1981) Cycling of nutrients and their utilization by plants in agricultural ecosystems. *Agro-Ecosystems* 7, 127–144.

McKenny, D.J., Drury, C.F. and Wang, S.W. (1997) Reaction of nitric oxide with acetylene and oxygen: implications for denitrification assays. *Soil Science Society of Ameria Journal* 61, 1370–1375.

Moolenaar, S.W., Lexmond, T.M. and van der Zee, S.E.A.T.M. (1997) Calculating heavy metal accumulation in soil: a comparison of methods illustrated by a case-study on compost application. *Agriculture, Ecosystems and Environment* 66, 71–82.

Oenema, O. and Roest, C.W.J. (1998) Nitrogen and phosphorus losses from agriculture into surface waters; the effects of policies and measures in the Netherlands. *Water Science and Technology* 37, 19–30.

Oenema, O., Boers, P.C.M., van Eerd, M.M., Fraters, B., van der Meer, H.G., Roest, C.W.J., Schröder, J.J. and Willems, W.J. (1998) Leaching of nitrate from agriculture to groundwater; the effects of policies and measures in the Netherlands. *Journal of Environmental Pollution* 102, 471–478.

PARCOM (1995) *PARCOM Guidelines for Calculating Mineral Balances.* Oslo and Paris Conventions for the Prevention of Marine Pollution, PRAM 95/7/6-E, Oviedo, 9 pp.

Romstad, E., Simonson, J. and Vatn, A. (eds) (1997) *Controlling Mineral Emissions in European Agriculture; Economics, Policies and the Environment.* CAB International, Wallingford, UK, 229 pp.

Sharpley, A.N. and Rekolainen, S. (1997) Phosphorus in agriculture and its environmental implications. In: Tunney, H., Carton, O.T., Brookes, P.C. and Johnston, A.E. (eds) *Phosphorus Loss from Soil to Water.* CAB International, Wallingford, UK, pp. 1–53.

Smaling, E.M.A. and Fresco, L.O. (1993) A decision-support model for monitoring nutrient balances under agricultural land use (NUTMON). *Geoderma* 60, 235–256.

Smaling, E.M.A. and Oenema, O. (1997) Estimating nutrient balances in agroecosystems at different spatial scales. In: Lal, R., Blume, W.H., Valentine, C. and Stewart, B.A. (eds) *Methods for Assessment of Soil Degradation.* Advances in Soil Science, CRC Press, New York, pp. 229–252.

Snedecor, G.W. and Cochran, W.G. (1967) *Statistical Methods.* The Iowa State University Press, Ames, Iowa, 593 pp.

Velthof, G.L. and Oenema, O. (1995) Nitrous oxide fluxes from grassland in the Netherlands: I. Statistical analysis of flux chamber measurements. *European Journal of Soil Scieence* 46, 533–540.

Wagenet, R.J. (1998) Scale issues in agroecological research chains. *Nutrient Cycling in Agroecosystems* 50, 23–34.

Technologies to Manage Soil Fertility Dynamics

J.J. Stoorvogel

Laboratory of Soil Science and Geology, Wageningen Agricultural University, PO Box 37, 6700 AA Wageningen, The Netherlands

ABSTRACT

As a result of inherent soil fertility as well as differences in land use history, soil fertility is extremely variable. Soil fertility is managed for several reasons: (i) to obtain high production levels; (ii) to maintain soil fertility; (iii) to safeguard environmental quality; and (iv) to obtain high quality products. The different objectives may lead to different perspectives on soil fertility management. The trade-offs between the different objectives and a priority setting by the stakeholders can be analysed to define a target soil fertility. A wide range of technologies is available for soil nutrient management, ranging from nutrient adding to nutrient saving. Using different technologies one can reach the target soil fertility. To select between the different technologies one should quantify the different soil nutrient flows and the derived nutrient use efficiency of the cropping system. Each of the technologies will influence one or more soil nutrient flows. An economic analysis in combination with crop growth simulation models can be used to determine the optimal strategy. In the economic analysis the sum of the costs for a possible reduction of yield, the cost of nutrient management, and the cost for possible environmental damage is minimized. Given specific soil fertility conditions, the objectives, and the socioeconomic environment, different technologies for soil fertility management can be selected from a large toolbox available to the farmers.

INTRODUCTION

Farmers and agricultural science have developed a wide range of management interventions to influence soil nutrient balances. These management interventions range from nutrient adding technologies, that is, nutrient additions through mineral or organic amendments, to nutrient saving technologies, that is, technologies that aim at an increased nutrient use efficiency through, for example, split application and agroforestry systems (Bationo et al., 1996; Smaling and Braun, 1996). Typically, one single technology does not solve nutrient-related problems and solutions have to be sought in a suite of technologies through integrated nutrient management. For example, when fertilizer is applied, technologies to prevent these nutrients from being lost to the environment should be applied at the same time. Despite a large toolbox with management interventions, soil nutrient management is still facing many problems. In marginal areas, soil nutrient depletion continues to affect soil fertility (e.g. Stoorvogel and Smaling, 1990; De Koning et al., 1997), whereas in high-input agriculture in Europe and the USA nutrient surpluses lead to environmental problems (Council for Agricultural Science and Technology, 1992; Van der Ploeg et al., 1995).

In the case of soil nutrient depletion, management interventions are necessary to sustain future agricultural production. The farming system determines the technologies to focus on. Subsistence farming may, in theory, have a more or less closed nutrient cycle. If, nevertheless, the system is depleting, nutrient saving technologies help to close the soil nutrient balance. In the case of market-oriented agriculture large quantities of nutrients will leave the agricultural system to urban areas or find their application in agroindustry for example. 'Natural' nutrient inputs, through deposition and nitrogen(N)-fixation, are not able to compensate for these losses. Therefore, market-driven agriculture needs an external source of nutrients.

Nutrient accumulation increases soil fertility and thus agricultural production. However, above certain levels, it will also result in an increased risk of nutrient emissions to the environment. Neutral soil nutrient balances do not necessarily imply that nutrient losses to the environment are low. Even with zero nutrient balances, high external input systems can lose large quantities of nutrients. It implies, however, that production is maintained and that the risk for nutrient emissions to the environment remains the same. Technologies to avoid nutrient accumulation and environmental contamination focus on neutral soil nutrient balances and high nutrient use efficiency. Nutrient use efficiency is here defined as the ratio between nutrients being used for crop production and the nutrients added to the system. With a high nutrient use efficiency and a neutral nutrient balance, contamination of the environment will be minimal.

Originally, much research focused on agricultural production (Colwell, 1994), but a gradual shift was made towards a long-term perspective, considering both current and future production as well as the environmental effects of nutrient emissions (Smyth *et al.*, 1993). Nowadays sustainable development strives for a high agricultural production of high quality agricultural products and maintaining soil and water quality. Maintaining soil quality implies soil nutrient balances are in equilibrium in the long run. From a practical, more pragmatic point of view, it can be questioned whether inherent soil nutrient stocks should and can be preserved. Clearing of tropical rain forests and subsequent agricultural land use, for example, liberate large quantities of nutrients (Juo *et al.*, 1995). Whereas not all these nutrients can be taken up by agricultural crops and/or secondary regrowth, and temporary fixation at the adsorption complex of the soil is also limited, the surplus is likely to be lost through leaching and denitrification. Management interventions to prevent these nutrient losses are costly and difficult to implement. However, with proper soil management, nutrient stocks stabilize at a fertility level that is lower but still acceptable for both agriculture and environment. The extent of tolerable soil nutrient depletion and accumulation should be well defined with threshold values specific to the situation defined by the soil nutrient stocks, crop requirements and the spatial and temporal scale. Scale issues are important as compensation in space and time takes place (Stoorvogel and Smaling, 1998). A depleting land use system, for example, is acceptable if followed by a fallow period where soil nutrient stocks are restored (for example in traditional slash and burn systems). An agricultural system with high emissions of nutrients to the environment may be similarly acceptable if surrounded by systems with low emissions.

This chapter starts with a discussion on the objectives of soil nutrient management. Why do we want to manage soil nutrients? It is followed by an inventory of the toolbox that is available to farmers to manage soil nutrients. Farmers will have to select from this toolbox. Finally long-term objectives of soil nutrient management will be discussed.

OBJECTIVES OF NUTRIENT MANAGEMENT

Soil nutrient management strives for different objectives:

- The maximization of agricultural production (in terms of actual production in the case of food security or in terms of net income in the case of market-oriented agriculture).
- Soil fertility maintenance.
- Environmental quality through the minimization of nutrient emissions to the environment.
- The production of high-quality agricultural products.

Maximize Agricultural Production

Agricultural production can be maximized through effective nutrient management. Maximum agricultural production will be reached when the supply of nutrients removes all nutrient limitations. In the case of the maximization of net returns, nutrients are supplied to the level where the cost of additional fertilization equals the corresponding increase of returns from additional production (De Wit, 1994). To reach maximum agricultural production, the proper technologies need to be selected that take care of the alleviation of all nutrient constraints. The application efficiency and the uptake efficiency need to be maximized to maximize the economic returns from the system.

Soil Fertility Maintenance

In sustainable agriculture one of the main aims is a steady state, where soil nutrient stocks are neither being mined nor accumulating. This implies a situation where, given specific spatial and temporal system boundaries, the inputs of nutrients equal the outputs of nutrients. Using different technologies, soil nutrient flows can be managed and, as a consequence, they can be used to stabilize losses or accumulation.

Environmental Quality

Especially in western high-input agriculture, high fertilization rates have led to the contamination of groundwater and surface water (e.g. Council for Agricultural Science and Technology, 1992). Increasingly, regulations aim to control the environmental effects of agriculture. Farmers need to increase the efficiency of their nutrient management to reduce these emissions.

High-quality Agricultural Products

Especially in western agriculture consumers and/or industry increasingly ask for high-quality agricultural products. The quality of many agricultural products is a function of nutrient management. In the sugar industry, for example, the sugar concentration in sugarbeet is more important than the total weight of the beets. Sugar concentrations are a function of soil nutrient management (Smit *et al.*, 1995). Similarly, the quality of barley for brewing malt is strongly related to fertilization (Russel, 1986). Another well-known example is the demand of consumers for 'organically grown' crops where, among other constraints, no chemical fertilizer is being used.

The Realization of the Objectives

In many cases the interests of the above objectives are conflicting. Figure 5.1 illustrates the complex interrelations between the different objectives as a function of fertilization. Each of the above objectives corresponds with certain optimal nutrient conditions in the soil during the growing season. Therefore, it is unlikely that all the objectives can be met simultaneously. As a result, the trade-offs are often a cause of debate (Viglizzo and Roberto, 1998). It is the task of soil fertility research to define the playing ground and the trade-offs between the different objectives. For example, to obtain a maximum yield, nutrient levels need to be so high that emissions to the environment cannot be prevented. The stakeholders need to determine the relative importance of the different objectives and determine a target soil fertility (see also Janssen, Chapter 2, this volume). The target soil fertility is the result of the inherent soil fertility, nutrient additions (from natural sources as well as from management) and nutrient use efficiency. The last factor can be subdivided into application efficiency, uptake efficiency, utilization efficiency and the harvest efficiency (van Noordwijk, Chapter 1, this volume).

THE TOOLBOX

Farmers around the world are equipped with a toolbox of alternative management practices that influence soil fertility directly or indirectly (see e.g. Uzo Mokwunye *et al.*, 1996; Braun *et al.*, 1997). The net inputs and outputs of nutrients of an agricultural system are listed in Table 5.1. Each technology will influence a number of soil nutrient flows. Fertilization, for

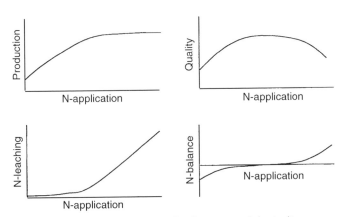

Fig. 5.1. Relationship between nitrogen fertilization and the indicators corresponding to the main objectives of soil nutrient management.

example, has a positive effect on N-fixation in nutrient limiting situations (see e.g. Jefing *et al.*, 1992). In Table 5.2 different technologies and their major effects on nutrient flows are listed. The agroecological conditions in combination with the socioeconomic setting and the objectives of nutrient management determine 'best' management.

External Nutrient Inputs

Mineral fertilizers

Probably the most widespread form of nutrient management is the application of mineral fertilizers. In most agricultural systems, natural nutrient

Table 5.1. Major nutrient inputs and outputs (Stoorvogel and Smaling, 1990).

Inputs		Outputs	
IN 1	Mineral fertilizer	OUT 1	Harvested product
IN 2	Organic fertilizer	OUT 2	Crop residues
IN 3	Deposition	OUT 3	Gaseous losses
IN 4	N-fixation	OUT 4	Leaching
IN 5	Sedimentation	OUT 5	Erosion

Table 5.2. Technologies and their effect on the different soil nutrient flows.

	Input				Output				
Technology	1	2	4	5	1	2	3	4	5
Mineral fertilizers	++				+				
Organic fertilizers		++			++		−		−
Fertilizer management					+		−−	−−	
Leguminous crops			++						
Non-symbiotic N-fixation			++						
Green manures		+						−−	
Agroforestry								−−	−
Crop improvement						−−	−		
Manure collection/storage		++						−	−
Livestock systems		+						−	
Erosion control									−−
Precision agriculture					+		−	−	−
Tillage								−	−
Pest management					+		−	−	−
Irrigation				+	+			−	
Regional-level solution	+	+			+				−

inputs through wet and dry deposition do not fulfil crop requirements. Therefore, the application of mineral fertilizers is required to alleviate nutrient constraints of the system. In general, mineral fertilizers do not contribute to an increase in soil nutrient stocks. Most of the nutrients are either taken up by the crop or lost through leaching and/or denitrification. Fertilizers in an insoluble form (e.g. rock phosphate), however, may contribute to the soil nutrient stock (Shapiro and Sanders, 1998).

Organic fertilizers
Increasingly the use of organic fertilizers (e.g. manure, crop residues) is propagated as an alternative and more sustainable form of soil amendment. The potential for organic fertilizers depends heavily on their availability. Typically, they are bulky and transport over long distances is economically not very attractive. Their use is therefore concentrated in mixed cropping systems where they originate from local sources. The use of organic fertilizers serves two purposes: the supply of essential nutrients for crop growth and the increase of organic matter contents in the soil. The latter results in an increase in the water-holding capacity of the soil, an increase in the soil nutrient stock, and prevents nutrients from leaching (see e.g. Zaccheo *et al.*, 1997).

Fertilizer Management

Simultaneously with fertilizer applications, farmers need to consider technologies that increase the efficiency of the application. A well-known measure is the split applications of mineral fertilizers to avoid high concentrations of mineral fertilizers in the soil that may be prone to leaching. Through split applications the efficiency increases by adapting fertilizer applications to match nutrient requirements of the crop with nutrient availability in the soil (Wilson *et al.*, 1998).

Although excess manure production has led to many environmental problems in Western Europe, in general organic fertilizers are scarce. It is, therefore, important that manure and crop residues, the two major sources of organic fertilizers, are stored and treated efficiently. For crop residue management, burning is generally considered to be the most non-sustainable, although in areas with serious weed problems it may be the only alternative. Incorporation in the soil is the most effective form of application of crop residues and manure. However, technical limitations and labour availability often do not allow for incorporation. With limited amounts of organic matter available, mulching of crop residues can be restricted to particular sites to obtain maximum profits (Lamers *et al.*, 1998).

Crop Selection

Leguminous crops

Especially in the tropics the socioeconomic environment seriously constrains the external input of nitrogen. At the same time an increasing number of farms is shifting to 'ecological farming'. Despite the fact that nutrients are exported through agricultural products, ecological farming does not use external sources of nutrients. The introduction of leguminous crops in the systems can increase the input of nitrogen through symbiotic N-fixation. This is one of the few technological solutions to replenish nitrogen exported from the system without using external inputs of nitrogen. However, to obtain high efficiency, it is necessary to inoculate seed with nitrogen-fixing bacteria and to alleviate other limitations (e.g. phosphorus restrictions).

Non-symbiotic N-fixation in paddy fields

Nitrogen-fixing organisms, such as blue-green algae, photosynthetic bacteria, and heterotrophic bacteria, grow in the surface layer of paddy fields under ponding conditions (Kohno et al., 1995). They fix gaseous nitrogen from the air. During the decomposition of these organisms readily available ammonium nitrogen will be produced. With proper management, including the inoculation with blue-green algae and photosynthetic bacteria, up to 30–40 kg N ha^{-1} $year^{-1}$ can be fixed (Roger and Kulasooriya, 1980; White, 1997).

Green manures

Green manures can be grown for different purposes. In most cases, green manures are cultivated to increase the soil organic matter content of the soil (e.g. McGuire et al., 1998). However, they can also be planted as a catch crop after harvesting the main crop. Soil mineral nitrogen after harvest is the major source of nitrogen pollution of groundwater in Western agriculture (Van der Ploeg et al., 1995; Booltink and Verhagen, 1997). Green manures play an important role by taking up this excess nitrogen and preventing it from leaching.

Agroforestry/deep rooting species

Most annual crops have a rooting depth of 30–90 cm. In general, nutrients below the rooting depth are lost from the system. However, if crops are intercropped with deep rooting species, the system is able to recuperate these nutrients and, thus, reduce the losses through leaching. Many agroforestry systems make use of these concepts and have shown positive results in terms of the recuperation of soil nutrients from deeper soil layers (Sanchez, 1995; Shepherd et al., 1996). However, Janssen (Chapter 2, this volume) claims that agroforestry only provides a solution under N-limiting conditions whereas deep-rooting species are not able to capture phosphorus, for example, at larger depths.

Crop improvement
Crop improvement may lead to new cultivars that are better adapted to specific climatic and soil conditions and as a result have higher uptake and harvest efficiencies. The development of cultivars with larger harvest indices has been an enormous improvement in terms of nutrient use efficiency, mainly through the higher harvest efficiency.

Livestock Management

Three main livestock systems can be distinguished: free range, mixed crop/livestock systems, and industrial livestock systems (De Haan *et al.*, 1997). The spatial distribution of these systems is governed by agroecological conditions. In West Africa, with relatively low population densities and large areas of fallow land, free-range systems dominate. In East Africa, with high population densities and little fallow land, mixed crop/livestock systems dominate. In capital-intensive European agriculture industrial livestock systems dominate. The different systems are strongly related to nutrient flows, mainly as a result of manure management.

Manure, however, is one of the few sources of nutrients in many cropping systems in the tropics. Crop nutrient requirements are in many cases restricted to the growing season, but manure production continues throughout the year. Sound manure collection as well as storage is extremely important to obtain a large quantity of high-quality fertilizer. In addition, storage systems strongly influence the nutritional value of manure (Tiquia *et al.*, 1998). The labour intensity and low quality of organic fertilizers increasingly calls for a combined application with mineral fertilizer (Mwangi, 1997).

Erosion Control

Stoorvogel and Smaling (1990) concluded that one of the major exports of nutrients is through erosion. Even with relatively small erosion rates that will not alarm any soil conservation expert, the relative importance of nutrient export was already significant. A large array of soil conservation practices exists ranging from soil tillage (e.g. contour ploughing) to intercropping and terracing. Soil conservation generally leads to a reduction of nutrient losses.

Precision Agriculture

Although traditionally farm management was not varied within fields, soil variation was known to occur at the field level resulting in large variations in soil nutrient flows. Detailed soil surveys and crop monitoring may lead to more precise fertilizer recommendations that are site-specific and more precise in terms of the timing of the applications. In this so-called precision agriculture (Robert et al., 1995), the fertilizer application will be specifically tailored to local soil and crop conditions avoiding an excess of nutrients and thus nutrient losses (Verhagen, 1997). Although the techniques are mostly being applied in high-tech Western agriculture, similar principles can be followed in low-tech, tropical agriculture as well (Bouma et al., 1995; Brouwer and Bouma, 1997; Stoorvogel, 1998).

Indirect Measurements

Tillage

Tillage can significantly change mineralization rates as well as the soil moisture regime (Jenkinson, 1990). As a result it may influence, for example, the size and composition of the soil organic matter pool and especially the labile fraction that determines productivity in the short term (Hu et al., 1997). In addition, tillage influences water movement and thus soil nutrient losses through leaching.

Pest management

Fertilizer recommendations are based on the crop requirements for non-limited crop growth. However, pests and diseases may seriously affect crop performance, leaving as a result a large quantity of readily available nutrients in the soil, which are likely to be leached before the next crop cycle. High uptake efficiencies can only be reached with proper pest management.

Irrigation

Soil nutrient leaching is directly linked to water movement in the soil. As a result, proper irrigation management can control nutrient loadings to aquifers (Diez et al., 1997). In addition, irrigation can lead to healthy crops that are able to uptake soil nutrients efficiently.

Solution at the Regional Level

Although solutions are often sought at the farm level, agricultural policies and incentives may be effective tools to influence farmers in their land allocation and management decisions (Gerner et al., 1996; Marsh, 1997;

Kuyvenhoven *et al.*, Chapter 6, this volume). Solutions to nutrient depletion and accumulation need to focus on economically feasible and socially acceptable technologies. In short, it is important that farmers see the need for change, otherwise adoption rates of alternative technologies are likely to be low.

Policy interventions range from the intensive margin, where the external inputs are being regulated, to the extensive margin where, through incentives for example, land allocation decisions are being influenced (Antle and Just, 1992).

At the intensive margin, the local or national government places restrictions on, for example, nutrient depletion. The Dutch government introduced a system to register N and phosphorus (P) balances at the farm level to deal with the eutrophication of soils and surface waters (Anonymous, 1995). With levies on exceeding critical levels for nutrient losses, the government hopes to significantly reduce the eutrophication of the environment (Oenema and Heinen, Chapter 4, this volume).

At the extensive margin, governments may influence land allocation and land management decisions of farmers through price policies and incentives. Land tenure, for example, is one of the requisites for farmers to invest in soil fertility management (Forster, 1993). If farmers do not have ownership or some other guarantee for future use of a particular field, they will not be inclined to invest in sustaining soil fertility and rather mine soil nutrients and leave the farm. Land ownership will therefore lead to more sustainable land management. Also institutional support through extension and credit plays an important role in land allocation and management decisions (Acheampong *et al.*, 1996). Improved quantitative techniques for land use planning using optimization in combination with simulation modelling permit the *ex ante* evaluation of agricultural policies and incentives (Antle and Just, 1992; Lauwers *et al.*, 1998).

HOW TO SELECT THE PROPER TECHNOLOGIES

Increasingly, well-defined fertilizer recommendation systems for tropical environments (e.g. Schnier *et al.*, 1997) are becoming available besides the existing ones for Western agriculture. These recommendations, however, deal with the application of chemical (or in some cases organic) fertilizer to maximize agricultural production (Posner and Crawford, 1992; Colwell, 1994). They cannot be used for the selection of technologies for proper soil nutrient management, whereas the other objectives, soil fertility maintenance, environmental quality, and the quality of agricultural products, are not included. To deal with, for example, environmental quality, integrated approaches are needed (Bouma, 1997; Webster 1997). Simulation models can play an important role in the evaluation of alternative management scenarios on, for example, production and water quality (Eckersten et al., 1996; Azevedo et al., 1997; Verhagen, 1997).

A standard procedure to determine the best integrated approach for soil nutrient management is lacking. It is crucial to make a thorough analysis of the cropping system in terms of the soil nutrient balance, that is, the different nutrient inputs into and outputs from the system (Smaling and Fresco, 1993; Smaling et al., 1996). At the same time the interaction between different nutrients needs to be considered (Janssen, Chapter 2, this volume; Benbi and Biswas, 1997). On the basis of the nutrient flow analysis, efficiencies can be calculated. The knowledge and understanding of the cropping system needs to be reflected against the original objectives.

The long-term goal is in many cases a steady state where neither accumulation nor depletion of nutrients occurs and where the different objectives are met to a certain degree. The options for integrated nutrient management strongly depend on the systems boundaries (Stoorvogel and Smaling, 1998). Do we need to restore soil fertility? Do we accept the actual state of the soil, or do we realize that maintaining current soil fertility is not feasible given the socioeconomic boundary conditions and do we consequently allow for a certain reduction in soil fertility?

Leaching of nutrients is a function of the applied amount of nutrients as well as crop management (e.g. Verhagen et al., 1995; Hack-ten Broeke and van der Putten, 1997). Fertilizer application does not necessarily enhance nutrient concentrations in seepage water, as it may also stimulate nutrient uptake by the vegetation (Holscher et al., 1997).

TOWARDS RANGES AROUND EQUILIBRIUM

Although soil nutrient balances are preferably in equilibrium, nutrient depletion or accumulation may be quite well acceptable for certain periods of time and on specific areas of land. In general, there is consensus that tolerable ranges for nutrient depletion and accumulation increase with larger spatio-temporal scales. Crop rotations, for example, preferably have a more or less balanced nutrient stock. However, specific activities within the rotation may deplete soil nutrient stocks. An extreme case is a rotational slash and burn system where nutrient accumulation takes place during a fallow period. In a subsequent (relatively short) period of crop production the nutrients that have been built up are depleted.

Nevertheless, these general theories are difficult to translate into practical threshold values for nutrient balances. Additionally, changes in soil nutrient stocks may not be very important if they are seen independently of the size of the nutrient stocks and crop requirements. Instead of thresholds for the rates of change, it might be more relevant to establish critical levels for soil nutrient stocks. Imbalances are accepted, but soil fertility levels under or above these critical levels are not allowed.

Techniques for nutrient saving and adding can keep soil fertility virtually at any level. At the same time, proper nutrient management may also give high yields on low fertility soils and losses to the environment can be prevented even in soils with extremely high nutrient concentrations. However, net returns may be low or even negative. From an economic point of view soil nutrient stocks should be kept between certain critical levels. With the implementation of the technologies and the definition of thresholds it is necessary to look for economically viable and socially acceptable alternatives. In the case of soil nutrient depletion, an economic analysis can determine whether the additional cost of cultivation at lower fertility levels is preferred to restoration and maintenance of high soil fertility levels. The analysis may provide us with a target soil fertility as defined by Janssen (Chapter 2, this volume). The target soil fertility is a function of the different objectives for, respectively, production level, soil fertility maintenance, soil and water quality, and the quality of agricultural products.

How do we determine the target soil fertility? Assume a cropping system where crop production is a function of nutrients supplied by soil and through fertilization. Additionally, costs for soil nutrient management can be subdivided into costs that directly lead to increased nutrient availability to the crop (C_f) and thus higher yields, and costs to maintain soil fertility at a specific level (C_{sfm}). Soil fertility can be maintained at any level, but within low soil fertility environments the costs to maintain fertility will also be lower. If soil fertility decreases, we can either accept a decrease in crop productivity with the coinciding costs (C_{pl}) or we can maintain crop productivity with an increasing cost of fertilization. Figure 5.2 shows three different scenarios for soil fertility management:

- Scenario A: lower fertility levels are not compensated for by additional fertilization. As a consequence, lower fertility levels lead to a reduction of crop productivity.
- Scenario B: production loss is partially tolerated. Production will drop but is partially compensated by increased fertilization.
- Scenario C: no production loss is tolerated. As a result increased costs will have to be made with lower soil fertility levels to maintain productivity.

Each scenario shows that, given its boundary conditions, C_{sfm}, C_f, and C_{pl} vary for different soil fertility levels. In this simplified example, the optimal soil fertility level will be located in the point where the sum of the costs, i.e. $C_{sfm} + C_f + C_{pl}$, is minimal. Soil fertility beyond this equilibrium point will lead to a situation where costs to maintain soil fertility are higher than the benefits of production. If soil fertility is lower than the equilibrium, crop production loss will be higher and will not be compensated by the lower costs of C_{sfm} and C_f. Each scenario will have its specific point where costs are minimal, indicating that policy objectives (e.g. food security or

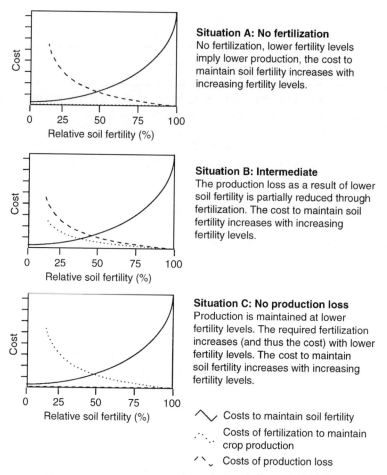

Fig. 5.2. Different scenarios for soil fertility management in relation to the different soil fertility levels.

low external input agriculture) determine the optimal soil fertility level. In reality, we have to deal with an analysis that incorporates the actual status of soil fertility. The transition from actual soil fertility to the target soil fertility should be incorporated in the analysis. Do we tolerate any rate of depletion until we are at a new, lower soil fertility level or does depletion have certain optimal ranges? Prices for agricultural inputs and outputs are likely to change in the future and so will the target soil fertility level, from an economic point of view. If the optimum changes to a higher soil fertility level, as a result of changes of input and output prices, the cost to restore soil fertility becomes important. In cases where the costs to restore soil fertility are large, one may have to consider how to avoid soil nutrient depletion in the beginning.

In cases where at high soil fertility levels nutrient emissions to the environment will take place, we will include the associated costs of pollution (C_p). These costs will increase with increasing soil fertility levels. The soil fertility, where the sum of the costs, $C_{sfm} + C_f + C_{pl} + C_p$, is minimal, corresponds to the optimal soil fertility level. Similarly, one can incorporate the costs associated with the reduced quality of agricultural products due to soil fertility management (C_{ql}). Again the total costs (= $C_{sfm} + C_f + C_{pl} + C_p + C_{ql}$) will be minimized.

The procedure will yield a target soil fertility. However, it does not include the current soil fertility and possible benefits or costs related to the transition from the current soil fertility to the target soil fertility.

CONCLUSIONS

- A large toolbox with alternative technologies is available for soil nutrient management ranging from nutrient adding technologies through external nutrient inputs to nutrient saving technologies that deal with nutrient use efficiency, crop selection, livestock management and precision agriculture.
- Soil nutrient management aims at the maximization of agricultural production, the maintenance of the soil nutrient stock, environmental quality and high quality agricultural products. Stakeholders will have to weigh the importance of the different objectives to come to a target soil fertility for the system.
- Successful nutrient management is the result of a set of farming practices aiming at a balanced soil nutrient stock that guarantees both current and future production. Changes in the socioeconomic environment require continuous adaptations of strategies and farming practices.

ACKNOWLEDGEMENT

The research of Dr Stoorvogel has been made possible by a fellowship of the Royal Netherlands Academy of Arts and Sciences.

REFERENCES

Acheampong, K., Gerner, H., de Jager, A., Honfoga, B. and Duivenbooden, N. (1996) Institutions and Support Services. In: Uzo Mokwunye, A., de Jager, A. and Smaling, E.M.A. (eds) *Restoring and Maintaining the Productivity of West African Soils: Key to Sustainable Development.* Miscellaneous fertilizer studies no 14. International Fertilizer Development Centre, Lomé, Togo, pp. 49–60.

Anonymous (1995) *Integral Memorandum on Manure and Ammonia Policy* (in Dutch). Ministry of Agriculture, Nature Management and Fishery, The Hague.

Antle, J.M. and Just, R.E. (1992) Conceptual and empirical foundations for agricultural-environmental policy analysis. *Journal of Environmental Quality* 21, 307–316.

Azevedo, A.S., Singh, P., Kanwar, R.S. and Ahuja, L.R. (1997) Simulating nitrogen management on subsurface drainage water quality. *Agricultural Systems* 55, 481–501.

Bationo, A., Rhodes, E., Smaling, E.M.A. and Visker, C. (1996) Technologies for restoring soil fertility. In: Uzo Mokwunye, A., de Jager, A. and Smaling, E.M.A. (eds) *Restoring and Maintaining the Productivity of West African Soils: Key to Sustainable Development*. Miscellaneous fertilizer studies no 14. International Fertilizer Development Center, Lomé, Togo, pp. 61–82.

Benbi, D.K. and Biswas, C.R. (1997) Nitrogen balance and N recovery after 22 years of maize-wheat-cowpea cropping in a long-term experiment. *Nutrient Cycling in Agroecosystems* 47, 107–114.

Booltink, H.W.G. and Verhagen, J. (1997) Using decision support systems to optimize barley management on spatial variable soils. In: Kropff, M.J., Teng P.S., Aggarwal, P.K., Bouma, J., Bouman, B.A.M., Jones, J.W. and van Laar H.H. (eds) *Systems Approaches for Agricultural Development*. Proceedings of the 2nd international symposium on systems approaches for agricultural development, Los Baños, Philippines, 6–8 December 1995. Kluwer Academic Publishers, Dordrecht, pp. 219–233.

Bouma, J. (1997) Soil environmental quality: a European perspective. *Journal of Environmental Quality* 26, 26–31.

Bouma, J., Brouwer, J., Verhagen, A. and Booltink, H.W.G. (1995) Site specific management on field level: high and low-tech approaches. In: Bouma, J., Kuyvenhoven, A. and Bouman, B.A.M. (eds) *Eco-regional Approaches for Sustainable Land Use and Food Production*. Proceedings of a symposium on eco-regional approaches in agricultural research, 12–16 December 1994, ISNAR, The Hague, pp. 453–473.

Braun, A.R., Smaling, E.M.A., Muchugu, E.I., Shepherd, K.D. and Corbett, J.D. (eds) (1997) *Maintenance and Improvement of Soil Productivity in the Highlands of Ethiopia, Kenya, Madagascar and Uganda: An Inventory of Spatial and Non-spatial Survey and Research Data on Natural Resources and Land Productivity*. AHI Technical Report Series no. 6, African Highlands Initiative, ICRAF, Nairobi.

Brouwer, J. and Bouma, J. (1997) *Soil and Crop Growth Variability in the Sahel: Highlights of Research (1990–95) at ICRISAT Sahelian Center*. Information Bulletin no. 49. ICRISAT, Andhra Pradesh and Agricultural University, Wageningen.

Colwell, J.D. (1994) *Estimating Fertilizer Requirements, a Quantitative Approach*. CAB International, Wallingford, UK.

Council for Agricultural Science and Technology (1992) *Water Quality: Agriculture's Role*. Task Force Report No. 120. Council for Agricultural Science and Technology, Ames, Iowa.

De Haan, C., Steinfeld, H. and Blackburn, H. (1997) *Livestock and the Environment, Finding the Balance*. European Commission Directorate-General for Development, Brussels.

De Koning, G.H.J., van de Kop, P.J. and Fresco, L.O. (1997) Estimates of subnational nutrient balances as sustainability indicators for agroecosystems in Ecuador. *Agriculture, Ecosystems and Environment* 65, 127–139.

De Wit, C. (1994) Resource use analysis in agriculture: A struggle for interdisciplinarity. In: Fresco, L.O., Stroosnijder, L., Bouma, J. and Van Keulen, H. (eds) *The Future of the Land. Mobilising and Integrating Knowledge for Land Use Options.* John Wiley & Sons, Chichester, pp. 41–55.

Diez, J.A., Roman, R., Caballero, R. and Caballero, A. (1997) Nitrate leaching from soils under a maize-wheat-maize sequence, two irrigation schedules and three types of fertilizers. *Agriculture, Ecosystems and Environment* 65, 189–199.

Eckersten, H., Jansson, P.-E. and Johnsson, H. (1996) *SOILN Model (ver 9.1) User's Manual, 3rd edition.* Swedish University of Agricultural Sciences Rep. 96.1, Uppsala.

Forster, N.A. (1993) Tenure systems and resource conservation in Central America, Mexico, Haiti and the Dominican Republic. In: Lutz, E., Pagiola, S. and Reiche, C. (eds) *Economic and Institutional Analyses of Soil Conservation Projects in Central America and the Caribbean.* World Bank Environment Paper number 8. The World Bank, Washington, DC, pp. 157–166.

Gerner, H., de Jager, A., Teboh, J.F., Bumb, B. and Dembélé, N.N. (1996) Economic policies and fertilizer market development. In: Uzo Mokwunye, A., de Jager, A. and Smaling, E.M.A. (eds) *Restoring and Maintaining the Productivity of West African Soils: Key to Sustainable Development.* Miscellaneous fertilizer studies no 14. International Fertilizer Development Centre, Lomé, Togo, pp. 35–48.

Hack-ten Broeke, M.J.D. and van der Putten, A.H.J. (1997) Nitrate leaching affected by management options with respect to urine-affected areas and groundwater levels for grazed grassland. *Agriculture, Ecosystems and Environment* 66, 197–210.

Holscher, D., Ludwig, B., Moller, R.F. and Folster, H. (1997) Dynamic of soil chemical parameters in shifting agriculture in the Eastern Amazon. *Agriculture, Ecosystems and Environment* 66, 153–163.

Hu, S., Coleman, D.C., Caroll, C.R., Hendrix, P.F. and Beare, M.H. (1997) Labile soil carbon pools in subtropical forest and agricultural ecosystems as influenced by management practices and vegetation types. *Agriculture, Ecosystems and Environment* 65, 69–78.

Jefing, Y., Herridge, D.F., Peoples, M.B. and Rerkasem, B. (1992) Effects of N-fertilization on N_2-fixation and N-balances of soybean grown after lowland rice. *Plant and Soil* 147, 235–242.

Jenkinson, D.S. (1990) The turnover of organic carbon and nitrogen in soil. *Philosophical Transactions of the Royal Society, London* 329, 361–368.

Juo, A.S.R., Franzluebbers, K., Dabiri, A. and Ikhile, B. (1995) Changes in soil properties during long-term fallow and continuous cultivation after forest clearing in Nigeria. *Agriculture, Ecosystems and Environment* 56, 9–18.

Kohno, E., Ogawa, Y. and Iwata, S. (1995) Productivity of paddy fields. In: Tabuchi, T. and Hasegawa, S. (eds) *Paddy Fields in the World.* JSIDRE, Tokyo, Japan, pp. 341–349.

Lamers, J., Bruentrup, M. and Buerkert, A. (1998) The profitability of traditional and innovative mulching techniques using millet residues in the West African Sahel. *Agriculture, Ecosystems and Environment* 67, 23–35.

Lauwers, L., Van Huylenbroeck, G. and Martens, L. (1998) A systems approach to analyse the effects of Flemish manure policy on structural changes and cost abatement in pig farming. *Agricultural Systems* 56, 167–183.

McGuire, A.M., Bryant, D.C. and Denison, R.F. (1998) Wheat yields, nitrogen uptake, and soil moisture following winter legume cover crop vs. fallow. *Agronomy Journal* 90, 404–410.

Marsh, J.S. (1997) The policy approach to sustainable farming systems in the EU. *Agriuculture, Ecosystems and Environment* 64, 103–114.

Mwangi, W.M. (1997) Low use of fertilizers and low productivity in sub-Saharan Africa. *Nutrient Cycling in Agroecosystems* 47, 135–147.

Posner, J.L. and Crawford, E.W. (1992) Improving fertilizer recommendations for subsistance farmers in West Africa: the use of agro-economic analysis of on-farm trials. *Fertilizer Research* 32, 333–342.

Robert, P.C., Rust, R.H. and Larson, W.E. (eds) (1995) *Site-Specific Management for Agricultural Systems*. ASA-CSSA-SSSA, Madison, Wisconsin.

Roger, P.A. and Kulasooriya, S.A. (1980) *Blue-green Algae and Rice*. IRRI, Los Baños, the Philippines.

Russel, G. (1986) *Fertilizers and Quality of Wheat and Barley*. Fertiliser Society, London.

Sanchez, P.A. (1995) Science in agroforestry. *Agroforestry Systems* 30, 5–55.

Schnier, H.F., Recke, H., Muchena, F.N. and Muriuki, A.W. (1997) Towards a practical approach to fertilizer recommendations for food crop production in smallholder farms in Kenya. *Nutrient Cycling in Agroecosystems* 47, 213–226.

Shapiro, B.I. and Sanders, J.H. (1998) Fertilizer use in semi-arid West Africa: profitability and supporting policy. *Agricultural Systems* 56, 467–482.

Shepherd, K.D., Ohlsson, E., Okalebo, J.R. and Ndufa, J.K. (1996) Potential impact of agroforestry on soil nutrient balances at the farm scale in East African Highlands. *Fertilizer Research* 44, 87–99.

Smaling, E.M.A. and Braun, A.R. (1996) Soil fertility research in sub-Saharan Africa: new dimensions, new challenges. *Communications in Soil Science and Plant Analysis* 27, 365–386.

Smaling, E.M.A. and Fresco, L.O. (1993) A decision-support model for monitoring nutrient balances under agricultural land use (NUTMON). *Geoderma* 60, 235–256.

Smaling, E.M.A., Fresco, L.O. and de Jager, A. (1996) Classifying and monitoring soil nutrient stocks and flows in African agriculture. *Ambio* 25, 492–496.

Smit, A.B., Struik, P.C. and van Niejenhuis, J.H. (1995) Nitrogen effects in sugarbeet growing: a module for decision support. *Netherlands Journal of Agricultural Science* 43, 391–408.

Smyth, A.J., Dumanski, J., Spendjian, G., Swift, M.J. and Thornton, P.K. (1993) *FESLM: an International Framework for Evaluating Sustainable Land Management: a Discussion Paper*. World Soil Resources Reports 73, FAO, Rome.

Stoorvogel, J.J. (1998) BanMan: A decision support system for banana management. In: Stoorvogel, J.J., Bouma, J. and Bowen, W. (eds) *Information Technology as a Tool to Assess Land Use Options in Space and Time*. Proceedings of an international workshop, 28 September – 4 October, 1997, Lima. Quantitative Approaches in Systems Analysis No. 16. The C.T. de Wit Graduate School for Production Ecology and DLO Research Institute for Agrobiology and Soil Fertility, Wageningen, pp. 13–22.

Stoorvogel, J.J. and Smaling, E.M.A. (1990) *Assessment of Soil Nutrient Depletion in Sub-Sahara Africa: 1983–2000*. 4 volumes. Report 28. The Winand Staring Centre for Integrated Land, Soil and Water Research, Wageningen.

Stoorvogel, J.J. and Smaling, E.M.A. (1998) Research on soil fertility decline in tropical environments: integration of spatial scales. *Nutrient Cycling in Agroecosystems* 50, 151–158.

Tiquia, S.M., Tam, N.F.Y. and Hodgkiss, I.J. (1998) Changes in chemical properties during composting of spent pig litter at different moisture contents. *Agriculture, Ecosystems and Environment* 67, 79–89.

Uzo Mokwunye, A., de Jager, A. and Smaling, E.M.A. (eds) (1996) *Restoring and Maintaining the Productivity of West African Soils: Key to Sustainable Development*. Miscellaneous fertilizer studies no 14. International Fertilizer Development Centre, Lomé, Togo.

Van der Ploeg, R., Ringe, H. and Muchala, G. (1995) Late fall site-specific soil nitrate upper limits for groundwater protection purposes. *Journal of Environmental Quality* 24, 724–733.

Verhagen, A. (1997) Site specific fertilizer application for potato production and effects on N-leaching using dynamic simulation modelling. *Agriculture, Ecosystems and Environment* 66, 165–175.

Verhagen, A., Booltink, H.W.G. and Bouma, J. (1995) Site-specific management: balancing production and environmental requirements at farm level. *Agricultural Systems* 49, 369–384.

Viglizzo, E.F. and Roberto, Z.E. (1998) On trade-offs in low-input agroecosystems. *Agricultural Ecosystems* 56, 253–264.

Webster, J.P.G. (1997) Assessing the economic consequences of sustainability in agriculture. *Agriculture, Ecosystems and Environment* 64, 95–102.

White, R.E. (1997) *Principles and Practice of Soil Science: the Soil as a Natural Resource*. Blackwell Science, Oxford, UK.

Wilson, C.E., Bollich, P.K. and Norman, R.J. (1998) Nitrogen application timing effects on nitrogen efficiency of dry-seeded rice. *Soil Science Society of America Journal* 62, 959–964.

Zaccheo, P., Genevini, P. and Ambrosini, D. (1997) The role of manure in the management of phosphorus resources at an Italian crop–livestock production farm. *Agriculture, Ecosystems and Environment* 66, 231–239.

Economic Policy in Support of Soil Fertility: Which Interventions after Structural Adjustment?

A. Kuyvenhoven, N. Heerink and R. Ruben

Development Economics Group, Department of Economics and Management, Wageningen Agricultural University, Wageningen, The Netherlands

ABSTRACT

Macroeconomic and sectoral policy measures significantly affect the conditions for decision-making at the farm household level; they have important, though often mixed, effects on agricultural development and the natural environment. Currency devaluation and output price reforms under structural adjustment in less developed countries (LDCs) create opportunities for improved nutrient management. But failing markets for labour, credit and other inputs (poverty-related), short time horizons, low farm incomes, insecurity of land use rights, and other typical features of low-income economies often prevent farmers from taking these opportunities, and may even lead to adverse effects on nutrient balances. In addition, cuts in public investments and input subsidies often have been detrimental to sustainable land use practices. Results of model simulations for southern Mali and Costa Rica show that soil mining practices can be economically efficient, even when more sustainable technology alternatives are available. In order to stimulate the implementation of sustainable nutrient management practices by farm households, economic reforms should include specific interventions to improve the functioning of markets by reducing the costs of transportation and other transaction costs, and improving access to credit, labour, information and soil fertility enhancing inputs. Where necessary, temporary price support may be used to facilitate the transition towards more sustainable practices.

INTRODUCTION

Negative nutrient balances are characteristic of most agricultural systems in sub-Saharan Africa and many systems in low-income countries elsewhere (e.g. Stoorvogel and Smaling, 1990; Van Lynden and Oldeman, 1997). Farm income levels are often too low to purchase enough mineral fertilizers and animal manure to compensate for the outflow of nutrients. Throughout the 1980s and 1990s, many of these low-income countries in Africa, Latin America and Asia implemented far-reaching macroeconomic and agricultural sector reforms to deal with short-term economic imbalances and structural distortions in their economies. Through their influence on the economic environment for farm household decision-making, these economic policy reforms have an important impact on agricultural development and on the related environment. In particular, changing crop and livestock production levels, the quantities of (imported) mineral fertilizers and animal manure used, and the profitability of soil conservation measures, can have a significant impact on nutrient balances.

The purpose of this chapter is to examine the effects of macroeconomic and other economic policy interventions on soil fertility in low-income countries, and to discuss appropriate policy interventions and incentives at the macro- and micro-level to encourage better land use practices. Because incomes are low and often uncertain and many markets function poorly or are even absent, the effects of policy measures on farm household decisions are likely to differ substantially from those in high-income countries. The resulting effects on the quality of the soil and other natural resources are often quite complex. Exchange rate devaluation, for example, immediately hurts rural households with little or no land through a rise in their cost of living. As a result, the quality of open access resources will deteriorate when basic food and firewood demand is going to be satisfied from increased use of marginal land and more deforestation. At the same time, households producing tradable agricultural commodities will face improved income and employment opportunities, which may induce them to shorten the fallow period but also enable them to spend more on fertilizers and soil conservation investments. After a certain time lag, higher agricultural product demand and lower real wages for rural labour may well result in hiring more casual labour in agricultural production. This may alleviate part of the resource degradation caused by the fall in the standard of living of landless households.

Macroeconomic imbalances (balance of payments deficits, fiscal deficits, inflation) can be outright detrimental to environmental conditions. Debt-ridden countries will hardly be impressed by the environmentally damaging practices of their export producers. Rather, such effects will be considered a reasonable cost of adjustment in much the same way as temporary unemployment. Austerity measures to curb unsustainable budget deficits tend to cut, in particular, subsidies and public investment (more so than public consumption), and may reduce expenditures on agricultural and environmental support programmes, research and investment alike.

Macroeconomic policy reforms constitute by no means sufficient conditions to induce sustainable land use and conservation measures for example. There are many reasons for this: (i) persistent poverty (of groups that do not benefit from the reforms) and few opportunities to transfer resources to the poor; (ii) structural imperfections in rural markets (high transportation and other transaction costs, access and information problems, absence of capital and insurance markets, lack of secure property rights); (iii) environmental 'goods' and 'bads' are not signalled in the market; (iv) many environmental goals have a trade-off in terms of income; and (v) environmental concerns are usually subordinated to growth and distributional issues in low-income countries. Specific policy interventions to safeguard natural resource use therefore remain indispensable, but should not be taken for granted in public decision making.

The following section briefly reviews the policy experiences in low-income countries with regard to agriculture and sustainability issues before and after structural adjustment, and presents an analysis of the impact of major economic policy instruments on sustainable land use and soil fertility. The question 'which economic policy interventions effectively support improved soil fertility management practices?' is then addressed. To this end, results of model simulations made with integrated bio-economic models of agricultural technology adjustment (for Costa Rica and southern Mali) are used. The chapter concludes with a discussion of how economic policy reform measures should preferably be devised in order to shape the conditions for improved soil nutrient management.

ECONOMIC POLICY REFORMS AND SOIL FERTILITY

General

During a substantial part of the post-war period, agriculture has not always been served well by policy-makers in less developed countries (LDCs), particularly in Latin America and Africa. Quantitative analyses of the effects of macroeconomic and sectoral policies show that: (i) rather than agricultural sector policies, exchange rate, trade and tariff policies have been the major determinants of agricultural prices and farmers' real incomes; (ii) agricultural price support, if any, has been insufficient to compensate for the strong price-depressing effect of macro-policies; and (iii) other compensating interventions (input subsidies, subsidized grain marketing) have often been regressive instead of progressive and were a significant burden for the government budget (World Bank, 1986; Krueger, 1995; Sadoulet and de Janvry, 1995).

During the 1970s and 1980s, many African and Latin American countries were confronted with increasing and unsustainable deficits on the balance of payments and government account, rising inflation and rapidly increasing foreign debts. When fundamental macroeconomic balances needed to be restored in the 1980s, agriculture's new sectoral policy orientation implied a change from a controlled but unprotected sector to a more free but moderately protected sector in line with other activities, without major claims on the budget (Norton, 1992, p. 52). Structural adjustment programmes (SAPs) sought to achieve this ambitious turn-round towards restoring equilibrium and enhancing efficient supply response of agricultural production. Most reforms therefore stressed the need to increase production and net export of agricultural products, to adjust the real exchange rate and to increase agriculture's productivity and efficiency.

After structural adjustment,[1] agriculture's contribution to overall welfare goals focuses on three major areas of policy concern: efficient pricing, higher productivity and shifts to higher-value crops. Proper agriculture policy will emphasize sound macroeconomic and trade policies to ensure correct price incentives, proper management of the natural resource base of agriculture, and an enabling structural and institutional policy to overcome access constraints to markets, input supply and technology. In general terms this means: (i) fewer price and market interventions; (ii) more emphasis on public expenditure for infrastructure, support services and institutional development; (iii) a shift from commodity to factor taxation; and (iv) appropriate and targeted interventions in the case of genuine market failures (i.e. environmental externalities, degradation of common resources, national goals like equity or food security that are not satisfied by markets). It is the market-failure argument that justifies fiscal subsidies to alleviate malnutrition, establishment of credit programmes (not subsidized interest rates), encouragement of positive environmental activities, or promotion of the transition to new technologies and input packages.

The potential environmental impact of policy measures taken during structural adjustment and affecting agriculture has only recently attracted attention (e.g. Sebastian and Alicbusan, 1989; Repetto, 1989; Pearce and Warford, 1993; Reed, 1993, 1996; Munasinghe and Cruz, 1995; Munasinghe 1996; Taylor, 1996). There is a wide consensus that the effects of macro and sector policies on natural resource use are usually mixed, because various measures have different and often opposite consequences. We will discuss some of the most important policy measures and their potential effects for sustainable land use at farm household and regional level in terms of nutrient management. In doing so, we will distinguish major nutrient inputs (mineral fertilizer, animal manure, green manure) and nutrient outputs (plant and animal production, runoff and erosion) that are influenced by farm household decision-making (see e.g. Smaling *et al.*, 1996). The policy measures[2] considered are the adjustment of agricultural output prices through price liberalization and reduction of export taxes, the reduction of agricultural input subsidies, devaluation of the exchange rate, public expenditure reduction and institutional reform.[3] Finally, we discuss interactions between different policy instruments.

Adjusting Agricultural Output Prices

Raising (relative) prices of agricultural products or ending discrimination against export crops[4] has far from conclusive effects on land use. Higher prices are expected to stimulate agricultural production, either through the cultivation of more land (extensification) and/or the use of more inputs (intensification). Where suitable agricultural land is unavailable and inputs are either unobtainable or too expensive, extensification is likely to occur through shortening of the fallow period or by clearing ecologically fragile land. In both cases, the outflow of nutrients from the soil will increase and nutrient balances deteriorate. This appears to be the case in large parts of sub-Saharan Africa (see e.g. Smaling *et al.*, 1996). When, on the other hand, production increases are achieved through the intensified use of mineral fertilizers and other modern inputs, nutrient balances may improve (depending on the net balance between outflows from increased production and inflows from increased use of mineral fertilizer and manure).

Besides an increase in agricultural production, a switch is expected to take place from subsistence crops to commercial crops (and between different commercial crops, when their relative prices also change) when discrimination against exports is reduced. Individual crops generally react more strongly to price changes than agricultural production as a whole, because it is easier to shift resources from one crop to another than it is to draw more resources into agriculture. The resulting changes in cropping patterns will have important consequences for soil erosion and nutrient balances, since some crops are more likely to exhaust the soil than others. Tree crops like coffee, cocoa, rubber, palm oil and bananas, provide a continuous root structure and canopy cover, and are more suitable for sloping terrain. Other commercial crops such as cotton and groundnuts, and staple food crops like cassava, yams, maize, sorghum and millet, leave the soil more susceptible to erosion. In West Africa, erosion rates for areas where tree and bush crops are grown are typically half to one-third of the rates where staple crops are grown (Repetto, 1989, pp. 71–72). Also, nutrient balances for areas under cultivation of commercial export crops (like cotton) are generally better maintained compared with local staple food crops (see e.g. Van der Pol, 1992). Pesticides that are used to annihilate the plagues that easily emerge in monocultures (which is often the case with commercial export crops), however, may eliminate useful organisms that contribute to a good physical and biological state of the soil.

The main objective of removing price discrimination against agricultural products is to raise farmers' incentives for improved input use and to increase prospects for different types of investments in levelling, terracing, draining, irrigating and other forms of land improvement. When prices are depressed, lower profitability reduces the demand for farmland and/or restricts the use of labour and other yield-increasing inputs. Because farmland cannot be massively shifted to other uses, land prices will be lower than they would be without price control. As a result, returns on investments in improved input use and land conservation are depressed. An increase in relative prices of agricultural products is therefore expected to stimulate anti-erosion investments and investments in soil fertility improvement (through fertilizer, rock phosphate, or other soil amendments).

A few caveats should be taken into account. Firstly, measures aimed at soil conservation and land improvement require labour inputs. When labour is scarce, conservation measures may not be undertaken. Secondly, security of land use rights is an important precondition for farm households to consider long-term investments in land improvement. When land use rights are ill-defined, higher prices will hardly enhance household investments in conservation projects, but result in excessive exploitation and cultivation of marginal lands. Similar effects may occur when price increases are expected to be temporary. Thirdly, the full benefits of investments in land improvement can only be reaped after several years. This holds especially for investments in soil fertility, the initial effects of which are usually limited. Moreover, farmers need time to learn new technologies and adapt them to local circumstances. Short time horizons, caused by poverty, may prevent farmers from undertaking such investments. Fourthly, the net effect of higher prices depends on the size of the marketed surplus. When farmers produce mainly for their own consumption, because of high transaction costs or to cover risk for example, higher agricultural prices are unlikely to have much impact on investments in land improvement. The effect of price liberalization on farm households' real income is dependent on their net supply or demand position in the food market. Higher farm gate prices will be favourable for net selling households, but negative expenditure effects arise for net buying households (Reardon *et al.*, 1988). Therefore, price increases can only result in positive incentives for improved input use or farm-level investment when income effects dominate expenditure effects.

It may be concluded that the effect of output price adjustments on soil nutrient balances is far from conclusive. Will crop switching or output expansion dominate? To what extent is agricultural supply response constrained by poor infrastructure and lack of institutions and services? Will intensification be feasible, and, if so, will fertilizer be over- or underused? If there are access problems with inputs, will extensification encroach on fragile lands? How different are cropping systems in terms of soil erosion and nutrient depletion? The answers to these questions determine to a large extent the resulting effects on nutrient balances.

Two case studies described in Munasinghe and Cruz (1995) may serve to illustrate some of these issues. In China (Jiangsu Province), market-oriented reforms led to higher prices for agricultural products and chemical fertilizer and an increase in the costs of labour (as a result of rapid industrialization). Application of crop and animal residues to the land, which is a labour-intensive activity, was discouraged. Demand for chemical fertilizer, however, increased. Due to the high population pressure, extensification is not an option. Land is still collectively owned, but land use rights are allocated to individual farm households. In some communities, the allocation takes place rather arbitrarily. It is argued that the resulting uncertainty about land rights encourages short-run profit maximization and exploitation of the land, at the expense of sustainability in production. In Ghana, on the other hand, the main source of agricultural production increases since the introduction of the SAP is the expansion of cultivated area rather than agricultural intensification. A 10% increase in cultivated land is estimated to increase output directly by 2.7%. Tribal land use customs (which allow village chiefs to reallocate land that is left idle) have not been adequately adjusted, resulting in a further reduction of fallow periods which already appear to be too short. This, in turn, leads to an estimated 2.5% loss in sustainable agricultural productivity. The resulting net effect of expanding cultivated area is still positive but only 0.2%.

Reduction of Input Subsidies

An important element of SAPs is the abolition or reduction of subsidies on agricultural inputs (fertilizers, pesticides, equipment), and the liberalization of the markets for these inputs. Underpricing of inputs tends to be frowned upon, because it encourages overuse – which harms the environment – and benefits large purchasers most – which intensifies inequalities.

Governments in many LDCs have traditionally pursued the dual and inconsistent objectives of keeping food prices low for urban consumers while simultaneously encouraging agricultural production (Krueger, 1995, p. 2538). Agricultural inputs like chemical fertilizer, pesticides, equipment and credit were subsidized in recent decades to modernize agriculture and to increase production in order to keep up with the growing food (and foreign exchange) needs of rapidly growing populations. What was not necessarily addressed were the constraints farmers faced in acquiring or adopting agricultural inputs. Was it insufficient knowledge, constrained access to credit, high transportation costs, the inability to take risk, unclear user rights, or unfavourable input/output price ratios? Clearly, a subsidy would not in all cases (though certainly in some) have been the first-best intervention to address these constraints.

A further complication is that it is not only overuse of inputs which is related to negative environmental effects (in which case extension, regulation and/or taxation are appropriate interventions), but underuse as well. When soil nutrient balances are negative, solutions must be sought in adding nutrients (through increased investment in mineral fertilizers, animal manure, or N-fixing species) or reducing nutrient losses (through erosion control, restitution of crop residues, or recycling of manure). Application of inorganic fertilizers may be required to enable the recovery of soil organic matter, and improve the efficiency of nutrient management. In large parts of Africa, use of mineral fertilizers has declined in recent years. Whereas the availability of fertilizers was often restricted prior to structural adjustment, it is currently the value–cost ratio that is the limiting factor. As a result of the fertilizer subsidy removal, value–cost ratios of fertilizer use in many food crops in West Africa have typically fallen below two, which is commonly regarded as a minimum value for application by farm households (Smaling *et al.*, 1996; Koning *et al.*, 1998).

A clear dilemma emerges here as illustrated by the positions of Sebastian and Alicbusan (1989) and Repetto (1989) who argue that the removal of input subsidies generally benefits the environment, while acknowledging that incentive problems arising from market failures (unsignalled soil depletion) should be corrected to induce farmers to respond to true costs and opportunities. Pearce and Warford (1993) and Munasinghe and Cruz (1995) later argue both positions. What matters again is the nature of the constraints to maintain soil fertility. Is it poverty, risk aversion, limited availability of or access to credit or information on better land use practices, high marketing and distribution costs of fertilizer due to poor infrastructural facilities and long transport distances, or limited possibilities to switch from chemical to organic fertilizers? Understanding the relevant constraints to maintenance of soil fertility is therefore essential if cost-effective incentive systems are to be established. Subsidies should be used with care, because they are a recurrent cost on the budget, poor targeting is common, and proper use of the subsidized item is not guaranteed due to reselling. According to Pinstrup-Andersen and Pandya-Lorch (1994), fertilizer subsidies are, at best, a temporary measure when costs of fertilizer are high and governments have not dealt with the conditions that lead to these high costs.

Currency Devaluation and Public Investment Reduction

Exchange rate devaluation has many effects of which two are especially relevant here. First, it changes the relative price of export and (domestic) food crops in favour of the former. This will reinforce the shift in cropping pattern induced by price liberalization and strengthen the effects on soil erosion and nutrient balances. Secondly, a devaluation makes imported inputs (fertilizer, pesticides, farm equipment) more expensive. As such it has the same effect as removing subsidies.

An important question is to what extent the relative price changes that result from currency devaluations can be sustained. In this respect, the different recent experiences in Ghana and Burkina Faso are instructive (see Kempkes, 1997). In Ghana, successive devaluations of its national currency, the *cedi*, increased domestic prices of tradables (e.g. cocoa, cotton) and chemical fertilizers, and strongly depressed relative prices of non-traded foodcrops (e.g. maize). As a result, the fertilizer/food crop price ratio has increased rapidly. In combination with the removal of fertilizer subsidies, this has led to further declines in fertilizer use from already very low levels (Bumb *et al.*, 1994). In Burkina Faso and other francophone African countries, the CFA franc (which has a fixed exchange rate with the French franc) was devalued to half its original value in January 1994. As a result, prices of agricultural exports (e.g. cotton) and fertilizer roughly doubled. Within 2–3 years' time, however, prices of food crops (millet, sorghum, maize) in Burkina Faso also increased by about 100%, probably as a result of an increased domestic demand for these non-traded crops.[5] As a result, the fertilizer/food crop price ratio has returned to its pre-devaluation level.

Reductions in public expenditure and institutional reform constitute core problems in structural adjustment as far as rural development is concerned. With agricultural and environmental issues usually low on the priority list of financial and economic decision-makers, severe budget cuts have hit expensive (and not always effective) input supply, rural credit, and research and extension systems. In many LDCs, attempts to switch public funds to more cost-effective support programmes and public rural investment have not succeeded (if they were seriously made). Input costs to producers have therefore risen, seriously threatening the long-term incentive effect of price reforms. Moreover, an important number of small farmers are also net food consumers, and rising consumer prices have affected them negatively when the positive income effects (caused by higher farm gate prices) are outweighed by negative expenditure effects. As a result, agricultural supply response has been retarded and environmental problems insufficiently addressed.

Policy Complementarities

Policy reforms and interventions considerably differ in their effect on soil fertility improvement. Effects can be reinforcing (price reforms cum improved provision of critical infrastructure and information) as well as neutralizing (exchange rate adjustment cum removal of fertilizer subsidies). Moreover, interventions in seemingly unrelated fields (better functioning labour markets, diversification into non-agricultural activities) may stimulate more sustainable intensification practices. The cases below briefly illustrate these points.

The consequences of rising output prices for soil quality management in the case of a deficient labour market have been analysed by Bulte and van Soest (1999). Raising the price of agricultural output increases marginal labour productivity in soil conservation investment, which justifies the cost of incorporating more labour. Without such a possibility (i.e. when no labour market is present), farmers may well, but need not, forgo allocating their own labour for conservation purposes and accept soil degradation. Hence, price reforms that provide for higher profitability in crop production are not always consistent with improved soil fertility management, if public efforts to reduce transaction costs in the labour market (and hence improve its functioning) are absent.

Cash cropping and off-farm earnings can relieve farmers' constraints to more sustainable food cropping practices, as argued by Reardon et al. (1997). When the household diversifies, more cash and better access to finance become available. This enables farmers to purchase more fertilizers, manure or other inputs, and hire labour to increase food production and invest in soil conservation and fertility improvement. Of course, the possibility to engage in off-farm activities is not available to all farmers. This reinforces the classic argument for the creation of labour-intensive, off-farm employment opportunities.

The effects of a devaluation and a simultaneous removal of fertilizer subsidies in Senegal and Burkina Faso are examined by Reardon et al. (1997). Fertilizer use in rainfed agriculture in these two countries decreased substantially when subsidies were removed and availability of credit declined. Soil quality decreased and farmers reacted by increasing (groundnut) seeding densities to maintain short-run output levels. Although the 1994 devaluation increased the profitability of groundnut cropping, both in absolute terms and relative to food crops, farmers continued to use little fertilizer. At the same time, area expansion devoted to groundnuts was considerable. The explanation for this continued low fertilizer use lies in liquidity constraints at the farm level and an unfavourable price ratio between groundnut and fertilizer prices, which was insufficiently restored after the devaluation.

Reporting on the effects of the 1994 devaluation of the CFA franc, Reardon et al. (1997) note that the profitability of irrigated rice in Mali rose as output prices went up faster than input costs.[6] Further expansion of production in this intensified system was held back, however, by lack of complementary public investment in infrastructure (irrigable land), extension and rehabilitation.

The importance of (reductions in) public investment is also evident from a study by Gibson (1996) on the devastating effects of uncontrolled exploitation of the fertile valleys and lowlands of Nicaragua by large commercial growers. Smallholders were forced to cultivate steeper terrains using technically retarded lowland technologies and applying almost no crop rotation. Large producers of cash crops managed their land considerably better than small ones, but public intervention improved soil management of the latter when extension, credit and foreign exchange brought small and medium-sized farmers up to minimum practices. In the present critical debt situation, room for public sector interventions has disappeared, and overuse of fertilizers and pesticides by those producing for the export market looms large.

As the previous discussion shows, the effects of SAPs on the environment are generally mixed. This has partly to do with the fact that many SAPs were only partially implemented, focused more on short-term macroeconomic stabilization than on structural transformations, and paid insufficient attention to redirecting and reformulating complementary public support and investment programmes. It should be kept in mind, however, that these policy reforms were not designed to primarily serve environmental goals. In this respect, there is an interesting analogy with the introduction of green-revolution type high-yielding varieties. Their main aim was to increase food production using land-saving and often labour-augmenting techniques. Without it, the environment would have suffered from extensification. With it, new environmental problems related to intensification arose, and new policies were subsequently designed to mitigate these effects and anticipate them better in the future. Macroeconomic reforms, by correcting harmful policy failures, can serve the environment; they cannot, however, cope with market imperfections related to environmental quality. To deal with such imperfections, collective action is required.

INTERVENTIONS IN SUPPORT OF SOIL FERTILITY

General

The efficiency and sustainability of agricultural production systems can be enhanced through various types of policy interventions. Reviewing country experiences with policies to support more sustainable (agricultural) development, Taylor (1996), however, cautions against undue optimism. Societal consensus on either environmental or economic questions is the exception rather than the rule, and little policy-relevant advice has been forthcoming. The macro-rule that rents generated by exhausting natural resources should be reinvested in reproducible capital is systematically violated.[7] The resulting environmental losses, including soil nutrient depletion, have serious repercussions for future productivity. Taylor (1996) mentions annual damage prevention and soil restoration outlays of 2–5% of GDP (similar to estimates in Pearce and Warford, 1993). Such figures are comparable with the payment obligations of many LDCs during the debt crisis in the 1980s. Preventing environmental damages of such a magnitude can be socially highly rewarding, and provides a strong efficiency argument to develop more sustainable land management practices that maintain soil fertility. Promotion of these practices requires: (i) agrotechnical research into the possibilities to reduce nutrient losses; and (ii) economic policies that make replacement of nutrients economically attractive to farm households.

In this section, we will specifically address the question, which (combination of) economic policy interventions effectively support efforts to maintain or improve nutrient and soil carbon balances? Simulation results from bio-economic models of farm households in Costa Rica and southern Mali will be used for this purpose. The focus of the analysis is on price policy and market development. Policy options considered include price increases for food and cash crops (through price liberalization or currency devaluation), higher prices for fertilizer and other inputs (through reduced subsidies or currency devaluation), a reduction of transaction costs (through investments in infrastructure), and an increased supply of credit (through institutional and organizational measures). We first discuss the importance of bio-economic models for analysing agricultural supply response to policy measures, and makes a general comparison of the model results obtained for Costa Rica and southern Mali. Then the consequences of distinguishing between different farm household types are shown. Responses of farm households are likely to differ between richer farm types with more available resources and poor farm types with fewer resources. Simulation results from the model for southern Mali will be used for this purpose.

Policy Measures and Supply Response in Costa Rica and Southern Mali

Most agroecological research on soil nutrient balances focuses attention on the monitoring of nutrient stocks and flows at different spatial and temporal scales and the financial costs and benefits of alternative technologies that could redress nutrient depletion (e.g. Van der Pol and Van der Geest, 1993; Smaling et al., 1996). By focusing on profitability only, these studies cannot yield adequate insights into farmers' decisions with respect to adjustment of their farming systems and the socioeconomic conditions that induce them to do so. The effectiveness of various policy instruments in inducing farm households to move towards modifications of their land use and factor allocation can be assessed within a supply response framework that takes into account: (i) the available resource endowments (land, labour, capital); (ii) the multiple objectives of the farm household (i.e. profit, risk, leisure); and (iii) the market conditions (prices, access, etc.). Supply response analysis enables an objective appraisal of the response reactions of farm households to different policy instruments (Nerlove, 1979; Bond, 1983). Incorporation of agroecological considerations into economic supply response models permits the identification of possible trade-offs between farm household welfare objectives (e.g. consumption possibilities) and sustainability criteria (e.g. nutrient and soil organic matter balances).

Recently, integrated bio-economic modelling procedures have been developed that permit a simultaneous appraisal of adjustments in technology choice (and the consequent changes in supply, income, and soil nutrient and organic matter balances) induced by: (i) making available the alternative technologies under unchanged market conditions, and (ii) adjusting the market conditions to enhance adoption of alternative technologies that could improve both household welfare and sustainable land use (Kruseman *et al.*, 1997; Ruben *et al.*, 1997).[8] Applications of bio-economic simulation models are available for the humid Atlantic zone of Costa Rica and for the semi-arid Sudano-Sahelian region of southern Mali. Different production techniques for arable cropping and livestock activities are defined that represent currently used practices, often with negative nutrient balances, as well as alternative practices that are technically feasible though not (yet) widely applied and maintain balanced nutrient flows and stocks (Hengsdijk *et al.*, 1996).[9]

Simulation results show that in both cases the final impact of policy instruments turns out to depend heavily on the way local factor and product markets function (Kuyvenhoven *et al.*, 1995). While labour is a constraining factor in both settings, the labour market in Costa Rica is functioning well in contrast to that of Mali. The market for non-factor inputs (fertilizers, biocides) is competitive in Costa Rica, while in Mali access to fertilizers is rationed. Access to rural credit is a limiting factor, especially in Mali.

Supply response reactions to policy instruments in Costa Rica and Mali are rather diverse. In Costa Rica, adoption of alternative, agroecologically sustainable production technologies depends most on market prices and related transaction costs; in Mali it mainly depends on access to factor markets. Improvement of nutrient (organic matter) balances as well as farm household welfare require a different set of policy instruments in each situation. While input price policies (e.g. fertilizer subsidies) are effective for this purpose in Costa Rica, improved access to factor markets is a more suitable policy instrument in Mali. Subsidies on fertilizers in Costa Rica have a positive effect on both household welfare and sustainability objectives (nutrient and organic matter balances) due to the prevailing high substitution elasticity between inputs. In contrast, in Mali, only income effects of fertilizer subsidies are recorded since, in the context of imperfectly functioning labour and commodity markets, subsidies appear to be an incentive for less efficient fertilizer use (Kuyvenhoven *et al.*, 1995). Infrastructure improvement to reduce transaction costs and improve the functioning of markets turns out to be an effective instrument in both situations.

Careful management of relative prices for inputs and outputs and of transaction costs is especially important to enforce technological change towards improving sustainability. Increasing output prices for all crops in Costa Rica induces partial substitution of actual by alternative technologies, causing strong income increases but hardly improving nutrient balances (Kruseman *et al.*, 1995). This occurs because unit fertilizer requirements of a 'sustainable' crop tend to be higher than those of the actual cropping activities. Decreasing fertilizer prices have a stronger effect on changes in technology and contribute to improved nutrient balances, while farm household income improves simultaneously. Therefore, fertilizer price subsidies turn out to be an appropriate instrument to induce the desired modification in land use in Costa Rica (Kruseman *et al.*, 1995, pp. 120–121).

Responses of Different Farm Household Types in Southern Mali

Different response reactions can be expected for various types of farm households, taking into account their incomes, their resource endowments (land, labour, equipment) and their competitive position on local markets for food (net buyers or sellers) and labour. A detailed analysis of agrarian policy options to enhance sustainable land use for different types of farm households has been made for southern Mali (Kruseman *et al.*, 1997; Ruben *et al.*, 1997). Using a standard classification of the CMDT (the cotton board in Mali), four different farm types are distinguished that differ in terms of available resources (land, livestock, equipment for animal traction).

Three different simulations were used to determine the size and direction of response multipliers to policy variables: (i) base scenario to define the initial production structure under current prices and with actual activities; (ii) technology scenario under current prices but including actual and alternative activities; and (iii) policy scenario with different modifications in prices and institutional constraints to assess the impact of different instruments on adjustments in land use and farm household welfare. In the third scenario, various price instruments are evaluated: output prices for selected cash crops and cereals, and input prices for fertilizers. Infrastructure (and market) development is evaluated in terms of changes in the transaction costs, that is, the margin between farm gate and market-producer prices. Another indicator of market development refers to access to rural finance, which increases the possibility for farm households to make in-depth investments (animal traction, perennial tree crops, etc.).

Table 6.1 summarizes the results of the base run. Farm type A has the largest resources of land and livestock and reaches the highest income, while farm type D has few resources and therefore has the lowest farm income. Shadow prices (not shown in the table) indicate that farm type B faces the strongest constraints in incorporating additional labour. All nutrient and carbon balances are negative and all farms, especially the smaller ones, suffer serious soil erosion. The smaller farm types C and D dedicate a higher proportion of their land to cash crops; their cereal balances are therefore negative.

Results of the introduction of alternative, sustainable cropping activities[10] under current prices ('technology scenario') are presented in Table 6.2. Nutrient and carbon balances show a partial improvement, while net returns increase by between 15 and 29%. Although alternative activities are technically more efficient, a significant number of current activities are still

Table 6.1. Base run: objective values, nutrient balances and initial production structure.

Indicator	Unit	Farm type A	Farm type B	Farm type C	Farm type D
Farm size	ha	17.8	10.1	5.8	3.3
Labour force	persons	11.8	5.7	3.9	2.5
Livestock	TLUs	25.1	3.9	0.9	0.2
Net revenue	FCFA per capita	81,493	61,419	36,546	26,848
Carbon balance	kg ha^{-1}	−1,127	−1,193	−1,323	−1,182
N-balance	kg ha^{-1}	−42	−48	−47	−44
P-balance	kg ha^{-1}	2	0	−2	−1
Erosion	Mt soil loss ha^{-1}	34	42	50	48
Cereal balance	kg per capita	133	246	−79	−118
Cereals area	ha	10.8	6.6	2.1	1.2
Cash crops area	ha	4.5	2.5	2.5	1.4
Cowpea area	ha	1.6	1.0	1.3	0.8

Cereal balance measures demand minus own production of cereals per capita.
TLUs, tropical livestock units; Mt, metric tonnes.
Source: Ruben *et al.* (1997).

Table 6.2. Response multipliers for the technology scenario (percentage change in the value of the goal indicators compared to the base run).

Indicator	Farm type A	Farm type B	Farm type C	Farm type D
Net revenue	18	28	29	15
Carbon balance	65	44	32	33
N-balance	59	48	29	22
Erosion	−28	−33	−18	−37
Alternative activities	61	53	33	21

Note: For positive nutrient and carbon balances, the percentage change should be higher than 100%.
Source: Ruben *et al.* (1997).

maintained as economically efficient options, especially on the smaller farm types C and D. This is explained by the fact that these farmers prefer high current returns (at the cost of soil mining) over higher future returns, due to their high time discount rates.

Farm types A and B can still increase their cash crop production as their food balance is already positive, while food-deficit farm types C and D give preference to improved cereal production. For all households there is a shift from actual to alternative technologies in cash crop (i.e. cotton and groundnut) production, since differences in costs are not very substantial while benefits in terms of nutrient and carbon balances are important. Farm types C and D maintain a high proportion of actual activities, since their higher time discount rate reduces the income effects of carbon and/or nutrient improvements.

Although making alternative production techniques available is essential to enhance resource use efficiency, economic incentives to foster their adoption could further contribute to improved soil management. Economic policy interventions were therefore considered that could simultaneously contribute to higher farm revenues and improved carbon balances ('policy scenario'). Simulation results for this scenario are shown in Table 6.3. The data in the table represent percentage changes (of net revenues and nitrogen and carbon balances) due to a 1% change in the policy instrument, and should not be compared with the results of Table 6.2.

Improvement of the nitrogen and carbon balances can be reached through modification of the relative price between cereals and cash crops, through adjustment of input costs or credit availability, or through a reduction of transaction costs. Modifications in output prices (and transaction costs) tend to be more effective than input price policies to improve both the soil balance and farm revenue. Fertilizer subsidies encourage the adoption of more sustainable cotton activities, but the potentially positive effect on the nitrogen and carbon balances is offset by the choice of more

Table 6.3. Response multipliers for different policy instruments (percentage change in goal indicators compared to the technology scenario, due to a 1% change in the instrument variable).

Indicator	Level	Food price increase	Cotton price increase	Transaction costs decrease	Fertilizer price decrease	Credit supply increase
Net revenue	Farm type A	0.5	0.5	0.3	0.1	0.1
	Farm type B	0.4	0.5	0.1	0.2	0.0
	Farm type C	0.4	0.6	−0.2	0.1	0.1
	Farm type D	0.9	0.2	0.1	0.2	1.0
Nitrogen balance	Farm type A	2.2	1.8	2.1	−0.1	0.9
	Farm type B	0.4	0.1	0.0	−0.2	0.0
	Farm type C	0.2	0.6	−0.5	−0.2	0.1
	Farm type D	0.0	0.0	0.0	0.0	−0.1
Carbon balance	Farm type A	1.1	0.1	0.6	−0.1	0.3
	Farm type B	−0.2	−0.1	−0.8	−0.2	0.0
	Farm type C	0.7	0.6	0.6	−0.3	0.2
	Farm type D	0.1	1.1	0.6	0.0	0.3

Source: Ruben *et al.* (1997).

soil-depleting cereal activities. In other words, the negative impact of fertilizer price subsidies on the carbon balance acts as a disincentive to improve input efficiency at the farm household level. For the farm type facing the strongest labour constraints, farm type B, output price increases result in a relatively small improvement of the nitrogen balance and a worsening of the carbon balance. Apparently, labour limitations inhibit the selection of more sustainable techniques.

Summarizing the results, two major conclusions stand out. First, technical options[11] for a sustainable intensification of cropping systems, both in agroecological and socioeconomic terms, are available. With the full knowledge of sustainable technology and in the absence of transition costs incurred in implementing these technologies, however, nutrient and carbon balances still remain negative. Farm household types with low incomes and relatively few resources in particular are less inclined to adopt sustainable technologies. Soil mining practices are thus economically efficient, given the farm household resources, goals and risk perception. Secondly, modification of output prices can be helpful to improve nutrient balances to a certain extent. Fertilizer subsidies may, however, lead to less efficient resource use. Structural policies addressing transaction costs and labour and financial markets in Mali seem to offer better prospects to enhance tradability and reinforce intersectoral growth linkages, permitting higher supply response and easier adoption of sustainable land use practices.

CONCLUSIONS

Policy measures taken under structural adjustment have not always achieved their main goals (i.e. restoration of macroeconomic equilibrium and better functioning markets), as a result of incomplete execution and other deficiencies. To the extent that they succeeded, these policies have created opportunities for improved nutrient management. However, because in most structural adjustment programmes agricultural support programmes, complementary public investments and subsidies were drastically curtailed, nutrient balances have deteriorated where these cuts were most severely felt. The positive effect of higher output prices may be lost due to rising costs for access to purchased inputs or higher prices for consumption goods. Coupled with the generally low aggregate supply response of agriculture (due to poorly functioning or failing markets for credit, labour and so on) and the short time horizons of peasants in low-income countries, the challenge is to devise policy interventions favourable to sustainable land management practices while avoiding unsustainable claims on the public budget.

To be more cost-effective, interventions need to be selective and better targeted to the issue at stake, and should explicitly address the question of whether farmers can and want to adopt sustainable, yet intensive, technologies that do not exhaust soil nutrients. Measures should therefore not only improve access to soil-fertility enhancing inputs and technology, but also create the right incentives for farm households to use these inputs efficiently. Low-income households, which often prefer food security and satisfaction of basic consumptive requirements over everything else, face liquidity constraints and are prone to risk, weigh the opportunity cost of their resources carefully. Policy instruments applied in other settings may not automatically induce them to higher use of external inputs or family labour.

Adoption of sustainable intensification practices is positively affected by diversification of activities into cash crop production (profitability is higher than in food crops, credit on input provision is usually available) and non-farm activities, which have a higher profitability, allow risk diversification, and create additional income that can effectively substitute for failing credit markets (Reardon et al., 1997). By the same token, yield-enhancing external inputs and conservation expenditures by farmers react positively to public investments in infrastructure (which lower transaction costs), the development of marketing channels, and effective provision of information based on agricultural research. Here a careful course should be followed to steer between the Scylla of fiscally untenable public outlays and the Charybdis of waning public support for structural policy.

Fertilizer is indispensable to maintain soil nutrient balances, but to subsidize its use across the board has become controversial. Bringing down transport and distribution cost, improving access to farmer-relevant agrotechnical research and extension, or making better use of locally available substitutes can provide forceful incentives to improve nutrient management and investment in soil conservation.

Economic theory with respect to market failures indicates that more is needed than just 'getting prices right' to achieve the goal of sustainable soil nutrient management. Farmers are faced with time lags in learning to use new, more sustainable technologies, in the adaptation of technologies to local circumstances, and in reaping the productivity benefits of the resulting investments in soil fertility. Soil fertility-enhancing techniques should preferably have a low investment component and provide early and more stable returns. Where necessary, temporary price adjustments may be needed to induce farmers to adopt such technologies, and to realize the long-run gains in productivity (Koning et al., 1998). Until these technologies become financially profitable, specific input subsidies or guaranteed minimum prices cum import duties[12] could be used to increase short-term profitability. The results of Table 6.3 indicate that output price modifications tend to be more effective than input price policies in this respect (provided labour availability is not a bottleneck). Price support should be properly targeted to products in which the country has a potential comparative advantage, and be gradually reduced over time. Moreover, its long-term benefits to society should be carefully weighed against the (more direct) costs to society in terms of consumer welfare and rent-seeking behaviour by, for example, traders in agricultural products.

NOTES

[1] A more extensive discussion of the results of SAPs, which have often been mixed as a result of incomplete reforms, failures to redirect public expenditures, or other causes, is outside the scope of this chapter. Recent African experiences are reviewed in Azam (1996) and Van der Geest (1994).

[2] Sebastian and Alicbusan (1989) show that World Bank supported SAPs in the 1980s were characterized by adjustments in agricultural producer prices (in 54% of the countries reviewed), subsidy removals (65%), export tax reductions (28%), currency devaluations (34%), changes in public expenditure programmes (44%) and institutional reforms (67%).

[3] Parts of these sections draw on earlier work by Heerink and Kuyvenhoven (1993) and Heerink et al. (1996).

[4] Liberalization of agricultural trade under the Uruguay Round and reform of the Common Agricultural Policy (CAP) of the European Union have basically the same positive effect on output prices of tradable agricultural products in LDCs. In addition, LDCs' access to export markets is expected to improve. See also Heerink et al. (1996) for a discussion of the effects of agricultural trade liberalization on soil degradation.

[5] Similar effects can be observed for the prices for livestock and beef, which are mainly traded at regional markets in francophone West Africa. Available recent evidence for Burkina Faso and Mali shows that prices appear to stabilize at a level roughly twice the pre-devaluation level (Moll and Heerink, 1998).

[6] It should be noted that only the immediate price changes resulting from the CFA devaluation are considered in the study by Reardon et al. (1997). To analyse the longer term impact of the devaluation, it is important to know whether the changes in the prices of export crops, food crops, and fertilizer are sustained, or price ratios return to their pre-devaluation levels (as seems to have happened in Burkina Faso; see above).

[7] At the microeconomic level, these rents proved to provide an important contribution to farm household income, estimated by Van der Pol (1992) at 40% of farmers' total revenues from agricultural activities in southern Mali.

[8] See also Deybe (1994) and Barbier (1996) for similar applications of bio-economic modelling procedures.

[9] Sustainable techniques include wider use of fertilizers in arable cropping, production of fodder crops to enable the intensification of livestock production, introduction of woody species into pastures, recycling of crop residues (stubble grazing, underploughing), and soil conservation measures (simple or tied ridging).

[10] See note 9 for a specification of the sustainable techniques that are considered.

[11] Through a better use of crop residues and higher efficiency of fertilizer application. Integration of livestock and cropping systems may contribute to the same goal, provided that additional fertilizers are applied to produce fodder crops.

[12] When food imports are not competitive with domestic production, as in many African countries, the rising costs of imports may force the country to devalue its currency. Resources will subsequently be reallocated from less tradable agricultural crops to export crops (Sadoulet and De Janvry, 1995, pp. 356–366). As argued in the chapter, such reallocations will often be beneficial for nutrient balances.

REFERENCES

Azam, J.P. (1996) The diversity of adjustment in agriculture. In: Ellis, S. (ed.) *Africa Now*. James Curry, London, and Heinemann, Portsmouth, pp. 136–154.

Barbier, B. (1996) *A Method of Bio-economic Modelling at the Micro-watershed Level for the Central-American Hillsides*. International Food Policy Research Institute (IFPRI), Washington, DC.

Bond, M.E. (1983) Agricultural supply response to prices in sub-Saharan African countries. *IMF Staff Papers* 30, 703–726.

Bulte, E. and van Soest, D. (1999) A note on soil depth, failing markets and agricultural pricing. *Journal of Development Economics* 58, 245–254.

Bumb, B.L., Teboh, J.F. Atta, J.K. and Asenso-Okyere, W.K. (1994) *Ghana Policy Environment and Fertilizer Sector Development*. International Fertilizer Development Center (IFDC), Muscle Shoals and Lomé.

Deybe, D. (1994) *Vers une Agriculture Durable: un Modèle Bio-économique*. CIRAD-URPA, Paris.

Gibson, B. (1996) The environmental consequences of stagnation in Nicaragua. *World Development* 24, 325–339.

Heerink, N. and Kuyvenhoven, A. (1993) Macroeconomic policy reform and sustainable agriculture in West Africa. In: Huijsman, B. and van Tilburg, A. (eds) *Agriculture, Economics and Sustainability in the Sahel*. Royal Tropical Institute, Amsterdam, pp. 11–31.

Heerink, N., Kuyvenhoven, A. and Qu, F. (1996) Policy issues in international trade and the environment with special reference to agriculture. In: Munasinghe, M. (ed.) *Environmental Impacts of Macroeconomic and Sectoral Policies*. World Bank, Washington, DC pp. 131–155.

Hengsdijk, H., Quak, W., Bakker, E.J. and Ketelaars, J.J.M.H. (1996) *A Technical Coefficient Generator for Land Use Activities in the Koutiala Region of South Mali*. DLV Report no. 5. Research Institute for Agrobiology and Soil Fertility (AB-DLO), Wageningen.

Kempkes, Y. (1997) Impact of structural adjustment on fertilizer use – the case of Ghana and Burkina Faso. MSc thesis, Wageningen Agricultural University, Wageningen, and International Fertilizer Development Centre (IFDC), Lomé.

Koning, N., Heerink, N. and Kauffman, S. (1998) *Integrated Soil Management and Agricultural Development in West Africa: Why Recent Policy Approaches Fail*. Wageningen Economic Papers no. 09–98.

Krueger, A. (1995) Policy lessons from development experience since the Second World War. In: Behrman, J. and Srinivasan, T.N. (eds) *Handbook of Development Economics. Volume 3B*. North-Holland, Amsterdam, pp. 2497–2550.

Kruseman, G., Ruben, R., Hengsdijk, H. and van Ittersum, M.K. (1995) Farm household modelling for estimating the effectiveness of price instruments in land use policy. *Netherlands Journal of Agricultural Science* 43, 111–123.

Kruseman, G., Hengsdijk, H., Ruben, R., Roebeling, P. and Bade, J. (1997) *Farm Household Modelling System for the Analysis of Sustainable Land Use and Food Security: Theoretical and Mathematical Description.* DLV Report no. 7. Research Institute for Agrobiology and Soil Fertility (AB-DLO), Wageningen.

Kuyvenhoven, A., Ruben, R. and Kruseman, G. (1995) Options for sustainable agricultural systems and policy instruments to reach them. In: Bouma, J., Kuyvenhoven, A., Bouman, B.A.M., Luyten, J.C. and Zandstra, H.G. (eds) *Eco-regional Approaches for Sustainable Land Use and Food Production.* Kluwer, Dordrecht, pp. 187–212.

Moll, H.A.J. and Heerink, N.B.M. (1998) Price adjustments and the cattle sub-sector in central West Africa. In: Nell, A.J. (ed.) *Proceedings of the International Conference on Livestock and the Environment, Ede/Wageningen, The Netherlands, 16–20 June 1997, World Bank, FAO, IAC.* International Agricultural Centre (IAC), Wageningen, The Netherlands, pp. 72–87.

Munasinghe, M. (ed.) (1996) *Environmental Impacts of Macroeconomic and Sectoral Policies.* World Bank, Washington, DC.

Munasinghe, M. and Cruz, W. (1995) *Economywide Policies and the Environment.* World Bank Environment Paper no. 10. World Bank, Washington, DC.

Nerlove, M. (1979) The dynamics of supply: retrospect and prospect. *American Journal of Agricultural Economics* 61, 874–888.

Norton, R.D. (1992) *Integration of Food and Agricultural Policy with Macroeconomic Policy: Methodological Considerations in a Latin American Perspective.* ESD Paper 111. FAO, Rome.

Pearce, D.W. and Warford, J.J. (1993) *World Without End.* World Bank and Oxford University Press, Oxford.

Pinstrup-Andersen, P. and Pandya-Lorch, R. (1994) Poverty and income distribution aspects of changing food and agricultural policies during structural adjustment. In: Heidhues, F. and Knerr, B. (eds) *Food and Agricultural Policies under Structural Adjustment.* Peter Lang, Frankfurt, pp. 479–492.

Reardon, T., Matlon, P. and Delgado, C. (1988) Coping with household-level food insecurity in drought-affected areas of Burkina Faso. *World Development* 16, 1065–1074.

Reardon, T., Kelly, V., Crawford, E., Diagana, B., Dioné, J., Savadogo, K. and Broughton, D. (1997) Promoting sustainable intensification and productivity growth in Sahel agriculture after macroeconomic policy reform. *Food Policy* 22, 317–327.

Reed, D. (1993) *Structural Adjustment and the Environment.* World Wide Fund for Nature, Earthscan Publications, London.

Reed, D. (1996) *Structural Adjustment, the Environment, and Sustainable Development.* World Wide Fund for Nature, Earthscan Publications, London.

Repetto, R. (1989) Economic incentives for sustainable production. In: Schramm, G. and Warford, J.J. (eds) *Environmental Management and Economic Development.* Johns Hopkins University Press, Baltimore, pp. 69–86.

Ruben, R., Kruseman, G., Hengsdijk, J. and Kuyvenhoven, A. (1997) The impact of agrarian policies on sustainable land use. In: Teng, P.S., Kropff, M.J., ten Berge, H.F.M., Dent, J.B., Langsigan, F.P. and van Laar, H.H. (eds) *Applications of Systems Approaches at the Farm and Regional Levels.* Kluwer, Dordrecht, pp. 65–82.

Sadoulet, E. and de Janvry, A. (1995) *Quantitative Development Policy Analysis.* Johns Hopkins University Press, Baltimore.

Sebastian, I. and Alicbusan, A. (1989) *Sustainable Development: Issues in Adjustment Lending Policies.* Environmental Department Divisional Paper 1989-6. World Bank, Washington, DC.

Smaling, E.M.A., Fresco, L.O. and de Jager, A. (1996) Classifying, monitoring and improving soil nutrient stocks and flows in African agriculture. *Ambio* 25, 492–496.

Stoorvogel, J.J. and Smaling, E.M.A. (1990) *Assessment of Soil Nutrient Depletion in Sub-Saharan Africa, 1983–2000.* Report 28. DLO Winand Staring Centre for Integrated Land, Soil and Water Research (SC-DLO), Wageningen.

Taylor, L. (1996) Sustainable development: an introduction. *World Development* 24, 215–225.

Van der Geest, W. (1994) Structural adjustment in sub-Saharan Africa. In: Van der Hoeven, R. and Van der Kraaij, F. (eds) *Structural Adjustment and Beyond in Sub-Saharan Africa.* James Curry, London, and Heinemann, Portsmouth, pp. 197–226.

Van Lynden, G.W.J. and Oldeman, L.R. (1997) *The Assessment of the Status of Human-induced Soil Degradation in South and Southeast Asia.* International Soil Reference and Information Centre (ISRIC), Wageningen.

Van der Pol, F. (1992) *Soil Mining: an Unseen Contributor to Farm Income in Southern Mali.* Bulletin 325 Royal Tropical Institute, Amsterdam.

Van der Pol, F. and Van der Geest, N. (1993) Economics of the nutrient balance. In: Huijsman, B. and van Tilburg, A. (eds) *Agriculture, Economics and Sustainability in the Sahel.* Royal Tropical Institute, Amsterdam, pp. 47–68.

World Bank (1986) *World Development Report 1986.* Oxford University Press, New York.

Integrated Smallholder Agriculture–Aquaculture in Asia: Optimizing Trophic Flows

J.P.T. Dalsgaard[1] and M. Prein[2]

[1]Danish Institute of Agricultural Sciences (DIAS), PO Box 3950, DK-8830 Tjele, Denmark; [2]International Center for Living Aquatic Resources Management (ICLARM), MCPO Box 2631, 0718 Makati City, Manila, Philippines

ABSTRACT

A nutrient modelling approach was applied to show how the combination of crops, trees, livestock, and fish, that is, integrated agriculture–aquaculture (IAA), helps in optimizing trophic flows in Asian rice-based agroecosystems. Integrated natural resource management can benefit the farm's nutrient balance sheet and nutrient use efficiency, while generating productive systems. In the current study, N-efficiency and economic efficiency of IAA systems can be more than doubled compared with rice monoculture systems. The IAA system represents one potential avenue towards ecologically balanced and sustainable forms of tropical smallholder farming. The key question remains how to engineer the appropriate socioeconomic environment for the evolution of such systems.

INTRODUCTION

The ecological records of tropical agriculture and aquaculture (fish farming) share many features (Dalsgaard *et al.*, 1995). Where agrochemical inputs are easily available, affordable and applicable these are often (too) liberally applied; irrigated farming of high-yielding rice varieties in Asia is a case in point. The short-term outcome is productive and profitable agricultural systems. The longer-term impact is often one of declining yields and returns, alongside ecological deterioration both on-farm and downstream, particularly in rainfed areas (Cassman and Pingali, 1995; FARM, 1996).

©CAB *International* 1999. *Nutrient Disequilibria in Agroecosystems* (eds E.M.A. Smaling, O. Oenema and L.O. Fresco)

Modern rice farming is witnessing declining agronomic efficiency, in terms of yield per unit input. At the cropping system level, trends of stagnant yields and increasing input requirements to maintain yields are detected. Underlying causes may include a general degradation of soil quality such as a reduction in essential micronutrients, and a decline in soil N supplying capacity (Cassman and Pingali, 1995).

Intensive aquaculture (i.e. producing >15 t ha^{-1} year^{-1}) is 7–31 and 3–11 times as polluting as semi-intensive systems (i.e. producing in the range of 1–20 t ha^{-1} year^{-1}), in terms of quantities of nitrogen (N) and phosphorus (P) released to the environment per kilogram of fish produced (Edwards, 1993). Intensive fish farms are also heavy users of antibiotics and disinfectants (Pullin, 1993). Stand-alone fish farms (i.e. without other enterprises of crop and livestock production) are risky ventures for smallholders and show a record of financial and environmental disaster in Africa and Asia (Cross, 1991; McClellan, 1991).

At the opposite end of the spectrum, in low-external-input smallholder farming, ecological constraints are also being encountered. Here, part of the problem is that of inadequate access to external nutrient inputs (Brummett, 1996; R.E. Brummett, 1999, unpublished observations). These are often needed to complement internal resources in order to revive the production system (Reijntjes *et al.*, 1992). Where nutrients and on-farm labour are available, examples of organic farming including integrated fish farming can develop (Little and Muir, 1987; IIRR and ICLARM, 1992; Symoens and Micha, 1995; Mathias *et al.*, 1997).

The development of balanced, productive and environmentally sound forms of tropical smallholder agriculture may well lie somewhere in between these two extremes. Integrated farming, or integrated natural resources management, where modest amounts of external inputs are used to supplement the (re)use of internal resources could be a solution for many tropical smallholders (Costa-Pierce *et al.*, 1991; Edwards 1993; Lightfoot *et al.*, 1993a; Prein *et al.*, 1995). Neither the route of agrochemical- and capital-intensive agriculture, nor that of resource underutilization, appears feasible. The alternative then, could be a middle way. This study presents a comparison of different rice-based farming systems using a mass-balance analytical framework to assess their comparative ecological and economic performance characteristics.

SEMI-INTENSIVE INTEGRATED AGRICULTURE–AQUACULTURE

One potential avenue towards ecologically balanced farming is the development of semi-intensive, integrated, agriculture–aquaculture (IAA) systems. Where the environment (e.g. water availability, soil texture, topography), tradition (waste reuse, fish consumption), experience, economic circumstances (e.g. declining farm productivity), and new opportunities (e.g.

access to markets) present fish farming as an option, IAA can emerge successfully (IIRR and ICLARM, 1992; Lightfoot *et al.*, 1993a; Brummett and Noble, 1995; Edwards and Little, 1995; Lightfoot and Pullin, 1995; Pullin and Prein, 1995; Prein *et al.*, 1996a, b). Some examples include the following:

- In Ghana and Malawi, farmers have established fish ponds in suitable locations in lowlands such as small inland valleys and depressions and operated these during the months following the rainy season, simultaneously growing an array of crops and vegetables (Prein, 1993; Ofori *et al.*, 1993; Brummett and Noble, 1995; Noble, 1996).
- In Vietnam, smallholder farmers have linked pig rearing, rice, vegetable and fruit growing with fish and freshwater prawn farming to become a profitable and highly productive system in the Mekong River and Red River deltas (Tran and Demaine, 1996; FAO-RAP, 1996; WES, 1997; Rothuis *et al.*, 1998a, b; D.K. Chung, H. Demaine, V.T. Pham, Q.D. Nguyen and T.H. Bui, 1999, unpublished observations).
- In Thailand, profitable fish farming ventures have developed utilizing chicken and other livestock manures and chicken slaughterhouse wastes purchased from other commercial farms (Edwards and Little, 1995). Yet these off-farm integration systems tend towards greater intensity due to their often peri-urban setting and ensuing constraints of required water quality and quantity (Edwards *et al.*, 1983; Engle and Skladany, 1992; Edwards, 1993; Little *et al.*, 1994).
- In China, rice–fish farming covered an area of 800,000 ha by 1988 with an average fish production of 180 kg ha^{-1}. The integration of fish culture in rice fields entails an increase in rice production by at least 10% which is attributed to weed removal, bioturbation and fertilization by the fish (MacKay, 1995; Mathias *et al.*, 1997).

Smallholder IAA is defined here as diversification of agriculture in the sense that aquaculture is developed as a subsystem on a farm with existing crops, trees or livestock subsystems, or a combination. An output from one subsystem in an integrated farming system which otherwise might have been wasted becomes an input into another subsystem resulting in a greater efficiency of output of desired products from the land/water area controlled by a farmer (Little and Muir, 1987; Edwards *et al.*, 1988). IAA systems in general are labelled *semi-intensive* as opposed to *extensive* systems relying exclusively on natural feed produced without intentional inputs; the *intensive* system depending on nutritionally complete feeds (and fertilizers), with farmed organisms deriving little or no nutrition from natural feed produced *in situ* (Edwards, 1993).

Semi-intensive fish farming has been a feature of the smallholder farmscape for centuries, particularly in parts of Asia, where the development potential for IAA is considerable (Ruddle and Zhong, 1983; Yan and Yao, 1989; Lightfoot *et al.*, 1992; Edwards, 1993; Edwards *et al.*, 1996; Roger, 1996; Mathias *et al.*, 1997). One of the secrets to the success of IAA lies in the development of a fine-tuned balance between the plant, livestock, fish and human subsystems, facilitating intensive (re)use and integration of resources across trophic levels (Schaber, 1997).

On a larger village or community scale, the IAA system is successfully practiced in China in situations of high population density (e.g. 1500 persons km^{-1}). Studies of these systems based on energy and nutrient flow modelling have shown that the amount of external nutrient inputs in the form of feeds and fertilizers can be halved through internal recycling and increase of primary production (e.g. grass or phytoplankton) on existing areas (see Bossel, 1987; Ruddle and Christensen, 1993; Guo and Bradshaw, 1993). Still, in spite of considerable efforts, these large-scale high-production systems derive approximately 80–90% of their nutrient needs from outside the farmed area in the form of imported feeds and fertilizers including all of their considerable fossil fuel and electricity needs.

RICE FLOODWATER ECOLOGY

The rice floodwater ecosystem is by nature a very productive and stable system. The flooding favours soil fertility and rice production by: (i) bringing soil pH to near neutral; (ii) increasing availability of nutrients; (iii) retarding soil organic matter decomposition and thus maintaining soil N fertility; (iv) favouring N-fixation; (v) suppressing outbreaks of soil-borne diseases; (vi) supplying nutrients from irrigation water; (vii) depressing weed growth; and (viii) preventing water percolation and soil erosion (Roger, 1996). Much of the natural fertility of rice wetlands can be attributed to cyanobacteria (blue-green algae), the major indigenous N-fixing agent in ricefield floodwater. Balance studies have indicated that this biological fixation alone contributes 15–50 kg N per crop ha^{-1} in unfertilized fields (Koyama and App, 1979).

Ecological studies on the submerged soils of lowland ricefields are scarce, however. As a result, our quantitative understanding of the nutrient pathways through this agroecological system under field and farm conditions is limited. Although individual N contributions via different agents of biological N-fixation (BNF) can be estimated more or less accurately, total BNF in a ricefield has not yet been estimated by measuring simultaneously the activities of the various components *in situ* (Roger and Ladha, 1992).

A review of the literature (App *et al.*, 1984; Roger and Ladha, 1992; Lightfoot *et al.*, 1993b; Roger, 1996) provides the following approximate values of component N fluxes into and out of an inorganically fertilized rice agroecosystem (Table 7.1). Individual estimates vary substantially, for example from a few to 80 kg N (27 kg N on average) in the case of photo-dependent fixation by cyanobacteria, depending on management and site conditions (Roger, 1996). Inorganic fertilizer application reduces BNF. Roger and Ladha (1992) present BNF figures of only 8 kg ha^{-1} per crop in

Table 7.1. Published values of component N fluxes into and out of inorganically fertilized rice agroecosystems. See text for sources of published data.

	N flux
In-fluxes (gains):	
Dry and wet atmospheric deposition	~1.5 kg N ha^{-1} year^{-1}
Run-on/deposition with irrigation water	~10 kg N ha^{-1} crop^{-1}
BNF:	
Associative fixation in the rice rhizosphere	~4 kg N ha^{-1} crop^{-1}
Heterotrophic fixation associated with rice straw	~2–4 kg N t^{-1} straw
Heterotrophic fixation in flooded planted soil associated with organic debris	~10–30 kg N ha^{-1} crop^{-1}
Photodependent fixation by blue-green algae (cyanobacteria)	~27 kg N ha^{-1} crop^{-1}
Out-fluxes (losses):	
~50–75% of biologically fixed N lost through ammonia volatilization and denitrification	
Erosion and runoff (negligible)	
Leaching/deep percolation (negligible)	

plots with broadcast urea and 12 kg with deep placed urea, as opposed to 20 kg in no-N plots. Besides having a depressive effect on BNF, inorganic fertilizer efficiency is, in itself, low in wetland ricefields. Only 20–40% of applied N is recovered by the crop depending on N source, management and agroecological conditions (Lightfoot *et al.*, 1993b). The remaining 60–80% is essentially lost from the rice production system in gaseous form. Long-term fertility plot studies have produced little evidence to suggest that N, P or potassium (K) fertilization affects total soil N content over time (App *et al.*, 1984). Losses through erosion/runoff and leaching/deep percolation are generally ignored or assumed to be negligible in lowland rice farming on fine-textured, heavy soils.

Fish (and other aquatic organisms such as frogs, snails, bivalves and crustaceans) are a natural, although increasingly scarce, ingredient of the rice floodwater agroecosystem and have been an important contributor to the nutrition of traditional rice farming households in Asia. Fish culture in ricefields has been observed to have several beneficial effects, including: (i) reduced N losses from the floodwater due to ammonia (NH_3) volatilization; (ii) improved physico-chemical properties of the soil (soil fertility); (iii) increased nutrient cycling and availability; and (iv) increased rice yield and N uptake (Roger, 1996). Many of these potentially beneficial effects of fish on rice, however, remain unquantified.

Fish grazing on aquatic biomass contribute through their faeces to increased nutrient cycling and availability to rice. In Hunan Province, China, organic matter, N, available P and K were all higher in rice/fish fields than in rice-only fields (Hunan Research Team, 1987, cited in Roger, 1996). Increases in N concentration in rice grain by 5% and N uptake by 10% were observed in fields where fish were introduced (Panda et al., 1987, cited in Roger, 1996). The same study also reported that increased P uptake in the presence of fish was observed in some experiments (the underlying circumstances are not clear) as well as higher iron uptake in rice–fish culture than in rice alone. The presence of fish may thus benefit rice yield in both quantitative and qualitative terms.

Fish grazing on algal biomass reduces water turbidity. As high algal density is associated with high pH values through carbon dioxide (CO_2) consumption, algal reduction helps lower the pH towards neutral, which in turn reduces N loss through NH_3 volatilization in the earlier stages of crop growth (Roger, 1996). Other beneficial effects of fish on rice include direct consumption of rice pests and weeds by fish. Detrimental effects of fish on rice are primarily associated with uprooting of seedlings by herbivorous fish (e.g. carp) when stocked too early.

ECOPATH MASS-BALANCE MODELS

In a preliminary study of the trophic ecological interactions and N flows in the rice floodwater ecosystem, Lightfoot et al. (1993b) compared two systems: rice monoculture and rice–fish integration. The integration of fish in rice paddies increased rice yield compared with control plots by up to 30% (10–15% on average), while at the same time producing up to 500 kg fish ha^{-1}. It is hypothesized that the greater efficiency in rice production is related to an increased production of detritus due to the presence of fish through their bioturbation activity and their production of excreta, contributing to the replenishment of soil microbial biomass – the most important actor in the ecosystem in terms of N cycling.

Initially developed for the modelling and analysis of aquatic ecosystems (Christensen and Pauly, 1992), the ECOPATH approach and software is now also being applied to agroecosystems. The software is available from ICLARM. This mass-balance framework was employed by Lightfoot et al. (1993b) as it provides a good basis for exploring the characteristics of nutrient flows and budgets in rice agroecosystems. Figures 7.1 and 7.2 convey a visual impression of how the structure of a trophic network within a rice-based agroecosystem changes as a monoculture rice farm is transformed into a 'fully-fledged' more diversified IAA farm combining rice, fish, livestock and trees. ECOPATH diagrams individual farm components as boxes and indicates their biomass, production and consumption parameter values and linkages to other components, including detritus which denotes the soil resource base. See Christensen and Pauly (1992) and Dalsgaard and Oficial (1998) for more details on how to generate, analyse and interpret flow networks within the ECOPATH framework.

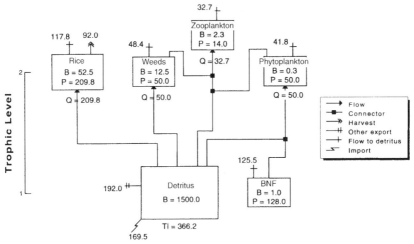

Fig. 7.1. ECOPATH flow diagram of a theoretical 1.0 ha monoculture rice farm (values in kg N ha^{-1} year^{-1}). B, average standing biomass; P, production; Q, consumption; TI, total input into detritus; BNF, bacterial nitrogen fixation.

In a comparative on-farm study of integrated and non-integrated rice farming, Dalsgaard and Oficial (1997) investigated N flows in four Philippine smallholder agroecosystems, including monoculture rice, diversified rice cultivation, and rice farming integrated with trees, livestock and aquaculture. The farms were monitored during one annual cycle and data collected on imports, recycled biomaterials, and harvested products for all compartments (rice, weeds, vegetables, fruit and multipurpose trees, bamboo, ruminants, poultry, pigs and fish). Table 7.2 lists important quantitative agroecological performance indicators for the four farms. The bioresource flow diagram in Fig. 7.3 illustrates the reuse of wastes and by-products within one of the integrated farms (Farm D). The N budget for the same farm is shown in Fig. 7.4.

This on-farm investigation showed that through integrated natural resources management, economically attractive, productive and (near) balanced systems can be generated and maintained, without resorting to large nutrient imports. It also showed that high application rates of inorganic fertilizers are not necessarily associated with a positive nutrient balance, but rather with high flows through the rice-based agroecosystem and high losses to the environment. The comparative modelling and analysis of the four smallholder farms emphasized gaps in our quantitative understanding of N fluxes through the rice agroecosystem: a positive balance in the order of 72 kg N ha^{-1} year^{-1} (Farm B, Table 7.2) suggests an impressive, but unlikely, build-up of soil N, supported neither by longer-term observations

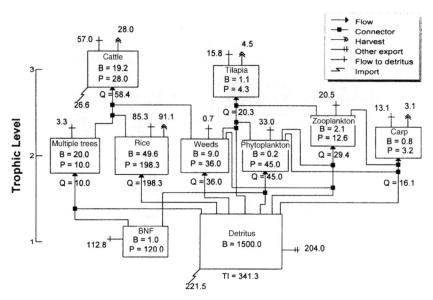

Fig. 7.2. ECOPATH flow diagram of a theoretical 1.0 ha integrated agriculture–aquaculture (IAA) farm system (values in kg N ha^{-1} year^{-1}). See Fig. 7.1 for details.

(App *et al.*, 1984) nor by long-term productivity trends (Cassman and Pingali, 1995); we are still very much in the dark as to the out-flows of nutrients through leaching and runoff. The former may be insignificant on waterlogged, well-puddled clay soils, whereas the latter will depend on water management and can be substantial where the surface water flow across a ricefield area is not carefully controlled after fertilizer applications.

DISCUSSION

The hypothetical models and actual case studies presented here indicate that one way to improve the balance, efficiency and impact of agricultural activities is through diversification and integration of farmed organisms across trophic levels. In our experience such integrated farming strategies benefit both the ecological and economic performance of smallholder farms, while reducing their off-site impact through better management and more intensive use of available soil and water resources.

Nitrogen is probably the most thoroughly researched macronutrient in the rice floodwater agroecosystem. Yet, our quantitative understanding of its inputs and outputs throughout this complex system remains fragmented.

Table 7.2. Agroecological performance indicators for four Philippine smallholder rice farm systems (from Dalsgaard and Oficial, 1997).

	Farm A High-fertilizer-input monoculture rice system	Farm B High-fertilizer-input diversified rice system	Farm C Low-fertilizer-input diversified and integrated rice system	Farm D Low-fertilizer-input diversified and integrated rice system
Surplus N (kg N ha^{-1} year^{-1})[a]	190	152	58	62
N balance (kg N ha^{-1} year^{-1})[b]	−2	72	1	−9
N efficiency[c]	0.19	0.17	0.40	0.38
N yield (kg N ha^{-1} year^{-1})[d,e]	43 (22)	45 (26)	39	33
Gross margin (US$ N ha^{-1} year^{-1})	~250	~750	~625	~600

[a]Lost from the farm system primarily in gas form (volatilized and denitrified N), and to a lesser extent through erosion/runoff.
[b]Expresses residual N not accounted for and thus assumed to be retained within the soil, as either available or unavailable N.
[c]Computed as the ratio of system N harvest over all N inputs (feeds, fertilizers, BNF, wet and dry atmospheric deposition, and run-on/sedimentation with incoming irrigation water).
[d]The productivity of the system expressed as total yield from all harvested products.
[e]Figures in brackets indicate yield with only one rice crop per year; on Farms C and D ricefields are left fallow during the dry season.

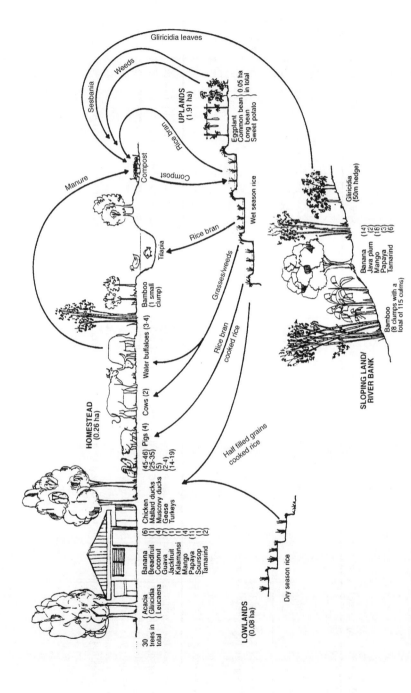

Fig. 7.3. Bioresource flow diagram of an integrated smallholder rice farm (Philippines). Values in parentheses next to enterprises are numbers of individuals (livestock or trees) unless otherwise specified (from Dalsgaard and Oficial, 1997).

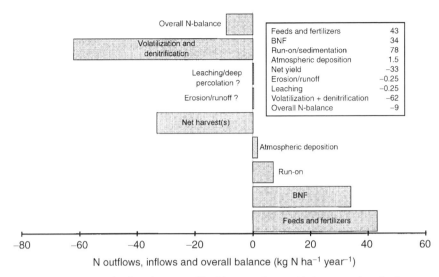

Fig. 7.4. Nitrogen budget for a smallholder rice farm with integrated agriculture–aquaculture (IAA) (Philippines); Farm D in Table 7.1.

This makes assessment and interpretation of nutrient budgets and flows at farm levels at best a difficult, and at worst a dubious, exercise.

A more complete understanding of N (and other nutrient) pathways through the rice floodwater agroecosystem requires complementary, *in situ* field studies at different spatial levels, below and above the farm scale. Plot studies can provide insights into the nutrient dynamics and availability within the soil-water resource base and nutrient exchanges across the soil-water interface. Catchment and water discharge studies give ideas of the magnitudes of nutrient loads in eroded soils, discharged water and sediments. Mass-balance studies can be developed as a first effort to identify the 'holes' in the systems – including gaps in our knowledge. More sophisticated (dynamic) modelling efforts will be needed to appreciate and quantify the physical and chemical processes behind the mass flows.

The overriding question remains, however, how to engineer the appropriate conditions for the evolution of such agroecosystems to occur on a wider scale. What are the technological, economic and/or social opportunities and constraints, which encourage or prevent farmers and farm communities from venturing into moderately intensive, integrated, nutrient balanced farming? Which tenure arrangements, market conditions, credit facilities and political incentives improve the acceptability and adoptability of IAA systems? What institutional mechanisms and communication channels within the researcher–adviser–farmer complex will further the development and spread of alternative, ecologically sound agricultural technologies and farming strategies? The transition towards IAA is knowledge and skill intensive and requires long-term commitment and entrepreneurially spirited farmers. IAA and learning to keep fish does not just mean handling a new, better variety under similar management conditions. Balanced, integrated natural resources management combines traditional and new knowledge systems and technologies to suit complex environments and find ways to utilize their heterogeneity and diversity in an ecologically sound manner.

REFERENCES

App, A., Santiago, T., Daez, C., Ventura, W., Tirol, A., Watanabe, I., De Datta, S.K. and Roger, P. (1984) Estimation of the nitrogen balances for irrigated rice and the contribution of phototrophic nitrogen fixation. *Field Crops Research* 9, 17–27.

Bossel, H. (1987) Systems study of ecofarming: the fish pond subsystem. In: Bruenig, E.F., Bossel, H., Welpel, K.P. Grossmann, W.D., Schneider, T.W., Wang, Z.H. and Yu, Z.Y. (eds) *Ecological-socioeconomic Systems Analysis and Simulation: a Guide for Application of System Analysis to the Conservation, Utilisation and Development of Tropical and Subtropical Land Resources in China*. Institute for World Forestry and Ecology, Hamburg-Reinbek and DSE/ZEL, Feldafing [Mitteilungen der Deutschen MAB Nationalkommission Nr. 24] pp. 170–203.

Brummett, R.E. (1996) The context of smallholding integrated aquaculture in Malawi: a case study for SubSaharan Africa. *Naga, the ICLARM Quarterly* 18(4), 8–10.

Brummett, R. (in press) Why Malawian smallholders don't feed their fish. *Naga, the ICLARM Quarterly*.

Brummett, R.E. and Noble, R. (1995) Aquaculture for African smallholders. ICLARM Technical Report no. 46, 69 pp.

Cassman, K.G. and Pingali, P.L. (1995) Intensification of irrigated rice systems: learning from the past to meet future challenges. *GeoJournal* 35(3), 299–305.

Christensen, V. and Pauly, D. (1992) ECOPATH II – a software for balancing steady-state ecosystem models and calculating network characteristics. *Ecological Modelling* 61, 169–185.

Costa-Pierce, B.A., Lightfoot, C., Ruddle, K. and Pullin, R.S.V. (eds) (1991) Aquaculture research and development in rural Africa. *ICLARM Conference Proceedings* 27, 52 pp.

Cross, D. (1991) FAO and aquaculture: ponds and politics in Africa. *The Ecologist* 21, 73–76.

Dalsgaard, J.P.T. and Oficial, R.T. (1997) A quantitative approach for assessing the productive performance and ecological contributions of smallholder farms. *Agricultural Systems* 55, 503–533.

Dalsgaard, J.P.T. and Oficial, R.T. (1998) Modeling and analyzing the agroecological performance of farms with ECOPATH. ICLARM Technical Report no. 53, Manila, the Philippines, 54 pp.

Dalsgaard, J.P.T., Lightfoot, C. and Christensen, V. (1995) Towards quantification of ecological sustainability in farming systems analysis. *Ecological Engineering* 4, 181–189.

Edwards, P. (1993) Environmental issues in integrated agriculture-aquaculture and wastewater-fed culture systems. In: Pullin, R.S.V., Rosenthal, H. and Maclean, J.L. (eds) *Environment and Aquaculture in Developing Countries.* ICLARM Conference Proceedings 31, pp. 139–170.

Edwards, P. and Little, D.C. (1995) Integrated crop-fish-livestock improvements in South-East Asia. In: Devendra, C. and Gardiner, P. (eds) *Global Sgenda for Livestock Research.* Proceedings of the Consultation for the South-East Asia Region, IRRI, Los Baños, the Philippines, 10–13 May 1995. ILRI (International Livestock Research Institute), Nairobi, pp. 71–87.

Edwards, P., Weber, K.E., McCoy, E.W., Chantachaeng, C., Pacharaprakiti, C., Kaewpaitoon, K. and Nitsmer, S. (1983) *Small-scale Fishery Project in Pathum Thani Province, Central Thailand: a Socioeconomic and Technological Assessment of Status and Potential.* AIT Research Report No. 158. Asian Institute of Technology, Bangkok.

Edwards, P., Pullin, R.S.V. and Gartner, J.A. (1988) Research and education for the development of integrated crop-livestock-fish farming systems in the tropics. *ICLARM Study Reviews* 16, 53 pp.

Edwards, P., Demaine, H., Ines-Taylor, N. and Turongruang, D. (1996) Sustainable aquaculture for small-scale farmers: need for a balanced model. *Outlook on Agriculture* 25(1), 19–26.

Engle, C.R. and Skladany, M. (1992) *The Economic Benefit of Chicken Manure Utilization in Fish Production in Thailand.* CRSP Research Reports 92–45. Pond Dynamics/Aquaculture CRSP, Oregon State University, Corvallis City.

FAO/RAP (1996) The Vietnam case: higher farmer income through integrated farm systems (a guide to low-input sustainable agriculture for the Asian farmer). Farm Management Technology Bulletin October 1996, FAO/RAP Publ, Bangkok.

FARM (1996) Annual progress report January to December 1995. FARM (Farmer-centred Agricultural Resource Management Programme). FAO/UNDP, Bangkok, 163 pp.

Guo, J.Y. and Bradshaw, A.D. (1993) The flow of nutrients and energy through a Chinese farming system. *Applied Ecology* 30, 86–94.

IIRR and ICLARM (1992) *Farmer-proven Integrated Agriculture-Aquaculture: a Technology Information Kit.* International Institute of Rural Reconstruction, Silang, Cavite, Philippines, and International Center for Living Aquatic Resources Management, Manila, 183 pp.

Koyama, T. and App, A.A. (1979) Nitrogen balance in flooded rice soils. In: *Nitrogen and Rice.* International Rice Research Institute, Los Baños, Laguna, Philippines, pp. 95–104.

Lightfoot, C. and Pullin, R.S.V. (1995) An integrated resource management approach to the development of integrated aquaculture farming systems. In: Symoens, J.-J. and Micha, J.-C. (eds) *The Management of Integrated Agro-piscicultural Ecosystems in Tropical Areas.* Proceedings of the International Seminar, Brussels, 16–19 May 1994, the Technical Centre for Agricultural and Rural Co-operation (CTA), Wageningen, and the Belgian Royal Academy of Overseas Sciences (ARSOM), Brussels, pp. 145–167.

Lightfoot, C., Costa-Pierce, B.A., Bimbao, M.P. and dela Cruz, C.R. (1992) Introduction to rice-fish research and development in Asia. In: dela Cruz, C.R., Lightfoot, C., Costa-Pierce, B.A., Carangal, V.R. and Bimbao, M.P. (eds) *Rice-fish*

Research and Development in Asia. ICLARM Conference Proceedings 24, pp. 1–10.
Lightfoot, C., Bimbao, M.A.P., Dalsgaard, J.P.T. and Pullin, R.S.V. (1993a) Aquaculture and sustainability through integrated resources management. *Outlook on Agriculture* 22(3), 143–150.
Lightfoot, C., Roger, P.A. and Cagauan, A.G. (1993b) Preliminary steady-state nitrogen models of a wetland rice-field ecosystem with and without fish. In: Christensen, V. and Pauly, D. (eds) *Trophic Models of Aquatic Ecosystems*. ICLARM Conference Proceedings 26, pp. 56–64.
Little, D. and Muir, J. (1987) *A Guide to Integrated Warm Water Aquaculture*. Institute of Aquaculture Publications, University of Stirling, 238 pp.
Little, D.C., Kaewpaitoon, K. and Haitook, T. (1994) The commercial use of chicken processing wastes to raise hybrid catfish (*Clarias gariepinus* × *Clarias macrocephalus*) in Thailand. *NAGA, the ICLARM Quarterly* 17(4), 25–27.
MacKay, K.T. (ed.) (1995) *Rice-fish Farming in China*. IDRC, Ottawa, 276 pp.
McClellan, S. (1991) Integrated systems: rethinking a panacea. *Ceres* 131(23), 2–25.
Mathias, J.A., Charles, A.T. and Baotong, H. (eds) (1997) *Integrated Fish Farming*. CRC Press, Boca Raton, Florida, 420 pp.
Noble, R. (1996) Wetland management in Malawi: a focal point for ecologically sound agriculture. *ILEIA Newsletter* 12(2), 9–11.
Ofori, J.K., Prein, M., Fermin, F., Owusu, D.Y. and Lightfoot, C. (1993) Farmers picture new activities: Ghanean farmers gain insight in resource flows. *ILEIA Newsletter* 9(1), 6–7.
Prein, M. (1993) Ponds transform the ecology of small farms in Ghana. *Naga, the ICLARM Quarterly* 16(2–3), 25. Photosection.
Prein, M., Lightfoot, C. and Pullin, R.S.V. (1995) ICLARM's approach to the integration of aquaculture into sustainable farming systems. In: Devendra, C. and Gardiner, P. (eds) *Global Agenda for Livestock Research*. Proceedings of the Consultation for the South-East Asia Region, IRRI, Los Baños, the Philippines, 10–13 May 1995. ILRI (International Livestock Research Institute), Nairobi, pp. 117–125.
Prein, M., Lightfoot, C. and Pullin, R.S.V. (1996a) Farmer-participatory research approaches towards agriculture-aquaculture integration for sustainable management of natural resources. In: Preuss, H.-J.A. (ed.) *Agricultural Research and Sustainable Management of Natural Resources*. Lit Verlag Münster-Hamburg, pp. 173–184.
Prein, M., Ofori, J.K. and Lightfoot, C. (1996b) Research for the future development of aquaculture in Ghana. *ICLARM Conference Proceedings* 42, 94 pp.
Pullin, R.S.V. (1993) An overview of environmental issues in developing-country aquaculture. In: Pullin, R.S.V., Rosenthal, H. and Maclean, J.L. (eds) *Environment and Aquaculture in Developing Countries*. ICLARM Conference Proceedings 31, pp. 1–19.
Pullin, R.S.V. and Prein, M. (1995) Fishponds facilitate natural resources management on small-scale farms in tropical developing countries. In: Symoens, J.-J. and Micha, J.-C. (eds) *The Management of Integrated Agro-piscicultural Ecosystems in Tropical Areas*. Proceedings of the International Seminar, Brussels,16–19 May 1994, the Technical Centre for Agricultural and Rural Co-operation (CTA), Wageningen, and the Belgian Royal Academy of Overseas Sciences (ARSOM), Brussels, pp. 169–186.

Reijntjes, C., Haverkort, B. and Waters-Bayer, A. (1992) *Farming for the Future: an Introduction to Low-external-input and Sustainable Agriculture*. Macmillan, London.

Roger, P.A. (1996) *Biology and Management of the Floodwater Ecosystem in Ricefields*. International Rice Research Institute, Los Baños, Laguna, Philippines, 250 pp.

Roger, P.A. and Ladha, J.K. (1992) Biological N_2 fixation in wetland rice fields: estimation and contribution to nitrogen balance. *Plant and Soil* 141, 41–55.

Rothuis, A.J., Nhan, D.K., Richter, C.J.J. and Ollevier, F. (1998a) Rice with fish culture in the semi-deep waters of the Mekong Delta, Vietnam: a socioeconomic survey. *Aquaculture Research* 29, 47–57.

Rothuis, A.J., Nhan, D.K., Richter, C.J.J. and Ollevier, F. (1998b) Rice with fish culture in the semi-deep waters of the Mekong Delta, Vietnam: interaction of rice culture and fish husbandry management on fish production. *Aquaculture Research* 29, 59–66.

Ruddle, K. and Christensen, V. (1993) An energy flow model of the mulberry dike-carp pond farming system of the Zhujiang Delta, Guangdong Province, China. In: Christensen, V. and Pauly, D. (eds) *Trophic Models of Aquatic Ecosystems*. ICLARM Conference Proceedings 26, pp. 48–55.

Ruddle, K. and Zhong, G. (1983) *Integrated Agriculture-Aquaculture in South China: the Dike-pond System of the Zhujiang Delta*. Cambridge University Press, London.

Schaber, J. (1997) FARMSIM: a dynamic model for the simulation of yields, nutrient cycling and resource flows on Philippine small-scale farming systems. MSc thesis, Institut für Umweltsystemforschung, University of Osnabrück, 114 pp.

Symoens, J.J. and Micha, J.C. (eds) (1995) *The Management of Integrated Agro-piscicultural Ecosystems in Tropical Areas*. Proceedings of the International Seminar, Brussels, 16–19 May 1994, the Technical Centre for Agricultural and Rural Co-operation (CTA), Wageningen, and the Belgian Royal Academy of Overseas Sciences (ARSOM), Brussels, 587 pp.

Tran, N.T. and Demaine, H. (1996) Potentials for different models for freshwater aquaculture development in the Red River Delta (Vietnam) using GIS analysis. *Naga, the ICLARM Quarterly* 19(1), 29–32.

WES (West-East-South Programme) (1997) *Ecotechnological and Socioeconomic Analysis of Fish Farming Systems in the Freshwater Area of the Mekong Delta 1996–1997*. College of Agriculture, Cantho University, 124 pp.

Yan, J. and Yao, H. (1989) Integrated fish culture management in China. In: Mitsch, W.J. and Jørgensen, S.E. (eds) *Ecological Engineering: an Introduction to Ecotechnology*. John Wiley & Sons, New York, pp. 375–408.

Improving Nutrient Management for Sustainable Development of Agriculture in China

Jin Jiyun,[1,2] Lin Bao[2] and Zhang Weili[2]

[1]*Beijing Office, Potash and Phosphate Institute of Canada (PPI/PPIC);* [2]*Soil and Fertilizer Institute, Chinese Academy of Agricultural Sciences, 30 Baishiqiao Road, Beijing 100081, China*

ABSTRACT

For several thousands of years before 1950, Chinese farmers used organic materials to maintain soil fertility in crop/soil systems, but at low productivity level. The low input and low output traditional production system cannot sustain Chinese society with increasing population. Therefore, after 1950, chemical fertilizer became important to increase crop production. China then went through four main stages in chemical fertilizer use, more or less following the law of the minimum: (i) nitrogen (N) fertilizer plus organic manure in the 1950s; (ii) N and phosphorus (P) fertilizers plus organic manure in the 1960s; (iii) N, P and potassium (K) fertilizers plus organic manure in the 1970s; and (iv) N, P, K, and micronutrients fertilizers plus organic manure in the 1980s.

In the last 45 years, chemical fertilizer use in China has contributed greatly to the sustained increase of crop production, and the maintenance and improvement of soil fertility. However, fertilizer use in China is still not balanced. Unbalanced use of fertilizers has resulted in depletion of soil K and some secondary and micronutrients in soils, and low fertilizer use efficiency. It is shown that more attention needs to be given to balanced fertilization and improved nutrient management to improve fertilizer use efficiency in China for both a sustained increase of agricultural production and improvement of environmental quality.

INTRODUCTION

China has to support over 22% of the world's population on less than 7% of the world's arable land. China has an annual increase of 15 million people and an average annual decrease in cultivated land of 350,000 ha, mainly due to industrialization of rural areas and other non-agricultural use. To support the large and ever-increasing population with limited and ever-decreasing cultivated land, China has to develop sustainable agricultural production systems which can: (i) ensure adequate supply of food, oil, fibre and other products to improve people's living standard and provide enough raw materials for related industrial developments; (ii) make the most efficient use of limited land resource and purchased external inputs; (iii) improve soil fertility and productivity; (iv) increase profitability of farming; and (v) improve environmental quality.

The unique situation and requirements mentioned above imply that Chinese agriculture must be one of high inputs, high outputs and high efficiency. Among the various agricultural input materials, fertilizer is the most expensive one. It has been reported that in general chemical fertilizer input (cost) accounts for 30–40% of the total external inputs (energy plus materials) in China (IBSRAM Newsletter). In some high-yielding regions, cost of fertilizers accounts for 50% or more of the total input costs (IBSRAM Newsletter). However, fertilizer is the most effective of the various external input materials used. It has been reported that 30–50% of the total crop production increase in China in the recent past was as a result of chemical fertilizers (Lin Bao and Jin Jiyun, 1991).

Because of the effectiveness of fertilizer use in increasing crop production, fertilizer consumption increased rapidly in China in the last 45 years. By 1995, the total nutrients input into cropping systems in China reached 53 million tonnes, with 33.18 million tonnes on chemical fertilizers and the rest in organic form (Li Jiakang and Lin Bao, 1996). China became number one in the world in total fertilizer nutrient consumption. With this level of nutrient input into the cropping system, the balanced supply of all essential crop nutrients for their efficient use, and close monitoring of the nutrient balance in the system becomes a critical and important task. This chapter provides an historic review of nutrient management in China, and the present and future outlook for the nutrient balance in Chinese agriculture.

ORGANIC AND INORGANIC NUTRIENT SOURCES IN CHINESE AGRICULTURE

China has a long tradition of using organic manures to maintain and improve soil fertility. According to historical records, the use of organic manure in China can be traced back to 3000 BC in the Shang dynasty. The common use of manure started in the Warring States period and Qin dynasty (221–206 BC), about 2000 years ago. Historically, manure use in China has contributed greatly to agricultural production and development of civilization. Three facts must be mentioned in order to fully understand how China was able to sustain a reasonable yield using only organic sources of nutrients: (i) in the early days, sufficient arable land was available, and people continuously opened new reasonably fertile land for agriculture as needed; (ii) the relatively low requirement by a slowly growing population with low living standards; and (iii) nutrients in lowland soils were supplemented by transport of organic materials from upland soils, resulting in heavy degradation and erosion of upland soils, and causing considerable environmental problems from which people still suffer today (Lin Bao et al., 1989).

The situation has now changed drastically. With the rapid population increase and improvement of living standards, demands on agriculture increased very quickly. Most arable land has now been opened for agricultural use, and the heavily degraded and eroded upland soils allow little transport of remaining nutrients to the lowland, especially with the increasing concern about soil erosion and environmental quality. Therefore, chemical fertilizers become a necessity to sustain soil fertility of the lowlands, improve those of upland soils, while maintaining a sustained increase in agricultural production.

Average yields of wheat and rice in China from 221 BC to the present (1995) are presented in Table 8.1. The increase in wheat yields from 793 to 1465 kg ha^{-1} covered a period of more than 2000 years in the low-input

Table 8.1. Average yields (kg ha^{-1}) of wheat and rice in China from 221 BC to present.

Dynasty	Year	Yield, kg ha^{-1}	
		Wheat	Rice
Qin	(221–206 BC)	793	—
W. Han	(206 BC–AD 24)	904	603
Song	(960–1279)	780	1560
Ming-Qing	(1368–1911)	1465	2930
PCR	1952	735	2408
	1965	1020	2940
	1980	1890	4133
	1986	3040	5338
	1990	3225	5805
	1995	3539	6020

Source: Liu Gengling (1988) and China Agricultural Yearbooks.

and low-output organic farming system, an average increase of 17 kg every 50 years. However, from 1952 to 1995 with modern agricultural production systems, wheat yields increased from 735 to 3539 kg ha^{-1}, an average increase of 65 kg every year. For rice, the increase from 603 to 1560 kg ha^{-1} spanned about 1200 years in traditional agriculture, an average increase of 40 kg every 50 years. But, with modern agricultural production systems, from 1952 to 1995, rice yields increased from 2408 to 6020 kg ha^{-1}, an average increase of 84 kg every year.

This historical perspective demonstrates that increases in grain yields since 1952 exceed those of any other period in China's history. There are many factors responsible for this accelerated growth of agricultural production, including development and use of high-yielding varieties, improvement of water management and pest control, and so on. However, one of the most important factors was the increased use of inorganic fertilizers.

Organic manure has been an important source of nutrients in China's history, and still is today. Before 1949, almost no inorganic fertilizer was used in China. Since then, with the continuous increase of crop production, plant nutrient supply in the form of organic manure increased considerably. From 1949 to 1995, the annual application of nutrients ($N + P_2O_5 + K_2O$) in organic manure increased from 4.28 million tonnes to 17.07 million tonnes (Table 8.2) (Li Jiakang and Lin Bao, 1996). However, at the same time,

Table 8.2. Nutrient[a] consumption and contribution of organic manure in China.

		Contribution of organic manure							
	Total nutrients[a]	Total nutrients		N		P_2O_5[a]		K_2O[a]	
Year	(Mt)	Mt	%	Mt	%	Mt	%	Mt	%
1949	4.29	4.28	99.9	1.62	99.6	0.79	100	1.87	100.0
1957	6.95	6.58	91.0	2.49	88.7	1.23	96.0	2.86	100.0
1965	9.13	7.37	80.7	2.93	70.8	1.38	71.5	3.06	99.9
1975	16.03	10.65	66.4	4.10	53.0	1.94	54.6	4.62	97.3
1980	24.00	11.31	47.1	4.16	30.6	2.06	41.8	5.09	92.8
1985	31.56	13.83	43.7	5.04	28.6	2.56	38.6	6.23	85.1
1990	40.89	16.10	36.4	5.26	23.2	2.67	29.2	8.17	80.1
1995	53.00	17.07	32.2	6.13	21.6	3.30	24.9	7.63	67.0

[a]To keep the nutrient units and the data consistent with the official statistics in China, N, P_2O_5 and K_2O were used for nitrogen, phosphorus and potassium nutrients, respectively, for nutrient consumption and application rate throughout this chapter, except where indicated otherwise.
Mt, million metric tonnes.
Source: Li Jiakang and Lin Bao (1996).

agricultural production increased very rapidly, and the nutrient requirement was far exceeding the amount of nutrients supplied in organic manure. Therefore, inorganic fertilizers were introduced in the 1950s, and their use increased very rapidly thereafter. As a result, the contribution of organic manure in total nutrient supply decreased with time, from almost 100% in 1949 to only 32.2% in 1995. Organic manure supplied 21.6% of total N input and 24.9% of total P_2O_5 input in 1995, but a large percentage (67.0%) of the total K_2O consumption was still from organic manure (Table 8.2).

With time, nutrient input as inorganic fertilizers became more and more important in agriculture production in China. Statistics indicated that in the 45 years from 1951 to 1995, total consumption of chemical fertilizers was significantly correlated with total grain production, and with total cotton production. Close correlation was also found between fertilizer application rate and grain and cotton yields (Lin Bao, 1991). Assuming that 80% of the total fertilizer was used for grain production and that 1 kg of applied nutrient increased grain production by 8 kg, it has been estimated that about 35% of the total grain production in the 5 years from 1986 to 1990 was due to use of chemical fertilizers (Lin Bao, 1991).

In the past 45 years, soil fertility management in China passed through different stages: the first stage of combined used of N fertilizer with organic manure before 1960; the second stage of combined use of N and P fertilizers with manure in the 1960s; and the third stage of combined use of N, P, K fertilizers, some micronutrients, and organic manure after middle 1970s (Jin Jiyun and Portch, 1991). In the 1960s, with development of high-yield crop varieties and other improvements in farming practices, N application increased crop yield significantly. At the same time, other nutrients in the soils were being depleted. Thus, soil P, the next most deficient plant nutrient in the soil, became a yield-limiting factor. Therefore, P fertilizer application became an important measure for further increases in crop yield. From the late 1960s and early 1970s onwards, with continuous increases of crop yield through increased use of N and P fertilizers, K deficiency was observed, first in southern China, but has gradually extended to the north. The change of fertilizer use efficiency and fertilization practices in China during these 45 years can be considered as a demonstration of the Law of the Minimum on a large scale. After the most severely deficient nutrient (N) was corrected by fertilization, the next deficient nutrient (P) limited crop yield.

From the 1950s to the 1980s, China's fertilizer use was far from balanced, with more N used than P and K. This situation has been slowly improved from the late 1980s. At present, China is the world's largest user of chemical fertilizers in total quantity. However, fertilizer use in China is still not well balanced, with less P and K used relative to N (Table 8.3).

The changes in nutrient input as mentioned above significantly affected the nutrient balance in China's agricultural production systems. Before the middle of the 1970s, the majority of plant nutrient input was still from organic sources (Table 8.2), and there was relatively low nutrient input as inorganic fertilizers. The total input could not balance the crop removal. As a result, negative balances were found for N, P and K, and all three major nutrients were subjected to depletion from soils (Table 8.4; Li Jiakang and Lin Bao, 1996).

Table 8.3. Fertilizer consumption in China, 1952–1995, in million metric tonnes.

Year	N (Mt)	P_2O_5 (Mt)	K_2O	Ratio N : P_2O_5 : K_2O
1952	0.078	—	—	—
1957	0.32	0.05	—	—
1965	1.33	0.61	0.003	1 : 0.46 : 0.002
1975	3.31	1.46	0.113	1 : 0.44 : 0.030
1980	9.42	2.88	0.39	1 : 0.31 : 0.041
1985	12.59	3.08	1.09	1 : 0.24 : 0.087
1987	13.89	4.84	1.23	1 : 0.35 : 0.089
1988	14.90	5.12	1.40	1 : 0.34 : 0.094
1989	16.20	5.70	1.67	1 : 0.35 : 0.103
1990	17.40	6.47	2.03	1 : 0.37 : 0.117
1991	18.48	7.19	2.39	1 : 0.39 : 0.129
1992	18.95	7.65	2.70	1 : 0.40 : 0.142
1993	19.94	8.61	2.97	1 : 0.43 : 0.149
1994	20.62	9.25	3.31	1 : 0.45 : 0.161
1995	22.24	9.95	3.76	1 : 0.45 : 0.169

Note: Estimate based on assumption that there is 30% N, 54% P_2O_5 and 16% K_2O in compound fertilizer.
Source: Agriculture Yearbooks.

Table 8.4. Nutrient balance in farm land in China (in 10,000 tonnes).

	Input						Output			Balance		
	Organic form			Inorganic form								
Year	N	P_2O_5	K_2O	N	P_2O_5	K_2O	N	P_2O_5	K_2O	N	P_2O_5	K_2O
1949	162	79	187	0.6	—	—	291	138	306	−129	−59	−119
1957	249	123	286	31.6	5.2	—	511	236	562	−246	−110	−276
1965	293	138	306	121	55	0.3	522	237	560	−169	−60	−254
1975	410	194	462	364	161	13	749	334	813	−157	−28	−338
1980	416	206	509	943	287	39	867	378	933	21	29	−385
1985	503	256	621	1259	408	109	1114	479	1208	19	63	−478
1990	526	280	693	1740	647	203	1279	559	1375	117	174	−478
1995	611	330	760	2224	995	376	1373	577	1455	350	449	−315

After 1980, with the increased use of N and P fertilizers, the balances for N and P gradually changed from negative to positive. But with inadequate supply of K fertilizer and large percentage of straw removal from the system, soil K depletion continued, causing rapid expansion of areas showing soil K deficiency and significant crop yield response to K fertilizer application.

From the above discussion, it is clear that both organic and inorganic fertilizers are vital nutrient sources for sustained development of agriculture in China. Therefore, attention is needed to ensure a combined and rational use of both organic and inorganic nutrient sources for a sustainable development of agricultural production in China.

NUTRIENT MANAGEMENT AND SUSTAINABILITY OF AGRICULTURE IN CHINA

Soil Nutrient Management and N Fertilizer Use Efficiency

Unbalanced use of fertilizers causes low yields, low nutrient use efficiency, low quality of products, and environmental problems. Thus, sustainability of agriculture is threatened.

Results obtained from a large number of field trials conducted by the National Network on Chemical Fertilizer Efficiency during the period from 1958 to 1962 and that from 1981 to 1983 are summarized in Table 8.5. It shows clearly that in the period from 1958 to 1962, application of N resulted in considerable increases in crop yields. However, by the early 1980s, N efficiency dropped significantly for all three grain crops. The most

Table 8.5. Changes in chemical fertilizer use efficiency with time.

Nutrient	Crop	Yield increase in kg per kg nutrient	
		1958–1963	1981–1983
N	Rice	15–20	9.1
	Wheat	10–15	10.0
	Maize	20–30	13.4
P_2O_5	Rice	8–12	4.7
	Wheat	5–10	8.1
	Maize	5–10	9.7
K_2O	Rice	2–4	4.9 (6.6 in south)
	Wheat	ns	2.1
	Maize	ns	1.6

Source: Lin Bao, 1991.

important reason was the unbalanced use of N fertilizers without adequate supply of P and K. This can be verified by the increases of P and K fertilizer use efficiency in the early 1980s compared with that in the late 1950s and early 1960s. The excess N, not taken up by crops, was probably leached out of the soil profile to groundwater, causing environmental problems (Zhang and Jin, 1996).

Research has indicated that balanced use of N fertilizer with P, K, and other needed nutrients can improve N fertilizer use efficiency. Data in Table 8.6 indicate that balanced use of P and K fertilizers with N in barley increased N utilization rate from 27.5 to 50.9%, and decreased N losses from 43.2 to 23.9% (Zhou Yimin, 1994).

Data in Table 8.7 indicate that increase of N rate from 60 to 120 kg N ha^{-1} alone without the combined use of K reduced rice paddy yields. In this case, additional N applied was not used by crops, but was probably leached out with irrigation water or volatilized, creating environmental problems. However, with an increase in the K rate, paddy yield was increased and N use efficiency was improved (Science and Technology Department of Ministry of Agriculture, 1991).

Table 8.6. Effect of P and K application on N utilization rate by barley.

Treatment	% of N uptake by crop	% of N remaining in soil	% of N lost
N	27.5	29.4	43.2
NP	44.2	25.6	30.3
NPK	50.9	25.3	23.9

Source: Zhu Yingmei (1994).

Table 8.7. Interactive effect of nitrogen and potassium on rice yields.

Treatment			
kg N ha^{-1}	kg K$_2$O ha^{-1}	Paddy yield (kg ha^{-1})	Yield increase (%)
60	0	3370	—
60	56	4834	43.4
60	112	5226	55.1
120	0	3084	—
120	56	4986	61.7
120	112	5598	81.5

Source: Zhejiang Provincial Soil and Fertilizer Institute, Zhejiang, China, 1991 (unpublished data).

Unbalanced Fertilization and Soil Nutrient Depletion

Unbalanced fertilization depletes soil nutrients not included in the fertilization programme. Results from a long-term field experiment with wheat–maize double cropping in Yutian County, Hebei province, indicated that for nine seasons, the crops removed a total of 711 kg of elemental K from the soil in plots without K application (NP treatment) (Table 8.8).

Even with K application both in chemical fertilizer form (NPK treatment) or in manure (NPKM treatment), the balance of K in the soil–crop system was still negative due to large removal of K by the high yields produced. As a result, the yield increase by K fertilization became more and more significant with time. In 1991, K application increased maize yield by 31–32%, while in the first maize crop in 1988 this was only 8–11% (Table 8.9).

Depletion of other plant nutrients from soils has also been observed in China. One of the interesting observations was sulphur (S) deficiency in the northeast region, where no S deficiency was expected or reported. No attention had been given to this plant nutrient until a soil nutrient survey was conducted with the cooperation of the China Program of the Potash and Phosphate Institute. This study discovered S deficiency in some paddy soils in this region. This was due to the shift of P sources from S-containing fertilizers, such as single super phosphate (SSP) (P_2O_5 15–18%, S 15–17%), to high analysis S-free fertilizers, such as diammonium phosphate (DAP), (18–46–0), and S-free compound fertilizers such as 15–15–15, since the late 1980s. In one field experiment in Jilin province, rice yield was increased by 28–36% by S application (Table 8.10; Zhang Kuan *et al.*, 1991).

Depletion of other secondary and micronutrients was also observed in the soil nutrient survey using the systematic approach as proposed by the China Program of Potash and Phosphate Institute. In this soil nutrient

Table 8.8. Balance sheet of soil K in a wheat–maize double cropping system for nine seasons, Yutian, Hebei, 1987–1991.

Treatment	K removed (kg ha^{-1})	K added (kg ha^{-1})	Balance (kg ha^{-1})
NP	711	0	−711
NPK	1357	841	−516
NPM	920	126	−794
NPKM	1592	967	−625

Notes: K, 112.5 kg K_2O ha^{-1} per season as KCl, M, 11.7 t manure ha^{-1}, only applied to wheat.
K added, the amount of elemental K added as both KCl and manure forms.
K removed, the amount of elemental K removed in crop harvest, including grain and straw.

Table 8.9. Yield responses of maize to K in a long-term field trial in Yutian, Hebei.

Year (crop)	Treatment	Yield kg ha^{-1}	%
1988	NP	7125	100
(Spring maize)	NPK	7724	108**
	NPM	7244	100
	NPKM	8027	111**
1991	NP	4809	100
(Summer maize)	NPK	6310	131**
	NPM	5070	100
	NPKM	6685	132**

Note: **Significant at 0.01 level.

Table 8.10. Effect of S application on rice yield in Jilin province.

Treatment	Yield (kg ha^{-1})	Yield increase by S application kg ha^{-1}	%
NPK	5745	—	—
NPKS	7680	1935	34

Note: N, 160 kg N ha^{-1}, P, 80 kg P$_2$O$_5$ ha^{-1}, K, 120 kg K$_2$O ha^{-1}.
Source: Zhang Kuan *et al.* (1991).

survey, soil samples were collected from major soil types throughout China. The samples were analysed for availability of all macro-, secondary- and micronutrients. Pot experiments with sorghum plants were designed with 13–16 treatments, including an optimum treatment (OPT). The OPT treatment was based on the soil testing and adsorption study, in which all essential nutrients were adjusted to a proper level and in a balanced manner. Missing element treatments from the OPT treatment were also arranged to verify the deficiency of the element in the soils. The relative dry matter yield of sorghum plants in each missing element treatment at harvest (taking the dry matter yield in OPT treatment as 100%) was used to evaluate the deficiency of the nutrient in the soil. Table 8.11 summarizes results of the sorghum pot experiments in 140 soils from 17 participating provinces, indicating the fairly widespread deficiency of S, calcium, magnesium, zinc, boron, molybdenum, copper and manganese (Jin Jiyun and Portch, 1991).

Table 8.11. Summary of soil nutrient survey of 140 soils from 17 provinces of China.

Treatment	No. of soils deficient (%)	Range of relative dry matter yield (%)	Mean of dry matter yield (%)	Location
−N	137 (98)	6.1–83.9	45.2	All 17 provinces
−P	126 (90)	8.5–89.7	39.6	All 17 provinces
−K	84 (60)	39.0–89.8	73.5	All 17 provinces
−Ca	20 (14)	2.2–89.0	52.8	GD,GX,SC,JX,HUB,LN
−Mg	25 (18)	34.0–89.7	74.4	GD,GX,SC,YN,GZ,SX,SD
−S	45 (32)	14.0–89.8	71.3	GD,GX,HUN,YN,HUB,HEB,SD,LN,JL,SX,GZ
−B	36 (26)	65.0–89.7	80.9	GD,LN,HUB,HUN,QH,HEB,GZ,YN,SX
−Cu	37 (26)	40.0–89.5	77.2	GD,JL,HUB,LN,HEB,QH,SC,JX,YN,SX,GZ
−Fe	17 (12)	46.0–87.5	79.4	SD,HEB,LN,JL,SX,GZ,QH,SC
−Mn	34 (24)	50.2–89.5	79.1	SD,LN,HEB,SX,QH,YN,GX,GD,GZ,SC
−Mo	28 (20)	38.7–89.4	79.5	LN,HEB,SX,ZJ,GD,HUB,HUN,GZ,YN,SC
−Zn	68 (49)	40.0–89.6	75.1	All 17 provinces

Notes: GD, Guangdong; JL, Jilin; LN, Liaoning; HL, Heilongjiang; HUN, Hunan; HUB, Hubei; YN, Yunnan; SC, Sichuan; SD, Shandong; JX, Jiangxi; HEN, Henan; HEB, Hebei; GX, Guangxi; ZJ, Zhejiang; SX, Shaaxi; GZ, Guizhou; QH, Qinghai.
Source: Jin and Portch (1991).

For sustained increase of agricultural production, it is necessary to identify all of these plant nutrient-limiting factors in soils and develop a balanced fertilization programme to systematically remove them as yield limiting factors. Otherwise, agriculture cannot be sustainable.

Soil Fertility Management and Environmental Quality

It is evident that overuse of N fertilizer will cause environmental problems, such as increased nitrate contents in ground and surface water, and excessive nitrate content in vegetables and other agricultural products. A survey of nitrate content in groundwater indicated that in the Beijing–Tianjin–Tangshan region, about half of the 102 groundwater samples had nitrate contents higher than the critical level of 50 mg NO_3 l^{-1}. In some extreme cases, nitrate content was as high as 300–500 mg l^{-1}. In general, nitrate content in groundwater is higher around the cities where vegetables are major crops. These usually receive more N, but relatively less P and K, compared with grain crops (Zhang and Jin, 1996).

According to a survey conducted in 1992, total N content in 25 lakes randomly sampled in China was higher than 0.2 mg l^{-1}, indicating that these lakes were reaching the eutrophic state in terms of N content (Zhu Yingmei, 1994). It is hard to estimate how much of the N in these lakes is from N fertilizer applied to farm land since many factors contribute to N accumulation in the water, such as industrial wastes, sewage, animal manure, etc. However, N applied in excess or in an unbalanced manner from farm lands and moved by leaching and/or surface runoff was undoubtedly one of the sources (Zhu Yingmei, 1994).

As discussed above and indicated in Table 8.6, application of N alone resulted in relatively low N use efficiency and great N losses. Part of the N loss was leached out of the soil profile to groundwater. However, addition of P and K fertilizers improved utilization of applied N, and reduced N losses and water pollution by N leaching.

Phosphate can be adsorbed very tightly by soil colloids. Therefore, little P can be leached out of the soil profile. However, bad soil management resulting in soil erosion, can carry some P from farm land to lakes and rivers, contributing to the eutrophic problem of surface water bodies. Therefore, the key to prevent the harmful influence of P fertilization on the environment is to control soil erosion, which is very serious in some locations of China. It has been reported that the total area in China suffering from water and soil erosion covers about 1.53 million square kilometres, accounting for 16% of the total territory in China. Soils lost from the loess plateau by erosion reach about 1.6 billion tonnes each year, containing about 40 million tonnes of nutrients (N + P_2O_5 + K_2O) (Yu Zhenrong *et al.*, 1993). It must be pointed out here that balanced fertilization can help re-vegetate eroded hillsides, therefore reducing erosion problems and helping to improve the environment. Researchers at IBSRAM reported on results from several countries in the sloping land network, that fertilizer alone (without any other conservation measure) reduces soil loss to 15 t ha^{-1} year^{-1} from 50 t ha^{-1} year^{-1} (IBSRAM Newsletter).

No harmful effect has been reported for K either in the water system or in the soil ecosystem. Overuse of K may cause induced deficiencies of other nutrients (magnesium, zinc, etc.) in soil and crop systems. However, this is not the case in China with generally low soil K content in most cultivated soils and very limited K application in relation to crop needs.

Proper Soil Fertility Management Improves the Soil Resource

Balanced use of chemical fertilizers is one of the key factors for proper soil management. It can increase crop yield and total biomass production, increase total organic matter in the soil–crop system, and eventually increase soil organic matter while improving soil properties and productivity (Mei Fangquan and Gu Shuzhong, 1993). In Quzhou county, Hebei province, soils were managed by increase of N and P fertilizer in a balanced manner (with the optimum ratio of N : P_2O_5 = 1 : 0.5) and return of crop straw and stubble to the soils. Soil organic matter content and N and P levels in 20 sites have been monitored since 1980. Results indicate that with comprehensive nutrient management (N and P fertilizer application plus straw and stubble return), soil organic matter content increased, on average, from 0.79% in 1980 to 1.30% in 1991. Total and available N and available P in the soils were also increased (Table 8.12) (Yu Zhenrong et al., 1993).

Sustainability and Profitability

Sustainable agricultural production must be profitable for the farmer. If farmers do not receive a reasonable profit for their labour, they will either abandon the land, or switch to a subsistence level of farming with reduced inputs. This type of subsistence farming is not sustainable because of plant nutrient depletion and, in the longer term, population increases on the same area of land. In China farmers are encouraged to produce high yields of high quality produce by high efficiency farming. Large area demonstrations have been established to visualize this principle. Economic analysis from one such area, a 15 t year^{-1} grain production project (produce a total of 15 tonnes grain in one year from two crops, wheat and maize) in Beijing region, revealed that high economic benefits were obtained in high-input, high-output multidisciplinary systems using integrated modern

Table 8.12. Change of soil organic matter and nutrients contents in top layer, Quzhou, Hebei.

Year	Organic matter content (%)	Total N (%)	Available N ($\mu g\ g^{-1}$)	Olsen-P ($\mu g\ g^{-1}$)
1980	0.79	0.051	33.3	3.8
1983	0.93	0.055	55.4	5.3
1984	0.94	0.072	40.4	6.4
1985	0.91	0.077	44.4	10.4
1986	1.08	0.066	42.1	10.8
1989	1.18	0.071	55.4	9.3
1990	1.24	0.074	51.3	9.8
1991	1.30	0.080	44.8	9.5

Average of 20 monitored sites.
Source: Yu Zhenrong et al. (1993).

technologies. The value/cost ratio was 1.80 (2.26 in some fields) from the 15 tonnes grain project area, while normal fields with lower yields had values between 1.57 and 1.74 (Table 8.13) (Wen Yupu, 1992).

PROBLEMS REMAINING IN NUTRIENT MANAGEMENT AND POTENTIAL FOR GRAIN PRODUCTION THROUGH IMPROVED NUTRIENT MANAGEMENT

From the above discussion, one can see that in the last 45 years China has made remarkable progress in improvement of chemical fertilizer use, and fertilizer application has played a vital role in the continuous increase of agricultural production in China. But there are many problems remaining. One related observation is that in the last 10 years, grain production did not increase with the increase of total fertilizer consumption at the same rate as before. From 1984 to 1994, chemical fertilizer consumption increased from 17.40 million tonnes to 33.18 million tonnes, a 90.7% increase in 10 years. However, the grain production was only increased from 407 million tonnes to 444 million tonnes, a 9.1% increase (China Agriculture Yearbooks). There are many factors associated with this; a significant portion of land was shifted from grain production to vegetable, fruit and other cash crops with more fertilizer use for higher profits. But, unbalanced use of fertilizer caused by associated fertilizer management problems contributed to the decrease of fertilizer use efficiency. There are many factors associated with this; the main factors are discussed below.

Unbalanced Supply of N, P and K Fertilizers

In general, Chinese farmers tend to use more N than P and K fertilizers, since the effect of N on plant growth can be more easily observed than

Table 8.13. Different yield levels and output/input ratios in wheat–maize cropping system, Beijing, 1990.

Yield (t ha^{-1})	Output (Yuan ha^{-1})	Input (Yuan ha^{-1})	Benefits (Yuan ha^{-1})	Output/input ratio
15.4	10,280	5,700	4,580	1.80
13.5–15.0	9,590	5,520	4,070	1.74
12.0–13.5	8,770	5,420	3,350	1.62
10.5–12.0	7,940	5,060	2,880	1.57

Source: Wen Yupu (1992).

P and K. This situation became very severe in recent years, especially in vegetable growing regions. The overuse of N caused decrease of N use efficiency, N loss and pollution to groundwater. A recent survey conducted by the Soil and Fertilizer Institute of the Chinese Academy of Agricultural Sciences in Beijing–Tianjin–Tangshan region, indicated that nitrate content in about 50% of the well water in 102 sites in 14 counties was above the critical level (Zhang and Jin, 1996).

Uneven Distribution and Use of Fertilizers by Regions

Statistics indicate that in economically developed regions, farmers use more fertilizers than in less developed regions. For example, in 1994, annual fertilizer use in nutrients in Fujian and Guangdong reached 838.5 kg ha^{-1} and 773.5 kg ha^{-1}, respectively, while in Heilongjiang and Gansu, the annual fertilization rate was less than 150 kg ha^{-1}.

Shift of Land from Grain Production to Vegetable and Other Cash Crops

Since farmers always look for high profit, in recent years, farmers have tended to grow more vegetable, fruit and other cash crops and use more fertilizers for those cash crops. Statistics indicate that from 1984 to 1994, the area for fruits was tripled, while the area for vegetable, melon and mulberry was doubled (China Agriculture Yearbooks). Although the government will continue to pay great attention to grain production in future, farmers will always try to use more fertilizers and other expensive input material to cash-making crops.

Deficiencies in Secondary and Micronutrients have not been Corrected in Time

With the increase of N, P and K fertilizer use and increase in crop production, secondary and micronutrients have become limiting factors in some regions. Recent surveys indicate that 51.1% of the arable land showed zinc (Zn) deficiency, 34.5% showed boron (B) deficiency, 21.3% showed manganese (Mn) deficiency, 28% showed S deficiency, 5.8% showed magnesium (Mg) deficiency and 29.5% showed calcium (Ca) deficiency. But only a small portion of the deficient land received corrective amounts of secondary or micronutrient fertilizer (Li Jiakang and Lin Bao, 1996).

Small-scale Operation of Farming and Low Education Level of Farmers are Retarding the Development of Balanced Fertilization

Because of the small-scale operation of farming and the low education level of farmers, it is difficult for Chinese farmers to adopt the improved farming practices which may be suitable for developed countries for increasing fertilizer efficiency. In fact, although great efforts have been made by the government to disseminate techniques in relation to soil testing and balanced fertilization, these techniques in practice are still seldomly applied by farmers. Therefore, it is very important to develop 'best management practices' for Chinese agriculture.

Attention is needed to overcome the above-mentioned constraints and to improve fertilizer use efficiency in China for a sustained increase of agricultural production and for improvement of soil productivity and environmental quality.

FUTURE OUTLOOK AND CONCLUSIONS

Although China has now become the world largest chemical fertilizer user in total quantity, the fertilization rate per unit planted area is only 160.5 kg of nutrient ha^{-1} of planted land in 1994. If the above-mentioned constraints can be removed, there is still good potential for a further increase of crop production through an increase of chemical fertilizer use.

It is estimated that chemical fertilizer consumption needs to be further increased, with a total of 42 million tonnes of nutrients expected to be used in the year 2000, 6.06 million more than the consumption in 1995. Assuming 70% of the increased use of fertilizer ($6.06 \times 70\% = 4.24$ million tonnes) is going to be used for grain crops and 1 kg nutrient can increase grain by 7 kg, the total grain increase from the increase in chemical fertilizer use will be 29.68 million tonnes.

Further improvement of fertilizer application, mainly balanced fertilization, will further increase grain production. It is estimated that with balanced fertilization and related scientific fertilization techniques, it is possible to improve chemical fertilizer use efficiency by 10–15%. Assuming balanced fertilization, 1 kg nutrient can produce 1 more kg grain, and assuming that 70% of fertilizers are used for grain crops, then $35.94 \times 70\% = 25.16$ million tonnes more grain could have been produced in 1995, and $42.00 \times 70\% = 29.4$ million tonnes more grain can be produced by 2000.

At present, China uses a large amount of chemical N fertilizer on soybean and groundnut, with a total planted area of about 200 million mu (13.33 million ha), wasting N resource and money. This N fertilizer could have been used for grain crops. Assuming 60 kg (0.06 tonne) N nutrient per ha could be saved from the legumes, a total of 13.33 × 60 = 800,000 tonnes N can be made available for grain crops. Assuming 1 kg N can produce an increase of 7 kg of grain, the grain increase from this saved N will be 5.60 million tonnes.

Adding the three components together, by 2000, total potential grain production increase could be 29.4 + 29.68 + 5.6 = 64.68 million tonnes, about 13.9% of the total grain production in 1995 (465 million tonnes).

To conclude, increasing fertilizer supply and consumption and improving fertilizer use efficiency by balanced fertilization has great potential for a further increase of grain production. But to realize this potential, great attention is needed to increase the supply of fertilizers, especially K and P, and to improve fertilization technology.

REFERENCES

China Agriculture Yearbooks (1980–1996) China Agriculture Press, Agricultural Publishing House, Beijing.

Asia land-management of sloping lands. International Board for Soil Research and Management *(IBSRAM) Newsletter* 19, 4.

Jin Jiyun and Portch, S. (1991) Summary of greenhouse plant nutrient survey studies of Chinese soils. In: Portch, S. (ed.) *Proceedings of the International Symposium on the Role of Sulphur, Magnesium and Micronutrients in Balanced Plant Nutrition.* Chengdu, People's Republic of China, April, pp. 209–215.

Li Jiakang and Lin Bao (1996) The role of chemical fertilizer in agricultural production in China. *International Symposium on Fertilizer in Agricultural Development,* Beijing, China. 15–18 October, China Agricultural Scientech Press, Beijing.

Lin Bao (1991) Make the most efficient use of fertilizers in increase crop production. In: *Soil Science in China: Present and Future.* Soil Science Society of China, Jiangsu Scientech Press, pp. 29–36 (in Chinese).

Lin Bao and Jin Jiyun (1991) The role of fertilizers in increasing food production in China: A historical review and prospects. In: *Balanced Fertilizer Situation Report-China*, 9 January, 1991, Beijing. Canpotex Ltd, Canada.

Lin Bao, Jin Jiyun and Dowdle, S. (1989) Soil fertility and the transition from low-input to high-input agriculture. *Better Crops International.* Potash and Phosphate Institute, June.

Liu Gengling (1988) Grain production and balanced fertilization in China. In: *Proceedings of the International Symposium on Balanced Fertilization.* Agricultural Publishing House, Beijing, pp. 23–27.

Mei Fangquan and Gu Shuzhong (1993) Analysis of the development of high-input and high-efficiency sustainable agroecosystem in China. In: *The International Symposium on Sustainable Agriculture and Rural Development*, Beijing, 25–28 May, 1993.

Science and Technology Department of Ministry of Agriculture (1991) *Potassium in Agriculture in Southern China.* Agriculture Publishing House, Beijing.

Wen Yupu (1992). Economic analysis of results from 15-tonne grain projects. *Better Crops International* December, 24–25.

Yu Zhenrong, Xin Dehui and Zai Zixi (1993) Improvement of soil fertility and nutrient cycling in the process of comprehensive control of salt-affected soil. In: *Proceedings of International Conference on Integrated Resource Management for Sustainable Agriculture*, Beijing, 5–13 September, 1993. Beijing Agricultural University Press, Beijing.

Zhang Kuan, Wu Wei, Wang Xoufang and Hu Heyun (1991) Effect of sulphur on the yield of rice in an alluvial paddy soil. In: Portch, S. (ed.) *Proceedings of the International Symposium on the Role of Sulphur, Magnesium and Micronutrients in Balanced Plant Nutrition*. Chengdu, People's Republic of China, pp. 176–177.

Zhang, W.L. and Jin, J.Y. (1996) Fertilizer use and agricultural development in 21st Century in China. In: *Proceedings of the Symposium on Agricultural Development in 21st Century in China*. Agricultural Scientech Press, Beijing, pp. 105–110.

Zhou Yimin (1994) Effects of fertilization on soil environment and vegetable quality in vegetable fields in Tianjin. In: *Proceedings of Symposium on Fertilization and Environment*. China Agricultural Scientech Press, Beijing, pp. 11–22.

Zhu Yingmei (1994) Fertilization and eutrophication of surface water. In: *Proceedings of Symposium on Fertilization and Environment*. China Agricultural Scientech Press, Beijing, pp. 40–44.

Nutrient Imbalances following Conversion of Grasslands to Plantation Forests in South Africa

R.J. Scholes[1] and M.C. Scholes[2]

[1] Division of Water, Environment and Forest Technology, CSIR, PO Box 395, Pretoria, 0001 South Africa; [2] Department of Botany, University of the Witwatersrand, Private Bag 3, WITS 2050, South Africa

ABSTRACT

This chapter discusses the changes in nutrient cycling that occur when native montane grasslands in South Africa are replaced by plantations of exotic fast-growing timber trees. South Africa has only a tiny area of indigenous forest. Therefore, for the past 60 years it has relied on plantations of pines and eucalypts to supply its timber and pulp needs. The plantations replace highland grasslands, in the wetter parts of the country frequently on highly weathered soils. The tree growth rate is high, permitting rotations of 10–25 years. There is concern that repeated removal of nutrients in the harvested wood will deplete the already low soil nutrient levels. Approximately 69% of the potassium and 45% of the calcium in the tree plus litter pool is removed per harvest. Up to 67% of the nitrogen in the tree plus litter pools is immobilized in the litter. Yield depression has been suggested to be occurring in third rotation plantations, but is not proven. The main body of plantations is downwind of a major industrial region and is subject to atmospheric deposition of nitrogen and sulphur. This adds to the acidification of the soils resulting from base cation uptake in the plantations. A decrease in pH of 0.3–1.0 units, relative to adjacent grasslands, occurs during the first 20 year rotation. There is uncertainty as to whether the foliar symptoms now apparent on pine needles are a consequence of cation deficiency, or a direct effect of acid deposition or high ozone levels.

INTRODUCTION

Most of South Africa is too arid to support forests. Only the eastern escarpment and seaboard receive more than 800 mm year^{-1}, which is sufficient for forests (Fig. 9.1). The rainfall is highly concentrated in summer, while the winter is dry and cool (because the altitude is mostly above 1500 m).

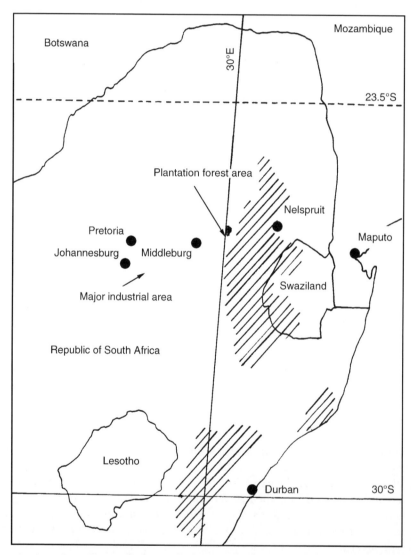

Fig. 9.1. About half of the afforestation in South Africa is located in the montane grasslands of the Mpumalanga province, north and west of Swaziland, and east of the major coal-burning industrial regions between Johannesburg and Middleburg.

This climatic combination naturally supports a grassland cover (Ellery *et al.*, 1991). The regular dry-season occurrence of grassland fires confines natural forests to the fire-protected positions in this landscape, which make up about 0.2% of the land surface (Geldenhuys, 1994).

The timber resources of the indigenous forest patches were exhausted soon after the arrival of European settlers. By the beginning of the 20th century, the demand for timber in the largely treeless interior of South Africa led to the establishment of plantations of non-indigenous trees in the highland grasslands. The first plantings were of black wattle (*Acacia mearnsii*), for fuelwood and tannin extract. Eucalyptus plantations to provide support timber in the gold mines followed. Large-scale afforestation with pines (mostly *Pinus patula*) began mid-century, in order to meet the demand for construction timber in the rapidly growing cities. Afforestation continued at a rate of about 25,000 ha year^{-1}, approximately half pines and half eucalypts, with the latter increasingly being used for the manufacture of paper pulp for export. The rate of new afforestation is currently around 10,000 ha year^{-1}, and the total planted area is 1.4 million ha, about 2% of the total land area of South Africa (Scholes *et al.*, 1996).

Plantation forests in South Africa consist of even-aged stands which are clear-felled when mature, and then replanted. The rotation periods are short by temperate forest standards, due to the relatively warm temperatures and year-round growth potential. Pine trees grown for saw timber are felled after 20–25 years. Eucalyptus trees grown for pulp are harvested after 10–15 years, and in some cases, after only 7 years. Most of the planted area is now in its second rotation, and many areas are in third rotation.

Plantation forestry has contributed significantly to the development of South Africa, both in rural and urban areas. The commercial forestry sector accounts for about 4% of the South African gross domestic product (a similar fraction to agriculture) and is an important source of foreign exchange. Further afforestation, however, is increasingly difficult. Bioclimatically, about 5% of the country could be afforested using currently available cultivars. If this were permitted, the last remnants of the species-rich montane grasslands – also under pressure from other forms of agriculture – would be lost (Scholes *et al.*, 1996).

The most compelling constraint to further expansion of the forests is the effect which the trees have on water resources, considered to be the most limiting natural resource in South Africa. The bioclimatically suitable area for afforestation coincides with the mountain catchments which supply a large fraction of the water used in agriculture, industry and homes throughout the country. The hydrological impact of plantations can be expressed as the additional evaporative loss relative to that of the natural vegetation it replaces. A series of long-term catchment afforestation experiments shows streamflow reduction in the range 300–600 mm year^{-1} (Lesch and Scott, 1993).

When the native grasses are replaced by deep-rooted, fast-growing trees, streamflow from the catchments is substantially reduced, especially once complete canopy cover is attained. Although the mean maximum evapotranspiration rate in midsummer from a stand of eucalyptus or pine trees is not very different from that of a grassland, the evergreen trees continue to use water throughout the winter, while the grasses are dormant. An Afforestation Permit System exists in South Africa to regulate the establishment of plantations. Permit decisions have largely been based on hydrological constraints, but biodiversity impacts are an increasingly important consideration.

Since the bioclimatically suitable higher rainfall areas often coincide with regions of highly weathered, nutrient-poor soils, the possibility of nutrient depletion through repeated harvest removals is of concern to the industry.

The purpose of this chapter is to examine the available information regarding nutrient balance within pine and eucalypt plantations in Mpumalanga province in the northeastern part of South Africa, which is the main centre for plantation forestry in South Africa (Fig. 9.1). In the first section, the nutrient budget of plantations in the region is compared with the budget for the natural system which they replaced. The second section examines the issue of soil acidification. The final section discusses the implications for sustainability of the plantation forestry industry.

THE NUTRIENT BUDGETS OF PLANTATIONS

There are indications of a reduced growth rate in third-rotation forests in Swaziland, which lies immediately south of Mpumalanga province, which have been attributed to nutrient depletion. As a consequence, fertilization has commenced in some forests there (Morris, 1992). As yet there are not enough growth rate studies in South Africa on compartments with a known growth history to be able to separate the effects of reduced nutrient supply from the effects of climate (such as the severe droughts of the early 1990s) and the growth enhancement resulting from improved tree breeding and silvicultural practice. It is only possible to interpret the significance of nutrient exports from plantation forests within the context of the complete nutrient cycle, including all the important inputs and outputs, and the size of the main nutrient pools. The key components are estimated below.

Nutrient Removal through the Harvesting of Wood

The nutrient contents of the stemwood, and nutrients removed in harvest, for the two main plantation species are given in Table 9.1. The quantity of material removed from the site depends on both the species planted, and

Table 9.1. Nutrient concentrations (mg kg^{-1}) in stemwood of *Pinus patula* (Morris, 1992) and *Eucalyptus grandis* (Noble and Herbert, 1989) and nutrients removed in harvest (kg ha^{-1} year^{-1}) calculated using the modal values of 83 t ha^{-1} of *P. patula* and 63 t ha^{-1} of *E. grandis*. For *P. patula* the standard deviation of the nutrient concentration is given in parentheses, and the range of nutrients removed. Only means are available for *E. grandis*.

	Pinus patula		Eucalyptus grandis	
Nutrient	Nutrient concentration (mg kg^{-1})	Nutrients removed in harvest (kg ha^{-1} year^{-1})	Nutrient concentration (mg kg^{-1})	Nutrients removed in harvest (kg ha^{-1} year^{-1})
Nitrogen	1100 (270)	3.6 (3.1–10.1)	2300	14.5 (13.6–55.4)
Phosphorus	220 (70)	0.7 (0.6–1.8)	70	0.4 (0.4–1.7)
Potassium	790 (540)	2.6 (2.2–7.4)	725	4.6 (4.3–17.5)
Calcium	540 (210)	1.6 (1.4–4.6)	700	4.4 (4.1–16.9)
Magnesium	110 (40)	0.3 (0.3–0.9)	215	1.4 (1.3–5.2)

on silvicultural practices. Plantations destined for saw timber make up about 25% of the total plantation area. They are mostly planted to pines at an espacement of about 3 m. The trees are thinned and pruned at 6 and 12 years. Thinnings are used for pulp, but small-dimension prunings stay in the forest. At around 20 years the trees are felled, and all the large-dimension timber is removed. The slash is allowed to decompose *in situ*, since burning it poses a fire hazard to surrounding plantations. Morris (1986) showed that up to 16% of the system nitrogen was volatilized if the slash is burned following thinning. A small amount of waste wood is removed for use as domestic fuelwood by nearby communities. Typical final yields of roundwood are in the region of 200 m^3 ha^{-1} (range 172–563), with a wet density of 0.88 metric tonnes (t) m^{-3} and a moisture content of 115%. Thus the amount of dry matter removed from the site once every 25 years ranges from 70 to 230 t ha^{-1}, with a modal value around 83 t ha^{-1}.

Plantations destined for pulp typically consist of eucalyptus hybrids, and are harvested without thinning or pruning after 10 years. They yield about 122 m^3 ha^{-1} of pulp timber (114–465), with a wet density of 0.86 t m^{-3} and a moisture content of 66%. Thus 63 (59–241) t ha^{-1} are removed every 10 years.

Multiplying these values by the nutrient content of the stemwood allows the mean annual cation export in the harvested timber to be calculated. Considerably more nutrients are removed when harvesting *Eucalyptus grandis* compared with *P. patula*. This is exacerbated by the short rotation times for *E. grandis*. Nitrogen, potassium and calcium are the nutrients that are removed in the greatest absolute amounts. Calcium is one of the main limiting nutrients in the system. *E. grandis* sequesters much more calcium than *P. patula*, leading to rapid soil depletion. These harvest losses would be considerably higher, especially for calcium, if bark was included, as the nutrient content of the bark is in most cases higher than in the stemwood. Efforts have been made in the last few years to remove as little bark from the sites as possible.

Losses through Soil Erosion

During the clearcutting operations, and for the 3 years that elapse until a forest cover is re-established, the soil is exposed to erosion. The plantations are often in mountainous areas, and the rainfall intensity is high. Although sediment loss from afforested catchments is generally low (lower even than grassland catchments which are grazed), in the year immediately following a harvest disruption they can be up to 100 t soil ha^{-1}. In the event of a forest fire (about 0.5% of the plantation area burns accidentally every year) soil losses can be even greater.

Inputs from Weathering of the Parent Material

The weathering rate is assumed to be 0.1 mm $year^{-1}$, based on estimates for similar material in Zimbabwe (Owen and Watson, 1979). This translates to a cation release rate of 1.8 kg calcium (Ca) ha^{-1} $year^{-1}$, 0.3 kg magnesium (Mg) ha^{-1} $year^{-1}$ and 1.9 kg potassium (K) ha^{-1} $year^{-1}$, or 0.163 mol (+) ha^{-1} $year^{-1}$.

The catchment approach has been applied to data on flow rates and water chemistry from the grass-covered research catchments at Mokubulaan, about 20 km southwest of the example site, where the parent materials are shales and quartzites. The estimated rate of production of cations is 0.39–0.86 mol (+) ha^{-1} $year^{-1}$. Note that the 'steady-state' assumptions may have been violated for this catchment due to the changes in precipitation chemistry which have occurred over the past 30 years, and are discussed in a later section.

Inputs from Atmospheric Deposition

There is a continuous input of cations through atmospheric deposition, both in the wet and dry form. Wet atmospheric deposition over the main plantation region in Mpumalanga province has been monitored at several stations over the period 1986–1996. The deposition rates for a number of sites in South Africa, as recorded in wide-mouthed receptors which are exposed

Table 9.2. Available estimates of annual volume-weighted wet deposition of H^+, SO_4^{2-}, NO_3^-, NH_4^+, Ca^{2+}, Mg^{2+} and K^+ in or close to afforested regions in Mpumalanga, and in other parts of South Africa. The period of collection is 1 year in four cases, 2 years in three cases and 7 years in three cases. The annual means and standard deviations are given.

Location		Mean annual deposition (kg ha^{-1} year^{-1})						
		H^+	SO_4^{2-}	NO_3^-	NH_4^+	Ca^{2+}	Mg^{2+}	K^+
Mpumalanga	Mean	0.37	16.0	7.4	2.1	3.6	0.7	2.1
(six sites)	SD	0.25	8.9	3.4	0.9	1.6	0.3	0.9
Other South Africa	Mean	0.23	23.0	2.9	3.1	7.7	2.3	4.4
(four sites)	SD	0.17	17.2	3.9	2.9	8.8	2.3	2.5

Sources: van Wyk (1990), Turner and de Beer (1993), Olbrich and du Toit (1993) and Turner (1993).

automatically with the onset of rain and are covered at other times, are given in Table 9.2. The origin of these cations, as estimated by isotopic analysis of airborne dust for a site 50 km east of Sabie (and 1000 m lower) is approximately 5% terrestrial (i.e. windborne soil particles, largely from ploughed croplands), 2% marine (the Indian Ocean is approximately 300 km to the east), 46% from biomass burning, and 47% industrial, from the coal-burning regions to the west (Piketh and Annegarn, 1994). The precipitation chemistry varies greatly between rainstorms, but integrated over the year amounts to 14.3 kg Ca ha^{-1} year^{-1}, 1.68 kg Mg ha^{-1} year^{-1}, 3.27 kg K ha^{-1} year^{-1} and 5.7 kg sulphur (S) ha^{-1} year^{-1} (Olbrich, 1995). The amount of dry deposition is more uncertain. It could be as large as, or even slightly greater than, the wet deposition. Olbrich (1995) and Zunckel *et al.* (1996) estimated the dry deposition to be 11.0 kg Ca ha^{-1} year^{-1}, 0.4 kg Mg ha^{-1} year^{-1}, 0.8 kg K ha^{-1} year^{-1} and 8.2 kg S ha^{-1} year^{-1}. This technique cannot be applied to nitrate, since it is directly absorbed by the foliage but, by inference, dry deposition is the major pathway for nitrogen (N) as well (approximately 10 kg N ha^{-1} year^{-1}).

Nutrient Immobilization in Litter

The litter layer in pine plantations above 1500 m in altitude has been observed to be accumulating to an unusual depth. In many places it exceeds 0.5 m in thickness. This poses a fire hazard, and a problem when planting seedlings for the next rotation. It also represents a large immobilization of nutrients. Table 9.3 shows the nutrient pools in a 42-year-old *Pinus patula* stand with an accumulation of 143 t of needle litter ha^{-1}.

Table 9.3. An approximate nutrient input–output (kg ha^{-1} year^{-1}) analysis for a landscape afforested with *Pinus patula* and *Eucalyptus grandis* in Mpumalanga. Negative values are losses of nutrients, positive values are gains.

Item	Nitrogen	Phosphorus	Potassium	Calcium	Magnesium
Inputs					
Primary weathering[a]	0.00	0.20	1.90	1.80	0.30
Atmospheric deposition[b]	6.00	0.08	4.08	7.13	0.19
Biological fixation[c]	0.50	0.00	0.00	0.00	0.00
Outputs					
Soil erosion[d]	−7.20	−0.56	−0.20	−0.80	−0.04
Leaching[e]	−0.10	0.00	−2.30	−5.30	−3.18
Wood harvest[f] *Pinus patula*	−38.5	−2.7	−3.4	3.1	−8.2
Litter immobilization[g] *P. patula*	−3.6	−0.7	−2.6	−1.6	−0.3
Balance *Pinus patula*	−42.9	−3.68	−2.52	−1.87	−11.23
Wood harvest, *E. grandis*	−14.5	−0.4	−4.6	−4.4	−1.4
Balance *E. grandis*	−15.3	−0.68	−1.12	−1.57	−4.13
Approximate pool size[h]	6480	830	436	1377	76

[a]Owen and Watson (1979). Data from granitic landscapes, Zimbabwe. The value for P is estimated by Professor M. Fey (Department of Geology, University of Cape Town); the range is 0.1–2.0 kg ha^{-1} year^{-1}, with low values on acid crystalline materials.
[b]Van Wyk (1990). Bulk deposition values at Sabie Forestry Research Centre.
[c]The rate is low because of the absence of symbiotic N fixers in these systems, and the acidity of the soil.
[d]Assuming a loss of 10 mm of soil per 20 year rotation. Almost all of this soil is lost during clear-felling. Topsoil data from Sabie Forest Research Centre, Christie (personal communication), Forestek, except for the total P data, which is from an infertile, sandy soil in the Northern Province (Scholes and Walker, 1993).
[e]Catchment data from Witklip, near Nelspruit, 1982–1987, 48% afforested.
[f]This chapter.
[g]Dames, J. and Scholes, M.C. (personal communication), Botany Department, University of the Witwatersrand. Pine plantation, Brooklands, near Sabie. A 22 cm thick litter layer (143 t ha^{-1}) had accumulated in 41 years. The litter layer is thicker in higher altitude sites and thinner at lower sites.
[h]Based on an oxisol at Sabie Forest Research Station, to 2 m depth. Only pools which are immediately available or mineralizable within a period of a few decades are included.

Litter accumulation under the natural grasslands is minimal with most of the litter decomposing within a year.

The main reason for the accumulation of the litter is thought to be climatic. Needle production continues at a high rate in the higher altitude sites, but needle decomposition is retarded. The mean annual temperature at 2000 m in Mpumalanga province (approximately the upper limit for pine plantations there) is around 15°C, and the mean minimum temperature of the coldest month is −1°C. While these temperatures are not low by comparison with the northern hemisphere regions where the pines originate (hence the continued high rate of growth and needle production), it is suggested that the varieties of decomposing fungi naturally present in South Africa do not function effectively to decompose needle litter at this temperature, or alternatively, the strains which have found their way from the northern hemisphere do not perform well under a combination of cool temperatures and winter aridity.

The Nutrient Balance Relative to the Soil Pool

Recognizing the uncertainties in some estimates, it is nevertheless clear that for certain elements, more is exported in timber products than is received through weathering and deposition. A measure of the significance of these imbalances can be gained by dividing them into the estimated pool of cations within the soil profile, giving a rough estimate of time to depletion (Table 9.3). Plant deficiency symptoms should appear well before depletion. Since these are highly weathered soils, the reserve of mineralizable material within the A and B horizons of the soil profile is assumed to be negligible. The pool size calculation is based on the cations extractable in 1 M ammonium bicarbonate. The soils are assumed to be 2 m deep on average (they vary from less than 1 m to up to 20 m). If the trees are obtaining the bulk of their cations from the saprolite layer in the carbon (C) horizon, the pool-size calculations will be underestimated. This does not invalidate the conclusion that the continued harvesting of timber without nutrient replacement is unsustainable in terms of nutrient supply, but does shift the time when deficiency symptoms should appear further into the future.

NUTRIENT BUDGETS FOR GRASSLANDS IN THE SAME REGION

The climax grassland ecosystems which occur in the region, and have for millennia, are assumed to be in nutrient equilibrium, although no long-term studies have been carried out to prove this. Total grass biomass (above and below-ground) in the grasslands is in the range of 4–8 t ha^{-1}, with a modal value of 6 t ha^{-1}. The nutrient contents of live grass are

7 (1.2–11.5) g N kg^{-1}, 0.81 (0.16–2.12) g phosphorus (P) kg^{-1}, 5.8 g Ca kg^{-1}, 0.55 g Mg kg^{-1} and 0.81 g K kg^{-1}. Small amounts of nutrients are immobilized in the litter layer, and as dead roots.

The cation input as a result of weathering is assumed to be equal to that estimated for plantations. Plantations may in fact have a greater access to cations from the C horizon, due to their greater depth of rooting. In the absence of data, it is assumed that the atmospheric deposition is the same as well, although dry deposition should be slightly higher in plantations than grasslands. There is likely to be a small input of nitrogen to the grasslands due to biological nitrogen fixation, of the order of 5 kg N ha^{-1} year^{-1}. Leguminous forbs with documented nitrogen fixing capacity constitute a small fraction of the herbaceous layer biomass. This input is approximately balanced by nitrogen losses due to 'pyrodenitrification', that is, the volatilization of nitrogen during the grassland fires which occur approximately once every 3 years.

THE CATION BALANCE AND SOIL ACIDIFICATION

Immobilization of basic cations in the tree biomass and plantation litter layer, and the net export of cations from the system in harvested wood, has led to a decrease in soil pH under the areas planted to exotic trees. There is simultaneously a concern that acid deposition, resulting mainly from the upwind burning of fossil fuels, will exceed the buffering capacity of the soil system, and threaten the sustainability of the plantations. Plantations are thought to be more susceptible to acidification than the grasslands which they replace, due to the multi-year needle longevity and high dependence of pines on mycorrhizae for phosphorus. The tall, rough plantation canopies are thought to be more effective in trapping dry acid deposition than the low, smooth grass cover.

Chlorotic bands on pine needles, similar to those described as symptoms of acidification in Europe, have been reported from several plantation estates in Mpumalanga. The chlorotic bands were found to have lower concentrations of magnesium (0.23 mg g^{-1}) and calcium (2.7 mg g^{-1}) than the healthy bands (0.35 mg Mg g^{-1} and 3.7 mg Ca g^{-1}), a marked breakdown in cuticle structure, and a higher incidence of fungal infection (Whiffler, 1990).

The two main sources of acidification are considered below.

Acid Deposition

The South African energy economy is highly dependent on coal, partly because of the vast reserves of this fuel, and partly because of strategic decisions to minimize dependence on other countries during the period of international isolation of South Africa. The main coal-mining and power-generation area is 50 km west of the plantation region. Nearly all of the ash in the power station emissions is trapped at source, but sulphur and nitrogen gases are not removed. Approximately 2×10^{12} g of S (as SO_2) and 0.9×10^{12} g of N (as NO_x) are injected into the atmosphere annually in a 30,000 km² block extending from Johannesburg in the west, to Middleburg in the East, and from Pretoria in the north to Sasolburg in the south (Wells et al., 1996). This quantity has been growing at about 3% per annum since the 1960s, and is projected to continue to rise at a similar rate.

The atmospheric circulation over southern Africa, particularly in the dry winter months, is extremely stable. Flue gas from the large power generating stations is injected at a height sufficient to avoid it being trapped in the ground inversion, so the ground-level concentrations of sulphur dioxide (SO_2) and ozone (O_3) (produced as a result of reactions between nitric oxide (NO), carbon monoxide (CO) and hydrocarbons) are generally within air-quality guidelines. The gases are trapped in a stable layer 1–6 km above the surface, where they convert to sulphuric acid (H_2SO_4) and nitric acid (HNO_3) as they circulate above the subcontinent. The industrial emissions are supplemented by emissions from vegetation fires in central Africa, especially during the winter months. The reaction products eventually wash out in the rain, or drift down as aerosols (Tyson et al., 1988), or are ejected into circulation systems off the subcontinent.

The observed rate of acid deposition in rainfall is similar to values reported for moderately industrialized areas of western Europe and the northwestern United States of America. It is reasonable to expect the acidification phenomena observed there after a century of exposure to acid deposition to present themselves in South Africa some time in the future. They would first become apparent on soils with a low buffer capacity, such as shallow soils, or those derived from quartzites.

The hydronium ion (H^+) accompanying the sulphate (SO_4^{2-}) and the nitrate (NO_3^-) in acid deposition, displaces cations from the soil profile into the groundwater. As a result, the soil pH drops, causing aluminium (Al^{3+}) (which is toxic to roots) to become more soluble. The loading of the ecosystem with nitrogen and sulphur is initially beneficial, since it acts as a fertilizer, but eventually the ecosystem can become nitrogen saturated, which causes changes in ecosystem nutrient retentivity, species composition (in natural systems) and ultimately loss of productivity. These effects are thought to be several decades away in South Africa. Direct impacts of acidic deposition on the growth rate and productivity of the plantation species and the grasslands are difficult to quantify. Long-term studies need to be initiated in order to apportion the additional acidity in the system to either deposition or decomposition of the pine litter.

Acidification due to Afforestation

A modified historic approach has been used to compare soil chemical properties between virgin grassland and adjacent forest plantations of *Pinus*, *Eucalyptus* and *Acacia* spp. (Du Toit, 1993). This approach assumes that the properties of the soil prior to afforestation were identical to those currently found under grassland immediately adjoining the plantation. A marked increase in soil acidity has been found to be associated with all the plantation species: lower values for soil pH (by 0.3–0.5 units), exchangeable base cations (by 1.6 cmol (+) l^{-1} on average) and acid neutralizing capacity (ANC, by 0.64 kmol ha^{-1} to 20 cm on average) were measured under plantations. The mean change in ANC for the solum (assuming that this is double the amount found in the top 20 cm) was equivalent to about 2.6 t calcium carbonate ($CaCO_3$ ha^{-1}). The greater than average acidification measured in black wattle (*Acacia mearnsii*) plantations compared with other species is attributed to its nitrogen fixing capacity, which provides an additional source of H^+. The rate of soil acidification, taking plantation age into account, correlated significantly with base status of the grassland soil, being greatest where the initial base status was high. If the trees died, and all litter decomposed *in situ*, the pH would theoretically return to its former value. In practice, cations are exported in the logs, so the pH continues to decline (at a lower rate) in subsequent rotations.

Comparing Acid Deposition and Acidification due to Plantations

The soil chemistry changes under plantations are similar, when compared over the same time span, to changes in European and North American forests, which have been attributed to atmospheric acid deposition. No systematic decline in the pH of soils under grasslands has been observed. Such a decline would be expected if atmospheric deposition were the principle agent of acidification, even allowing for different deposition rates between the rough plantation surfaces and the smooth grasslands. The failure to observe acidification in grassland soils is partly because a complete and rigorous survey, comparing known sites over time, using standardized techniques or archived samples, has not been undertaken. The information to date has mostly come from pairwise comparisons of adjacent plantation compartments and grassed areas on the same original soil.

A study of the distribution of the leaf banding symptoms in pine plantations failed to reveal a geographical pattern which would clearly implicate acid deposition of an industrial origin. At present it is not known whether the banded needles represent a direct symptom of poor air quality (for instance, high ozone, SO_2 or acidity), an indirect symptom caused by reduction of calcium and magnesium availability to the roots, or is totally unrelated to cation deficiencies (Whiffler, 1990).

THE SUSTAINABILITY OF PLANTATION FORESTRY IN SOUTH AFRICA

In the long term, plantation forestry in South Africa will be unsustainable unless the nutrient losses due to harvest and soil erosion are balanced by nutrient additions through fertilization. Similarly, the acid load imposed by harvest and atmospheric deposition will have to be balanced by liming. It is not clear how soon nutrient imbalances will lead to significant reductions in productivity at an industry-wide scale; some problems on shallow, inherently acid and infertile soils are likely within the next decade as many plantations enter their third or higher rotation.

With the shortening of rotation length, plantation forestry is becoming more like crop agriculture and less like traditional forestry. Cereal crop farmers have long accepted the need for replacement fertilization if they are to maintain high crop yields. Some South African plantations are experimenting with leguminous intercrops during the establishment phases of forest plantations. The benefits are additional nitrogen due to biological nitrogen fixation, weed control due to cultivation operations, and good community relations, since the landless poor have an opportunity to grow crops. Nitrogen is probably not the main limiting factor in the long term, due to the atmospheric inputs, and legume crops contribute to acidification.

Fertilization would add a few per cent to the cost of tree growing. Assuming balancing quantities, application at planting only and current costs of fertilizers and transport, the cost would be approximately $10 ha^{-1} in addition to the $600 ha^{-1} establishment, weeding, thinning and harvesting costs on a typical 20 year pine sawtimber rotation. Tree-growing operations currently operate on a very small profit margin. The timber industry is highly vertically integrated; the same few companies which own and manage the bulk of the plantations also control sawmilling, pulp mills, wood products and marketing. For strategic reasons, the bulk of the profit is taken in the processing steps. The pulp market, in particular, is highly volatile and cannot be strongly influenced by South African producers, and constrains log prices at some times. However, failure to reinvest in soil fertility will ultimately lead to declining productivity.

Atmospheric deposition is an externalized cost, largely from the energy generation sector, which is partially borne by the forestry industry. It is likely to increase by a few per cent per year for the next two decades, when the next generation of coal-burning power stations will come on stream. The state-owned electricity supply utility argues that retrofitting the existing equipment with flue-gas desulphurization technology will be prohibitively expensive and will divert resources from more deserving social and environmental causes. At present South Africa has an excess of installed generation capacity, so new plants are not planned in the near future. Integrated pollution control equipment, coupled with probable improvements in thermal efficiency, have the potential to reduce emissions when new plants are commissioned, if there is sufficient political pressure to require their implementation.

The other dimensions of forest sustainability, notably the increasing pressure on the forest industry with respect to water resources and the conservation of biodiversity, are likely to be less amenable to technical solutions. If the plantation area is reduced or geographically redistributed, then the problem of rehabilitation of formerly afforested land for other purposes will become a pressing issue. At present, experience in this regard is restricted to conversion of forests to profitable, high-intensity uses (such as sugar-cane or fruit trees) and small areas of streambank rehabilitation. The large-scale regeneration of biodiverse, extensive grasslands, with their associated fauna, has not been attempted.

REFERENCES

Du Toit, B. (1993) Soil acidification under forest plantations and the determination of the acid neutralizing capacity of soils. MSc thesis, The University of Natal, Pietermaritzburg, South Africa.

Ellery, W.N., Scholes, R.J. and Mentis, M.T. (1991) An initial approach to predicting the sensitivity of the South African Grassland Biome to climate change. *South African Journal of Science* 87, 499–503.

Geldenhuys, C.J. (1994) Bergwind fires and the location pattern of forest patches in the southern Cape landscape, South Africa. *Journal of Biogeography* 21(1), 49–62.

Lesch, W. and Scott, D.F. (1993) A report on the effects of afforestation and deforestation on streamflow and low flows at Mokobulaan. Unpublished contract report, FOR-DEA 620, to the Department of Water Affairs and Forestry. CSIR Division of Forest Science and Technology, Pretoria.

Morris, A.R. (1986) Soil fertility and long-term productivity of *Pinus patula* in the Usutu Forest, Swaziland. PhD thesis, The University of Reading, Reading, UK.

Morris, A.R. (1992) Dry matter and nutrients in the biomass of an age series of *Pinus patula* plantations in the Usutu forest, Swaziland. *South African Forestry Journal* 163, 5–11.

Noble, A.D. and Herbert, M.A. (1989) Estimated nutrient removal in a short-rotation *Eucalyptus grandis* crop on a Fernwood soil. *Annual Report*. Institute for Commercial Forestry Research, University of Natal, South Africa, pp. 139–145.

Olbrich, K.A. (1995) Research on impacts of atmospheric pollution and environmental stress on plantations. Unpublished contract report, FOR-DEA 874, to the Department of Water Affairs and Forestry. CSIR Division of Forest Science and Technology, Pretoria, South Africa.

Olbrich, K.A. and Du Toit, B. (1993) *Assessing the Risks Posed by Air Pollution to Forestry in the Eastern Transvaal, South Africa*. Confidential report to Eskom, Division of Forest Science and Technology Report FOR-C 214, CSIR, Pretoria, South Africa.

Owen, L.B. and Watson, J.P. (1979) Rates of weathering and soil formation on granite in Rhodesia. *Soil Science Society of America Journal* 43, 160–166.

Piketh, S.J. and Annegarn, H.J. (1994) Dry deposition of sulphate aerosols and acid rain potential in the Eastern Transvaal and the Lowveld regions. *Annual Proceedings of the NACA Conference*, Western Cape branch of the National Association for Clean Air (NACA), Cape Sun Hotel, Cape Town, South Africa.

Scholes, R.J. and Walker, B.H. (1993) *An African Savanna: Synthesis of the Nylsvley Study*. Cambridge University Press, Cambridge, 306 pp.

Scholes, R.J., Smith, R.E., Van Wilgen, B.W., Berns, J., Evans, J., Everard, D., Scott, D.F., Van Tienhoeven, M. and Viljoen, P.J. (1996) The costs and benefits of plantation forestry: case studies from Mpumalanga. Unpublished contract report, FOR-DEA 939, to the Department of Water Affairs and Forestry. CSIR Division of Forest Science and Technology, Pretoria.

Turner, C.R. (1993) A seven year study of rainfall chemistry in South Africa. *Annual Proceedings of the NACA Conference*, Dikhololo Game Lodge, Brits, South Africa.

Turner, C.R. and De Beer, G.H. (1993) Rain quality measurements in the escarpment region during 1992 to 1993. Eskom Technology Research and Investigations Report: TRR/S/93/118/rw, TRI, Rosherville.

Tyson, P.D., Kruger, F.J. and Louw, C.W. (1988) *Atmospheric Pollution and its Implications in the Eastern Transvaal Highveld*. South African National Scientific Programmes Report No 150, Pretoria, South Africa.

Van Wyk, D.B. (1990) Atmospheric deposition at selected sites in the mountain catchments of South Africa. *1st IUAPPA Regional Conference on Air Pollution: Towards the 21st Century*, Vol. 1. 24–26 October, National Association for Clean Air, CSIR Conference Centre, Pretoria, South Africa.

Wells, R.B., Lloyd, S.M. and Turner, C.R. (1996) National air pollution source inventory. In: Held, G., Gore, B.J., Surridge, A., Tosen, G.R., Turner, C.R. and Walmsley, R.D. (eds) *Air Pollution and its Impacts on the South African Highveld*. Environmental Scientific Association, Cleveland, South Africa. pp. 3–9.

Whiffler, J. (1990) The effect of acid rain on the morphology, anatomy and nutrient composition on the needles of *Pinus patula*. Honours project, Botany Department, The University of the Witwatersrand, Johannesburg, South Africa.

Zunckel, M., Turner, C.R. and Wells, R.B. (1996) Dry deposition of sulphur on the Mpumalanga highveld: a pilot study using the inferential method. *South African Journal of Science* 92, 485–491.

Nutrient Balances at Field Level of Mixed Cropping Systems in Various Agroecological Zones in Mozambique

10

P.M.H. Geurts, L. Fleskens, J. Löwer and E.C.R. Folmer

National Institute for Agricultural Research, Maputo, Mozambique; Sponserf 2, 6413 LS Heerlen, The Netherlands

ABSTRACT

Mozambique's agricultural production has been greatly affected by the civil war that ended in 1992 and a series of ravaging droughts, resulting in a large food deficit. The 1995 Agricultural Policy aims at attaining food self-sufficiency and production for the export market. There is a need to increase the area of cultivated land and its productivity in a sustainable manner. Nutrient balance studies assess the vulnerability of land use to chemical soil degradation and suggest interventions that contribute to sustained agricultural production. This chapter presents the results of a nutrient balance study of mixed cropping systems in various agroecological zones in Mozambique. The study was carried out in the 1995/96 season on smallholder fields, in arid to sub-humid dry southern Mozambique, with extensive areas covered by sandy soils, and in sub-humid dry northern Mozambique, in fields related in a catena, situated in an undulating landscape. In the field in arid southern Mozambique, basically, water controls the long-term nutrient balance and hence sustainability of rainfed agriculture. Nutrient availability is the main limiting factor of land productivity in all other fields. In sparsely populated semi-arid southern Mozambique shifting cultivation is still a feasible nutrient management option to sustain the nutrient stocks of the sandy soils. However, in coastal southern Mozambique, population pressure has reached levels that do not allow natural soil fertility regeneration by means of long fallows. Here it is crucial to include leguminous crops in the crop system, leave crop residues in the field, avoid grazing and employ nutrients from external sources. The potential of manure, bat guano and P fertilizers is discussed. In northern Mozambique, in Nampula, land use is intensive and both the shifting and the fallow system are inadequate to sustain the nutrient stocks through fallowing. It is crucial to make efficient use of all locally available nutrients by including leguminous crops, avoiding burning, returning crop residues to the field and employing effective soil conservation measures. Despite these measures, nutrient depletion rates during cultivation are considerable, calling for the use of external nutrients. Due to the absence of considerable numbers of livestock, manure is not an option and inorganic fertilizers need to be applied to sustain production.

INTRODUCTION

In Mozambique, land use patterns and productive capacity of the rural population have largely been affected by the civil war (1980–1992) and several severe droughts, resulting in a large food deficit. A third of the population fled to the urban centres and their surroundings (Abrahamsson and Nilsson, 1994). It was here where agricultural activities were concentrated, soils became over-utilized, fallow periods shortened and eventually disappeared, and field sizes decreased. Neither fertilizers nor manure were added, causing rapid soil fertility depletion. Meanwhile, large parts of the rural area experienced a war-induced, involuntary long-fallow. After the peace agreement (1992), the original pre-war situation has been partly restored: from 1994 people started to return to the rural areas, clear their fields and produce their own food again, reducing the dependency on food imports and benefiting from the soil fertility accumulated during the long and unplanned fallow.

The 1995 Agricultural Policy, recognizing the large food deficit, stressed the rehabilitation of agricultural production capacity and productivity, through sustainable use of natural resources. In order to promote sustainable land use, it is essential to demarcate areas prone to degradation and to provide sound and attractive agricultural practices that improve or conserve the production resources.

Folmer *et al.* (1998) assessed the chemical soil degradation of Mozambique at national level, following the methodology of Stoorvogel and Smaling (1990) and found annual depletion rates for cultivated fields of 33 kg nitrogen (N), 7 kg phosphorus (P) and 25 kg of potassium (K) per cultivated hectare. Areas with highest annual depletion rates (120 kg N, 25 kg P and 95 kg K ha^{-1}) were encountered in the high rainfall areas (>1000 mm annually), with relatively high soil fertility as well as high erosion rates (soil loss around 25 t ha^{-1} $year^{-1}$). These areas are mainly located in the provinces of Nampula, Manica, Tete and Niassa.

At field level, in various agroecological zones fields were sampled to determine the in- and outflows of the macronutrients N, P and K during one full cropping season (i.e. 1995/96) in order to assess the extent of nutrient depletion, identify its main causes and suggest interventions that contribute to sustained agricultural production.

MATERIALS AND METHODS

Agroecology and Soil Fertility

In southern Mozambique five fields were selected along a west–east transect, to include the main soil and climatic classes. Soils predominantly have a sandy texture and a strong rainfall gradient exists along the west–east transect, with both rainfall quantity and reliability increasing from west to east (Table 10.1). The rainfall reliability is expressed as the probability that a growing period of 110–120 days ($P_{GP\,>\,120\,days}$) can be expected (Westerink, 1995). In northern Mozambique, three average fields were constructed, located on the interfluve (field 6), middle slope (field 7) and valley bottom (field 8) of the Nakatone catchment: the fields are related in a catena, a representative pattern of most of Nampula and Cabo Delgado provinces (Serno, 1995).

Farms consist of several fields that are often at a considerable distance from each other (2–8 km), but without any nutrient transfer between them, and thus considered independent entities. Therefore, fields were sampled and not complete farms. From each farm one field was selected and randomly sampled around harvest time in order to establish the nutrient stocks and flows to and from the sampled fields during one full cropping (i.e. 1995/96) season. The cropping season sampled is part of a land utilization cycle consisting of a period of cultivation and fallow. The land utilization

Table 10.1. Location, agroecological conditions of sampled fields.

Field	Southern Mozambique					Nakatone catchment		
	1	2	3	4	5	6	7	8
Latitude South	22°58′	21°58′	22°2′	22°12′	22°8′	15°10′		
Longitude East	32°7′	34°7′	34°12′	34°47′	35°7′	39°17′		
Climate	Arid	Semi-arid		Sub-humid dry		Sub-humid dry		
Soil (FAO)	Arenosol	Luvisol	Arenosol		Luvisol	Lixisol	Lixisol	Arenosol
Altitude (m asl)	204	179	157	115	35	420		
Mean rainfall (mm year^{-1})	492		596		831	1137		
$P_{GP\,>\,120\,days}$	0.07	0.50	0.50	0.70	0.70	1.0		

intensity (R value) is defined as the period of cultivation divided by the land utilization cycle (Ruthenberg, 1976) and varies with agroecological condition and population pressure. Viable shifting cultivation systems have R values of below 0.33. No actual data on the land utilization intensity could be obtained because the civil war affected land use in the rural and peri-urban areas greatly and only after the peace agreement, from 1994, did people start to return to their original fields, taking up their normal pre-war cultivation practices and land utilization intensities. Therefore, land utilization intensities as practiced before the war were assumed to apply (FAO/UNDP, 1983).

In all fields soil type, slope and signs of erosion were determined. All fields sampled in southern Mozambique have sandy topsoils, with clay contents below 13% and kaolinite as the main clay mineral, low to very low N_{total} contents and very low available phosphorus levels. Soils of fields 4 and 5 overlie limestone, explaining the alkaline soil reaction, the higher ECEC and exchangeable K compared with the soils of fields 1–3. In the catchment of Nakatone, a gradient of increasing soil fertility can be observed from the higher positions on the catena to the valley bottom (Table 10.2). The soils on the interfluves have a sandy topsoil layer of 30 cm, in contrast to the soils on the middle slopes and valley bottoms with a sandy cover of 70 and > 100 cm respectively.

Table 10.2. Land utilization intensity, slope, flooding and topsoil fertility of sampled fields.

Field	Southern Mozambique					Nakatone catchment		
	1	2	3	4	5	6	7	8
Land utilization intensity (R)	0.05–0.17	0.06–0.23	0.50–0.63			0.60–0.80		1
Cultivation (year)	1–3	1–3	4–5			3–4		∞
Fallow (year)	15–20	10–15	3–4			1–2		0
Slope (%)	1.5	1.5	0.5	0.5	0.1	6.2	8.0	2.7
Cumulative annual flooding (days)	10	0	0	0	0	0	0	80
Crop residues	Grazed	Left, cotton burnt	Left	Burnt	Left	Left		
Sand % topsoil	87	96	89	86	89	78	83	87
Clay % topsoil	11	1	10	10	1	13	12	6
pHw	6.5	6.9	5.9	8.2	8.0	6.1	6.7	7.1
OM (%)	0.7	0.6	0.5	0.8	2.5	1.7	1.0	4.9
N_{tot} (%)	0.04	0.04	0.03	0.05	0.09	0.07	0.04	0.12
P-Olsen (ppm)	1	1	2	3	3	4	4	13
K (cmol(+) kg^{-1})	0.19	0.18	0.08	0.41	0.81	0.19	0.32	0.29
ECEC (cmol(+) kg^{-1})	4.7	3.6	0.7	12.8	16.9	5.8	8.7	16.7

Interviews

To obtain information on flooding, land utilization cycle and agronomic practices that relate to soil fertility management and nutrient movements of one complete cropping season, a structured interview was held with each farmer on the basis of a formal questionnaire.

Sampling

In southern Mozambique the selected fields were randomly sampled for crop density and yield of produce and residues. Sample plot size was 100 m². Plant tissue samples were collected for laboratory N, P and K analyses of produce and residues. At the same time, samples were taken from topsoil (0–20 cm) to establish the nutrient stocks, and where possible from the garbage heap and kraal manure.

In northern Mozambique in the Nakatone catchment (650 ha), data were collected along two transects of 2–2.5 km (summit to summit), perpendicular to the direction of the main drainage channel. Of sections of 50 m, primary data were collected on maximum slope, erosion intensity, crop density, fraction cultivated and soil cover index. At specific sections, topsoil samples, crop cuttings and plant samples of produce and crop residues were taken from sample plots of 1–16 m² dependent on crop density. Representative average fields in terms of soil and crop system were constructed at the interfluve, middle slope and valley bottom of the catena in the Nakatone catchment.

Calculations and Estimates

Both IN and OUT nutrient flows are divided into several components. Inputs into the system or field, occur through:

- application of inorganic fertilizers (IN 1);
- application of organic fertilizers (IN 2);
- wet deposition (IN 3);
- biological N-fixation (IN 4);
- sedimentation through irrigation or flooding (IN 5).

Outputs from the system or field, are caused through:

- harvest of produce (OUT 1);
- removal of crop residues (OUT 2);
- leaching of N and K (OUT 3);
- gaseous losses of N (OUT 4);
- water erosion (OUT 5).

With regard to the estimation and calculation of the nutrient in- and outflows the methodology of Stoorvogel and Smaling (1990) was mostly used. The relevant Land Water Classes are LR (low rainfall, rainfed land), PR (problem rainfed land, mean annual rainfall < 1200 mm) and NF (naturally flooded land) for fields 1, 2–7 and 8 respectively.

Application of organic fertilizers (IN 2)

Grazing of crop residues left in the field results in a (partial) return of crop residue nutrients to the fields. It is assumed that 80% (8 + 72) of the crop residues or crop residue nutrients effectively pass through the animals (OUT 2), with 8% of the total crop residue nutrients being retained in the animal and 72% excreted (Stoorvogel and Smaling, 1990). Cattle were always kept at the kraal at night-time and it is therefore assumed that animals spend 12 out of 24 hours in the kraal with 50% of the manure being dropped in the kraal, and 50% in the field (IN 2).

Wet deposition (IN 3)

Neither primary data of this component measured at the sites nor data measured at sites with similar climatic conditions within Mozambique are available. Transfer functions using an empirical quantitative relation between mean annual rainfall (MAR in mm) at the site and amount deposited were applied to estimate the wet depositions of N, P and K (Smaling and Fresco, 1993). Mean annual rainfall was used, as no full year actual rainfall data of 1995/96, recorded at the sites, are available. Dry deposition was not taken into account.

Biological N-fixation (IN 4)

The biological N-fixation is brought about by symbiotic and non-symbiotic N-fixation. Leguminous crops were visually checked for the presence of nodules and, if growth stage permitted, for the activity of the nodules (pink nodule interior). Measurements of the contribution of N-fixation to the N nutrition of the crops are not available. It was therefore assumed that in leguminous crops 60% of the total demand for nitrogen is supplied by symbiotically fixed nitrogen (Stoorvogel and Smaling, 1990). In lowland rice systems, the percentage of 80% given by Stoorvogel and Smaling (1990) was lowered to 60%, due to poor water level control in the Nakatone valley bottom. Symbiotic N-fixation was set to zero if no nodules or unfavourable soil conditions were found.

No primary data were available on non-symbiotic N-fixation. Therefore, the values provided by Stoorvogel and Smaling (1990), based on Land Water Class were used.

Sedimentation through irrigation or flooding (IN 5)

An allowance for the calculation of the sedimentation component was made for fields 1 and 8 that were reported to experience flooding (Table 10.2).

The sedimentation factor is set to 0.05 and 0.40 for field 1 and 8 respectively and multiplied by [IN (1 + 2 + 3 + 4) – OUT (1 + 2 + 3 + 4 + 5)] to estimate the contribution of sedimentation to this nutrient inflow.

Harvested produce (OUT 1) and crop residues (OUT 2)

Cassava (*Manihot esculenta*) is a crop with a cycle that can extend to a period over 1 year. As the nutrient balance was calculated on a year basis, productions per hectare of tubers and residues were standardized to a period of 1 year. Cowpea (*Vigna unguiculata*), a multipurpose crop, is a source of vegetables (young leaves), grain and green manure. However, no recordings of the amounts of leaves picked were made at the sites. Instead, leaf and grain yield data, collected under similar conditions (unpublished data), were used.

It is assumed that during crop residue grazing, 80% (72 + 8) of the residues or residue nutrients is effectively removed (OUT 2) by the animals, with 8% of the total crop residue nutrients being retained in the animal and 72% excreted (Stoorvogel and Smaling, 1990) in the form of manure.

Leaching of N and K (OUT 3)

Downward movement of dissolved nutrients applies especially to N and K, whereas the mobility of phosphorus is very low. Losses of phosphorus through leaching are therefore considered negligible and set to zero.

The transfer functions of Smaling and Stoorvogel (1990) yielded negative values for N and K leached. Therefore, the equations derived from Van den Bosch (1994) were used instead. The amount of N leached is assumed to be dependent on mineral soil N, fertilizer N, rainfall and clay content. The mineral soil N is calculated by multiplying the mineralization rate by the total amount of N in the plough layer (0–20 cm). The mineralization rate is set to 2% year^{-1}. For the sandy topsoils a bulk density of 1.5 t m^{-3} is assumed. The amount of K leached is thought to be related to exchangeable K, fertilizer K, rainfall and clay content. In the case of burnt crop residue (ashes), all potassium present in the crop residue is considered as fertilizer in the calculations (equation 10.1).

For clay < 35%:

$$N_{leached} = (2.1 \times 10^{-2} \times MAR - 3.90)/100 \times (0.02 \times N_{stock}) \quad (10.1)$$
$$K_{leached} = (2.9 \times 10^{-4} \times MAR + 0.11)/100 \times (1170 \times K_{exch} + K_{burned})$$

with $N_{leached}$ in kg N ha^{-1} year^{-1}, N_{stock} in kg N ha^{-1} (= N_{tot} (in %)/ 100 × bulk density (in t m^{-3}) × 2 × 10^6), $K_{leached}$ in kg K ha^{-1} year^{-1}, K_{burned} in kg K ha^{-1} year^{-1}, K_{exch} in cmol(+) kg^{-1} soil.

Gaseous losses of N (OUT 4)

Nitrogenous gaseous losses (OUT 4) do occur as a consequence of denitrification, volatilization of ammonia and burning. The transfer function used by Smaling and Fresco (1993) to estimate denitrification and volatilization yields negative values for the lower mean annual rainfall sites. Therefore, the classes of Stoorvogel and Smaling (1990) that affect losses through denitrification and volatilization of ammonia were used and adapted to the specific conditions of the fields sampled. The class values given by Stoorvogel and Smaling (1990) were multiplied by 1.5 when pHw >7 and for the 'naturally flooded' Land Water Class, the class value is multiplied by 0.5 when submerged conditions only prevail in part of the year. If burning of crop residues is practised, all nitrogen present in the residues is assumed to be lost (Stoorvogel and Smaling, 1990).

Erosion (OUT 5)

Erosion research in Mozambique is practically absent and no data are available of soil and nutrient losses through erosion. Nutrient loss through erosion depends on total soil loss and nutrient content of eroded soil. Soil loss was estimated using the Universal Soil Loss Equation (USLE) (Wischmeier and Smith, 1978), which estimates only rain erosion and does not allow negative values, such as deposition, therefore overestimating erosion (Bergsma, 1996). Values for the six factors of the USLE equation have been either estimated or observed. The R factor was estimated using the transfer function suggested by Roose (1975): $R_{factor} = 0.5 \times MAR$. The K-factor was obtained using soil analytical data of the topsoil of the sampled fields as an input in the nomographs given by Morgan and Davidson (1986). The slope is the maximum slope (in %) measured in the sampled plots and transect sections. The length of slope is fixed to 100 m in the sample plots in southern Mozambique because all fields were 2–3 ha and flat to almost flat. In the Nakatone catchment an estimate of the length of slope is obtained using the average field length along the transect of sections belonging to the same physiographic position. The C-factor has been estimated using actual field observations of the soil cover index in the month of harvest and estimates for the other months of the year on the basis of the crop calender, development stage, residue management, relative crop density (RCD) and literature data (Bergsma, 1996). The RCD was introduced as a support tool to estimate soil cover and is defined as the sum of the relative densities, that is, the ratios between actual density under mixed-cropping and the optimal density under sole cropping of the individual crops that make the mixed cropping system (Geurts *et al.*, 1997). Finally, the P-factor was established on the basis of the management practices of the farmers obtained through interviews and field observations.

Nutrient Stocks

The nutrient stocks are calculated by multiplying the content of each nutrient in the topsoil by the weight of the topsoil. The nitrogen stock could be calculated by using the measured N_{total} content. No primary data were available of the P_{total} and K_{total} contents of the topsoil. Some data were available of P_{total} contents of soils, being a routine analysis in Mozambique before 1975. A selection of topsoil data was made with kaolinite as the predominant clay (Casimiro, 1969) and within this group P_{total} was found to correlate well with clay percentage ($R^2 = 0.63$; significant at the 1% probability level). Through the transfer function, $P_{total} = 0.00155 \times $ clay (with P_{total} in % P and clay in %), an estimate of the P_{total} of the soils of the sampled field was obtained.

Neither K_{total} data are available of the soils of the sampled fields, nor data of similar soils in other locations within Mozambique. Therefore, the K_{total} percentages given by Stoorvogel and Smaling (1990) were used, with fields 1–7 and 8 set to 0.04 and 0.08% K respectively.

RESULTS

Nutrient Flows

Inorganic fertilizers (IN 1) were not used in any of the fields (Table 10.3). In southern Mozambique, despite the fact that farmers of fields 1–4 possess manure and recognize its plant nutritional value, none of the farmers applied manure (IN 2) to the sampled fields. The animals stay overnight in the kraal where annually 700–750 kg of dry cattle manure accumulates per head of cattle (Geurts et al., 1997). Its quality with regard to N and K is reasonable, but the manure is very poor in phosphorus (Table 10.4). During the day the animals graze or browse on natural pastures. Only in field 1, did the farmer allow his livestock to feed on the crop residues, returning with the animal droppings 1.3 kg N, 0.3 kg P and 6.7 kg K ha^{-1} to the field. Farmers of northern Mozambique have no tradition in keeping cattle due to the occurrence of trypanosomiasis. Therefore, farmers neither possess nor have access to manure.

All farmers keep a garbage heap, but in none of the fields was garbage applied. The quality of the garbage is very poor (Table 10.4). The garbage heap is regularly burnt and contains kitchen refuse, processed produce residues and considerable amounts of sand originating from homestead cleaning. The amount of garbage is limited; however, no estimate of the annual production could be obtained. Bat guano is a valuable, local source of nutrients (Table 10.4) and is found in caves in the area of Vilanculos situated in the coastal belt of southern Mozambique. This deposit is estimated at 30,000 t (Ledder, 1994).

Wet deposition (IN 3) provided the fields annually with small but essential quantities of nutrients: 3–5 kg N, 0.5–0.8 kg P and 2–3 kg K ha^{-1}. Particularly in situations where neither inorganic nor organic fertilizers are applied, wet deposition is the main, if not the only source of P and K.

Table 10.3. N, P and K flows of the eight sampled fields.

Field code	IN components (%)					Σ IN (kg ha⁻¹ year⁻¹)	OUT components (%)					Σ OUT (kg ha⁻¹ year⁻¹)
	1	2	3	4	5		1	2	3	4	5	
Nitrogen												
Southern Mozambique												
1	0.0	14.6	35.5	45.7	4.2	8.7	51.1	15.1	6.2	16.0	11.7	18.8
2	0.0	0.0	63.1	36.9	0.0	5.4	42.0	0.0	7.0	35.6	15.4	29.5
3	0.0	0.0	22.3	77.7	0.0	15.3	74.3	0.0	6.1	14.9	4.7	33.6
4	0.0	0.0	11.4	88.6	0.0	29.9	15.8	0.0	6.6	69.7	7.9	70.2
5	0.0	0.0	66.9	33.1	0.0	6.0	50.0	0.0	6.8	12.5	30.7	59.8
Nampula												
6	0.0	0.0	10.3	89.7	0.0	45.9	50.8	0.0	6.3	3.8	39.1	133.2
7	0.0	0.0	6.4	93.6	0.0	73.7	70.9	0.0	4.6	4.8	19.7	103.7
8	0.0	0.0	14.8	54.5	30.7	31.9	41.1	0.0	30.9	12.9	15.0	46.5
Phosphorus												
Southern Mozambique												
1	0.0	25.7	38.3	0.0	36.0	1.3	86.3	7.3	0.0	0.0	6.4	10.5
2	0.0	0.0	100.0	0.0	0.0	0.6	98.2	0.0	0.0	0.0	1.8	3.7
3	0.0	0.0	100.0	0.0	0.0	0.6	84.4	0.0	0.0	0.0	15.6	4.0
4	0.0	0.0	100.0	0.0	0.0	0.6	54.9	0.0	0.0	0.0	45.1	2.8
5	0.0	0.0	100.0	0.0	0.0	0.7	95.6	0.0	0.0	0.0	4.4	5.4
Nampula												
6	0.0	0.0	100.0	0.0	0.0	0.8	44.6	0.0	0.0	0.0	55.4	31.3
7	0.0	0.0	100.0	0.0	0.0	0.8	65.6	0.0	0.0	0.0	34.4	21.0
8	0.0	0.0	46.2	0.0	53.8	1.7	79.0	0.0	0.0	0.0	21.0	3.1
Potassium												
Southern Mozambique												
1	0.0	68.4	20.7	0.0	11.0	9.8	44.3	49.3	0.8	0.0	5.6	30.2
2	0.0	0.0	100.0	0.0	0.0	2.2	56.4	0.0	6.9	0.0	36.7	9.6
3	0.0	0.0	100.0	0.0	0.0	2.2	87.1	0.0	3.4	0.0	9.4	17.3
4	0.0	0.0	100.0	0.0	0.0	2.2	36.6	0.0	28.4	0.0	35.1	9.9
5	0.0	0.0	100.0	0.0	0.0	2.6	51.4	0.0	10.2	0.0	38.4	16.5
Nampula												
6	0.0	0.0	100.0	0.0	0.0	3.1	74.0	0.0	1.1	0.0	25.0	92.6
7	0.0	0.0	100.0	0.0	0.0	3.1	82.7	0.0	1.6	0.0	15.7	100.4
8	0.0	0.0	43.5	0.0	56.5	7.1	63.7	0.0	10.6	0.0	25.7	14.1

All leguminous crops showed nodulation, although it was not always possible to verify effective nodulation. The total amount of biologically fixed nitrogen and its relative contribution to the total nitrogen inflow were considerably increased by the inclusion of leguminous crops into the crop system, from 2 kg N ha^{-1} year^{-1} and approximately 35% (non-symbiotic N-fixation only, in fields 2 and 5) to 69 kg N ha^{-1} year^{-1} and 89% (field 7) respectively (Table 10.3), making biologically fixed N, especially symbiotically fixed N, the most important source of nitrogen in the absence of external inorganic and organic fertilizers.

Table 10.4. Chemical and physical characterization of nutrient resources.

Nutrient resource	DM (%)	Bulk density (t fresh m^{-3})	% of DM		
			N	P	K
Cattle manure	82	0.78	1.99	0.11	1.38
Garbage	85	0.97	0.47	0.09	0.17
Bat guano	88	—	1.88	2.36	0.57

Some sedimentation (IN 5) was observed in fields 1 and 8. The contribution to the total nutrient inflow to the field was 0.4 kg N, 0.5 kg P and 1.1 kg K ha^{-1} year^{-1} in field 1 and 10 kg N, 0.9 kg P and 4 kg K ha^{-1} year^{-1} in field 8. In field 8, in the valley bottom, sedimentation is a particularly important source and replenisher of nutrients, providing more than 30% of the N and more than 50% of the P and K.

Harvested produce (OUT 1) is by far the largest nutrient drain (Table 10.3). Nutrient extraction by harvested produce in the Nakatone fields is notably higher than that in the southern Mozambique fields.

All crop residues were left in the field, except for field 1 in arid southern Mozambique, where the farmer allowed his livestock to feed on the crop residues. The net effect of day time grazing of crop residues, without kraal manure recycling to the field, leads to nutrient losses of 1.5 kg N, 0.5 kg P and 8.2 kg K ha^{-1}.

Losses of N and K due to leaching (OUT 3) are generally small, in the range 1–14 kg N ha^{-1} and 0.2–3 kg K ha^{-1} or generally 5–10% of the total outflow of both N and K. Burning of crop residues increases the susceptibility of potassium contained in the residues to leaching, as can be seen in field 4 in particular. This effect could hardly be traced in the OUT 3 values of field 2 because of the low cotton biomass (Table 10.5). Nitrogen leaching in the Nakatone valley bottom field was high because of the high N level of the topsoils (Table 10.2).

Nitrogen losses due to denitrification are small, 2–8 kg N ha^{-1} or 4–17% of the total nitrogen outflow. Burning of crop residues increases the nitrogenous gas losses to 11 and 49 kg N ha^{-1} in field 2 and 4 respectively, bringing the relative contribution of N gaseous losses to the total N outflow to 36 and 70%.

In the southern Mozambique fields, estimated nutrient losses due to water erosion (OUT 5) were considerably smaller than those on the interfluve and middle slope Nakatone fields. In the southern Mozambique fields these nutrient losses amounted to 2–18 kg N ha^{-1}, 0.1–1.3 kg P ha^{-1} and 2–6 kg K ha^{-1} as compared with 20–52 kg N ha^{-1}, 7–17 kg P ha^{-1} and 16–23 kg K ha^{-1} obtained on the interfluve and middle slope Nakatone fields. Nutrient losses in the Nakatone valley bottom are similar to those obtained on the flat southern Mozambique fields.

Table 10.5. Crops, crop densities and yields of produce and residues of the sampled fields.

	Southern Mozambique					Nampula		
Field	1	2	3	4	5	6	7	8
Crop	Actual 1995/96 cropping system					Average cropping system		
1	Maize	Maize	Maize	Maize	Maize	Cassava	Cassava	Rice
2	Cowpea	Cotton	Sorghum	Groundnut	Sorghum	Cowpea	Cowpea	
3	Melon		Cowpea	Cowpea		Maize	Bambara	
4							Maize	
RCD (%)	51	42	54	48	38	66	99	94
Crop (produce (kg dry matter ha^{-1}))								
1	821	614	305	331	1,461	3,658	4,980	1,608
2	53	211	363	230	381	616	898	
3	1,690		529	39		2,559	166	
4							1,477	
Crop (crop residues (kg dry matter ha^{-1}))								
1	1,129	1,241	381	1,375	670	909	1,237	5,355
2	14	396	2,739	1,532	675	1,711	2,496	
3	423		138	10		2,021	549	
4							861	
All crops (total biomass (kg dry matter ha^{-1}))								
	4,130	2,462	4,455	3,517	3,187	11,474	12,664	6,963

Notes: Produce components: bambara groundnut, pods; cassava, tubers; cotton, cotton seed; cowpea, pods + picked leaves; groundnut, pods; maize, ears; rice, paddy; melon, fruits; sorghum, panicles. Crop residue components for all crops except rice: stems + leaves; rice, stems + leaves + empty heads.

Nutrient Balances and Stocks

In all fields the N, P and K nutrient balances were negative, implying that all fields experience nutrient depletion during the cultivation period (Table 10.6). In southern Mozambique, the nutrient depletion during the 1995/96 season was 10–54 kg N, 2–9 kg P and 7–20 kg K ha^{-1}. Within the southern Mozambican fields, there was a clear distinction between fields 1–3 and 4–5 in the nitrogen balance, primarily caused by the practice of burning (OUT 4) on field 4 and the high nutrient extraction with the harvested produce (OUT 1) and erosion (OUT 5) in field 5. In Nampula, on the interfluve and middle slope fields nutrient depletion was highest: 30–90 kg N, 20–30 kg P

Table 10.6. Nutrient balances and stocks of the eight sampled fields (kg ha^{-1}).

Field code	Σ IN	Σ OUT	Balance	Stock
Nitrogen				
Southern Mozambique				
1	8.7	18.8	−10.0	1200
2	5.4	29.5	−24.1	1200
3	15.3	33.6	−18.3	900
4	29.9	70.2	−40.3	1500
5	6	59.8	−53.8	2700
Nampula				
6	45.9	133.2	−87.2	2100
7	73.7	103.7	−30.1	1200
8	31.9	46.5	−14.7	3600
Phosphorus				
Southern Mozambique				
1	1.3	10.5	−9.2	492
2	0.6	3.7	−3.1	23
3	0.6	4	−3.4	474
4	0.6	2.8	−2.2	455
5	0.7	5.4	−4.8	46
Nampula				
6	0.8	31.3	−30.5	933
7	0.8	21.0	−20.3	567
8	1.7	3.1	−1.4	441
Potassium				
Southern Mozambique				
1	9.8	30.2	−20.4	1245
2	2.2	9.6	−7.4	1245
3	2.2	17.3	−15.1	1245
4	2.2	9.9	−7.7	1245
5	2.6	16.5	−13.9	1245
Nampula				
6	3.1	92.6	−89.6	1245
7	3.1	101.4	−98.3	1245
8	7.1	14.1	−7.0	2489

and 90–100 kg K ha^{-1}. The valley bottom, however, showed depletion rates much like those experienced in the southern Mozambican fields: 15 kg N, 2 kg P and 7 kg K ha^{-1}.

Topsoils are sandy and organic matter contents generally low (Table 10.2). Consequently, nutrient stocks are expected to be low and estimated at 900–3600 kg N, 20–900 kg P and 1200–2500 kg K ha^{-1}.

DISCUSSION

Field 1 in Arid Southern Mozambique

The favourable weather in the 1995/96 season, the accumulated soil fertility during the war-forced-fallow (period 1982–1994) and crop failure in the 1994/95 season due to a serious drought allowed a very good crop in the 1995/96 season. In this season the nutrient stocks were reduced by 10 kg N ha^{-1}, 10 kg P ha^{-1} and 20 kg K ha^{-1}. The climatic conditions, basically water availability, control the long-term nutrient balance or sustainability of the cropping systems in this zone. Breman's statement that 'only in the desert (arid areas) water is the dominant limiting factor and that in the semi-arid tropics, nutrient availability is generally the most limiting factor' (Breman, 1990) is in line with the above findings. A successful harvest is only expected once every 14–15 years (Table 10.1), with crop failures in most years, without any or hardly any nutrient being extracted by harvested produce. During these years, a small positive balance is possible, allowing the regeneration of soil fertility lost in a successful season, provided the farmer prevents livestock from grazing on the crop residues. Grazing of the field's crop residues is detrimental to the field's soil fertility if the animals are kept overnight in the kraal, without recycling the manure to the field. Crop residue grazing affects the soil K status of the field, a consequence of the comparatively high K content of stems and leaves. The farmer does not use manure. In addition, he does not possess sufficient manure, as at present (1996) he holds only one head of cattle. During the war, the farmer lost all his livestock.

Fields 2 and 3 in Semi-arid Southern Mozambique

In fields 2 and 3, generally, the moisture availability allows a reasonable harvest once every 2 years. In a year of crop failure, the P and K in- and outflows of field 2 and the N, P and K in- and outflows of field 3 are balanced, with OUT 1 set to 0 and assuming somewhat lower figures for IN3, IN4, OUT3, OUT4 and OUT5. However, the N balance of field 2 continues to be 10 kg N ha^{-1} negative. This is due to the absence of a leguminous crop in the cropping system and the phyto-sanitarian practice of burning cotton residues.

During a successful season like 1995/96, the soils of fields 2 and 3 lost 18–24 kg N, 3 kg P and 7–15 kg K ha^{-1}. Over a period of 4 years (Table 10.2), with two successful harvests, the total depletion is to be likely in the range of 30–40 kg N, 5 kg P and 10–25 kg K ha^{-1}. Reducing the N outflow by leaving the cotton residues unburnt in the field is not a feasible soil fertility management option, for phyto-sanitarian reasons. The inclusion of a leguminous crop in the cropping system of field 2 at densities corresponding to RCDs of 10–15%, able to fix biologically around 10 kg of N year^{-1} ha^{-1}, can reduce the N depletion from 40 to 20 kg ha^{-1}. Through the system of shifting cultivation, still a feasible nutrient management option in the sparsely populated parts of southern Mozambique, a fallow period of around 10–15 years is required to compensate for the remaining depletion experienced during the cultivation period, assuming a positive balance of 2 kg N, 1 kg P and 1 kg K ha^{-1} year^{-1} during fallowing (Stoorvogel and Smaling, 1990). These fallow periods (Table 10.2) are reported to be practised in southern Mozambique (FAO/UNDP, 1983).

The farmers of fields 2 and 3 possess livestock, 10 and 9 head of cattle respectively, making available to each of these farmers, around 7 t of dry cattle manure or 140 kg N, 8 kg P and 97 kg K. The nutrients in manure originate mainly from the natural pasture land and are redistributed by livestock to the kraal. The manure available to the farmers could sustain the soil nitrogen stock of around 6 ha, the soil phosphorus stock of 3 ha and the potassium stock of approximately 10 ha, reducing the need for long-term fallowing. The produced manure is very poor in phosphorus due to the poor P content of the natural pastures growing on the P deficient Arenosols of southern Mozambique. The manure quality could be improved by the addition of small quantities of P fertilizer. The reality, however, is that manure is hardly used due to: (i) its bulkiness, complicating transportation and application; (ii) increased weed infestation; and (iii) increased crop susceptibility to drought.

Fields 4 and 5 in Sub-humid Dry Southern Mozambique

The P and K depletion of fields 4 –5 was similar to that of fields 1–3 in the arid and semi-arid climate zones: 2–5 and 8–14 kg ha^{-1} respectively. The N depletion in fields 4 and 5, however, was considerably larger, 40–54 kg N ha^{-1}, than that experienced in fields 1–3 and mainly the result of crop residue burning in field 4, and in field 5 due to the absence of a leguminous crop (IN 4), a relatively high extraction through harvest (OUT 1) and erosion (OUT 5). Burning was responsible for nearly 70% of the total N outflow. The cropping system in field 4 had a high proportion of leguminous crops (Table 10.5, RCD = 0.19) and without crop residue burning a balanced nitrogen in- and outflow would have been obtained, showing the importance of a leguminous crop to the maintenance of the N stock. On the other hand, in field 5 no leguminous crop was included. The inclusion of leguminous crops at a RCD of 0.15 could reduce the negative balance by 20 kg N ha^{-1} to 35 kg N ha^{-1} in field 5. To counterbalance the remaining 35 kg N ha^{-1} in the absence of organic and inorganic fertilizer, a fallow period of around 15 years is required. Similar periods (10–15 years) are required to compensate for the net loss of potassium. However, in the coastal belt of southern Mozambique, with population densities reaching 70 persons km^{-2} (Pililão, 1989) and land utilization intensities of 0.5–0.6, the system of shifting cultivation is becoming an increasingly unsuitable management option to sustain the soil nutrient stocks, especially those of N and K. Here, manure and bat guano, locally available and renewable nutrient resources, could be applied to cropping systems that include leguminous crops and exclude the practice of burning. Rates of 1500–2500 kg ha^{-1} are required to sustain the macronutrient soil stocks. Another nutrient management option is the application of mineral fertilizers. Nitrogen and potassium fertilizer have rarely been shown to be beneficial due to their high solubility and consequently high susceptibility to leaching in sandy soils (Chaguala and Geurts, 1996). In addition, in the case of crop failure, these fertilizers are lost, requiring annual reapplication. Fertilizer phosphorus, however, has been shown to be a feasible option for the smallholders cultivating the sandy soils of sub-humid dry southern Mozambique because of its good initial and residual values (Chaguala and Geurts, 1996). To maintain the soil phosphorus stock, small annual doses in the range of 10 kg fertilizer P ha^{-1} are required.

Fields 6–8 of the Nakatone Catena

In northern Mozambique, rainfall reliability is generally higher than in southern Mozambique and total crop failure due to moisture stress is not expected. Total biomass in the Nakatone catchment fields was at least twice that in southern Mozambique fields (Table 10.5). Total P and K inflows were similar to those obtained in southern Mozambique, but the nitrogen inflow was considerably higher, owing to the higher leguminous crop biomass and densities, corresponding to RCD values of 0.11–0.29. The N, P and K outflows of the interfluve and middle slope fields of the Nakatone catchment (field 6 and 7 respectively) were two to three times higher than those of the southern Mozambique fields, principally caused by the larger harvest extraction and erosion losses. The cultivation of the cassava-based cropping systems on the interfluves and middle slopes drains potassium in particular, annually 90–100 kg K ha^{-1}, and is expected to produce highly potassium-deficient soils in the near future, requiring considerable K input from mineral sources.

The estimated annual soil loss was estimated at 37 and 25 t ha^{-1} on the interfluve and middle slopes, respectively. Farmers can decrease the rate of soil loss by maintaining a good soil cover, using optimal crop densities and more adequate conservation measures. Assuming a reduction in soil loss of 50%, annual depletion rates of 20–60 kg N, 15–25 kg P and 75–90 kg K ha^{-1} should still be anticipated on the interfluves and middle slopes. Shifting and even fallow systems are unsuitable nutrient management options to restore soil fertility as fallow periods are too short (Table 10.2) due to land scarcity in the Nakatone catchment. Owing to the absence of significant numbers of livestock, manure is not available to farmers for soil fertility correction and grazing of crop residues is not practised. Furthermore, crop residues were not burnt on the Nakatone fields and leguminous crops already form an important component of the cropping system. Therefore, to achieve sustainability, the application of chemical fertilizers is required in rates equal to the annual depletion rates. Nitrogen and phosphorus fertilizer rates of 20–60 kg N and 15–25 kg P ha^{-1} proved to be economically feasible. K fertilization, however, has so far rarely shown economic responses (Geurts, 1997). At present, available K levels of these interfluve and middle slope soils vary between 0.2 and 0.3 cmol(+) kg^{-1}, close to the critical level of 0.2 cmol(+) kg^{-1} (Landon, 1984). Monitoring of the K in- and outflows and soil status of cassava-based cropping systems in particular, is required.

The macronutrient inflows and outflows in the Nakatone valley bottom field resemble those in the southern Mozambique fields, except for the nitrogen inflow, which was found to be higher, owing to sedimentation caused by annual flooding. A small negative balance was estimated for the valley bottom and in agreement with field observations of slight erosion signs. Fields are continuously cropped, mainly with lowland rice, and due to the absence of locally available and renewable nutrient resources, fertilization of the valley bottom fields with 10–20 kg N ha^{-1} is required to balance the N outflow. The application of phosphorus fertilizers is not recommended, owing to the balanced phosphorus in- and outflows and high available soil P (13 ppm P-Olsen). Despite this depletion, no potassium fertilization is recommended as no economic responses to potassium fertilizer are expected due to sufficiently high available soil potassium levels (0.3 cmol(+) kg^{-1}) and comparatively small annual K depletion rates (5–10 kg K ha^{-1} year^{-1}), relative to the soil K stock (2500 kg K ha^{-1}).

It is important to note that the phosphorus balances of most fields, except for those on Lixisols, are only slightly negative with the inflows and outflows being small and almost in equilibrium. This is no surprise; looking at the very low and deficient levels of available soil phosphorus (Table 10.2), low phosphorus stock (Table 10.6) and low biomass production (Table 10.5), and one can conclude that these fields have reached sustainability at the lowest possible level. Fallow periods of only 3–5 years suffice to compensate for the P depletion of one cropping season.

Shortcomings

The results are based on observations of cultivated fields of only one cropping season. This cropping season forms part of the cultivated period which in turn is part of the land utilization cycle. Nutrient depletion in one season is no indication of an unsustainable situation and data of the full cycle, including fallowing, are required to judge if land use is sustainable in the long term. Only the inflows and outflows 1 and 2 could be quantified through direct field measurements. The other flows were established indirectly through transfer functions, assumptions or qualitative information collected during interviews. To improve the understanding of the natural processes and management practices influencing nutrient flows, the long-term nutrient balance and hence sustainability of land use systems, a full land utilization cycle should be followed and all flows determined through direct field measurements. However, taking into consideration the Mozambican research infrastructure with limited personnel, financial means and materials, the methodology followed in this paper provides a useful tool to improve the soil fertility management of land use systems in Mozambique.

CONCLUSIONS

In field 1 in arid southern Mozambique, rainfall amount and reliability only allow extensive land use with crop failures in most years and water availability being the main limiting factor. Livestock is an important component of the farming systems in arid and semi-arid regions, providing food, traction and manure. Kraal manure, however, is hardly used and grazing of the crop residues of field 1 resulted in a nutrient transfer to the kraal and a net loss of K in particular. Water availability in the other fields allows rainfed agricultural output levels that call for specific soil fertility management to sustain production.

Climatic conditions in semi-arid southern Mozambique permit a successful crop once every 2 years. Leguminous crops that effectively fix nitrogen are crucial in the maintenance of the soil nitrogen stock. Provided that leguminous crops are included and crop residues incorporated (where possible), nutrient depletion in fields 2 and 3 experienced during the cultivation period can effectively be restored by the system of shifting cultivation, still a feasible nutrient management option in this sparsely populated region, using fallow periods of 10–15 years. Manure, although available and its nutritional value recognized, is hardly used.

Fields 4 and 5 are situated in the densely populated coastal belt of southern sub-humid, dry Mozambique with land utilization intensities of 0.5–0.6. Here, shifting cultivation is inadequate to compensate for the nutrient depletion during the cultivation period. Burning of crop residues leads to large nitrogen losses and should be avoided. After maximizing nutrient input and minimizing losses through judicious management of the field system nutrients, there is still a need for the application of external nutrients. Nutrients from organic sources are locally available, like manure and bat guano, and rates of 1500–2500 kg ha^{-1} are required to balance the nutrient inflows and outflows. The plant nutritional value of the garbage heap is too poor to be of use for soil fertility improvement. Inorganic N and K fertilizers have not proved to be effective in the sandy soils of southern sub-humid, dry Mozambique. Phosphorus fertilizer, however, is a feasible management option to the smallholder to correct the phosphorus balance of the flat sandy soils.

In northern Mozambique, on the interfluve and middle slope fields of the Nakatone catchment, highest nutrient depletion rates were recorded, caused by high biomass output and erosion. Owing to high population pressure, shifting and fallow systems are incapable of sustaining the nutrient stocks. It is, therefore, crucial that farmers make effective and efficient use of all available nutrient resources. Due to the presence of trypanosomiasis, livestock has never been a tradition in this area and consequently farmers have no access to manure. To turn the production systems in field 6 and 7 into sustainable ones, nitrogenous and mineral fertilizers need to be employed. Rates of 20–60 kg N and 15–25 kg P ha^{-1} proved to be economically feasible in these agroecological conditions. Although crop responses to K fertilization rarely proved to be economically attractive, the cultivation of the cassava-based cropping systems on the interfluves and middle slopes annually drains large quantities of potassium (90–100 kg K ha^{-1}), and is expected to produce highly potassium deficient soils in the near future, requiring considerable K input from mineral sources. Fields in the Nakatone valley bottom are continuously cropped and, due to the absence of locally available and renewable nutrient resources, fertilization of the valley bottom fields with 10–20 kg N ha^{-1} is required. Neither P or K fertilizer is recommended due to low P and K depletion rates and high available P and K soil levels.

Soil phosphorus is a serious limiting factor in fields 2–7 as is shown by the small negative balance, the small in- and outflows, low available soil phosphorus and low phosphorus stocks, and one can conclude that these fields have reached sustainability at the lowest possible level.

REFERENCES

Abrahamsson, H. and Nilsson, A. (1994) *Moçambique em Transição. Um Estudo da história de Desenvolvimento Durante o Período 1974–1992*, 1st edn. Padrigu and CEEI-ISRI, Maputo, Mozambique, 365 pp.

Bergsma, E. (1996) *Terminology for Soil Erosion and Conservation*. Prepared for Sub-commission C, Soil and Water Conservation of the International Society of Soil Science. ISSS, ITC, ISRIC, Wageningen, The Netherlands

Breman, H. (1990) No sustainability without external inputs. In: *Beyond Adjustment, Sub-Sahara Africa*, Africa Seminar, Maastricht, The Netherlands.

Casimiro, J.F. (1969) *Os Solos de Algumas Unidades Experimentais do IIAM*. IIAM. Comunicações 41. Lourenço Marques (Maputo), Mozambique.

Chaguala, P. and Geurts, P.M.H. (1996) *Adubação Orgânica e Mineral Numa Rotação Num Arenossolo no Sul de Moçambique (1950–1965)*. Série Terra e Água do INIA. Comunicação 85. Maputo, Mozambique, 18 pp.

FAO/UNDP (1983) *Land Resources Inventory and Study of the Population Supporting Capacity of the Coastal Zone of the Districts of Manjacaze and Zavala, Gaza and Inhambane Province, Mozambique*. FAO/UNDP Project, Land and Water Use Planning. AGOA/MOZ/75/011 Field Document 47. Maputo, Mozambique.

Folmer, E.C.R., Geurts, P.M.H. and Francisco, J.R. (1998) *Assessment of Soil Fertility Depletion in Mozambique*. Agriculture, Ecosystems and Environment 71, Issues 1–3 (1993), 161–169.

Geurts, P.M.H. (1997) *Recomendações de Adubação Azotada e Fosfórica para Culturas Alimentares e Algodão em Moçambique*. Série Terra e Água do INIA, Maputo, Mozambique, Comunicação 88, 66 pp.

Geurts, P.M.H., Fleskens, L., Löwer, J. and Folmer, E.C.R. (1997) *Field Level Nutrient Budgets of Mixed Cropping Systems in Mozambique*. INIA, Maputo, Mozambique, 45 pp.

Landon, J.R. (1984) *Booker Tropical Soil Manual*. Booker Agriculture International, London, UK, 474 pp.

Ledder, H. (1994) *Notícia Expicativa da Carta de Jazigos e Ocorências de Minerais não Metálicos na Escala 1 : 1 000 000*. Maputo, Mozambique.

Morgan, R.P.C. and Davidson, D.A. (1986) *Soil Erosion and Conservation*. Longman Scientific Technical, Essex, UK.

Pililão, F. (1989) *Moçambique. Evolução da Toponimiae da Divisão Territorial 1974–1987*. Dinageca, Maputo, Mozambique.

Roose, E.J. (1975) *Erosion et Ruisellement en Afrique de l'Ouest: Vingt Anées de Mesures en Petites Parcelles Expérimentales*, Cyclo. ORSTOM, Adiopodoumé, Ivory Coast.

Ruthenberg, H. (1976) *Farming Systems in the Tropics*. Clarendon Press, Oxford.

Serno, G. (1995) *Os Solos dos Distritos de Ribaue, Lalaua, Mecumburi, Muecate, Nampula, Monapo, Nacaroa, Namapa e Memba, Provincia de Nampula*. Série Terra e Agua do INIA, Maputo, Mozambique Comunicação 77a + 77b.

Smaling, E.M.A. and Fresco, L.O. (1993) A decision support model for monitoring nutrient balances under agricultural land use (NUTMON). *Geoderma* 60, 235–256.

Stoorvogel, J.J. and Smaling, E.M.A. (1990) *Assessment of Soil Nutrient Depletion in sub-Saharan Africa: 1983–2000*. Winand Staring Centre, Wageningen, The Netherlands.

Van den Bosch, H. (1994) *A Decision Support System for Sustainable Use of Soil Macronutrients in Kenyan Farming Systems*. Working paper, DLO Winand Staring Centre, Wageningen, The Netherlands.

Westerink, R.M. (1995) *Evaluation of Length of Growing Period and Crop Growing Possibilities in Mozambique*. Nota Técnica 76. INIA, DTA, Maputo, Mozambique.

Wischmeier, W.H. and Smith, D.D. (1978) *Predicting Rainfall Erosion Losses – A Guide to Conservation Planning*. Agricultural Handbook 537. US Department of Agriculture, Washington, DC, 58 pp.

11 Nutrient and Cash Flow Monitoring in Farming Systems on the Eastern Slopes of Mount Kenya

J.N. Gitari, F.M. Matiri, I.W. Kariuki, C.W. Muriithi and S.P. Gachanja

Kenya Agricultural Research Institute, Regional Research Centre – Embu, PO Box 27, Embu, Kenya

ABSTRACT

Land degradation is threatening the basis of many farming communities in East Africa. In order to determine farm households' perception of soil nutrient depletion and to acquire comprehensive knowledge of the dynamics of the farming systems such as nutrient flows and economic performance, a 1 year monitoring activity was conducted in 1995 in five different land use zones (LUZs) in the Embu and Mbeere districts of eastern Kenya. The monitoring included primary data collection such as quantities and prices of inputs and outputs of crop and livestock activities, soil samples, measurement of nutrient contents of farm products, labour and farm assets. Consultation of existing and relevant secondary data such as soil maps and agroclimate data were included in the monitoring process. A Farm-NUTMON accounting tool was developed to calculate nutrient flows, balances and economic performance indicators. Nutrient monitoring results indicate a net negative N-balance at farm level in all LUZs, except LUZ 5. In the latter, browsing of cattle and goats results in a concentration of nutrients from the surrounding pastures. Leaching is the major source of N loss especially in high rainfall zones such as LUZ 1 and 2. For P and K the results show a mixture across the LUZs. Farm types which predominantly grow food crops show a trend of strong nutrient mining at plot level, due to low external inputs of nutrients because of high fertilizer and low crop product prices. Economic performance shows a large variation between farms within a LUZ and between LUZs. Households with higher cash flows appear to have more positive nutrient balances due to application of chemical fertilizers.

INTRODUCTION

In Kenya, the agricultural sector generates the lion's share of gross national income and its performance is of paramount importance to the national food supply. The main determinants of the performance of the agricultural sector are weather conditions and soil fertility. As far as the latter is concerned, Stoorvogel and Smaling (1990) showed that agroecosystems in sub-Saharan Africa (SSA) lost approximately 22, 2.5 and 15 kg ha^{-1} year^{-1} of N, P and K respectively between 1982 and 1984. Hence, nutrient mining of agricultural soils is a real menace and results from crop production with low levels of nutrient inputs coupled with poor nutrient conservation practices. Sustainability of the agroecosystems is multifaceted involving agronomic, ecological, social as well as economic factors. In the highlands of East Africa, traditional soil fertility practices cannot be maintained under conditions of mounting population growth and land scarcity. Thus, land degradation is now threatening the very basis of the farming communities (Stahl, 1993).

To determine nutrient and cash flow status in eastern Kenya, a sample of 15 farm families spread over five land use zones (LUZs) in Embu and Mbeere districts were monitored over a period of 1 year (two cropping seasons in 1995). The objectives were: (i) to determine farm households' perception of soil nutrient depletion and related constraints and potentials; (ii) to acquire a comprehensive knowledge of a farm system and the dynamic functionality of its internal and external nutrient flows in a spatial and temporal context; (iii) to quantify nutrient flows and balances of existing farming systems at different spatial scales; (iv) to quantify the economic performance of existing farming systems at different spatial scales; (v) to identify, on-farm test and evaluate with stakeholders relevant integrated nutrient management (INM) technology options; and (vi) to identify and evaluate with stakeholders relevant policy instruments enabling INM-technology adoption.

THREE-LEVEL CHARACTERIZATION OF THE AREA

Districts

Embu and Mbeere districts have an area of 2805 km^2 and are located on the eastern slopes of Mount Kenya. The elevation ranges from 760 to 2070 m asl. Rainfall varies between 640 and 2000 mm year^{-1} and falls in two distinct periods (March–July and October–December). The districts can be classified as 4.7% humid, 3.3% sub-humid, 12.7% semi-humid, 16.4% transitional and 62.9% semi-arid (Jaetzold and Schmidt, 1983). Major LUZs in both districts were defined using secondary data, satellite images and expert knowledge. These are Tea/Dairy (1), Tea/Coffee/Dairy (2), Coffee/Maize (3), Tobacco/Food crops (4), and Livestock/Shifting cultivation (5) (Fig. 11.1).

The predominant soil types in the major LUZs of Embu and Mbeere districts are shown in Table 11.1. The soils and agroecology of the area are greatly influenced by the presence of Mount Kenya (Jaetzold and Schmidt, 1983).

According to the 1989 national population census, the population of Embu and Mbeere districts was 370,138 people, growing at a rate of 3.7% per annum. The average population density is 132 persons km^{-2}, but most

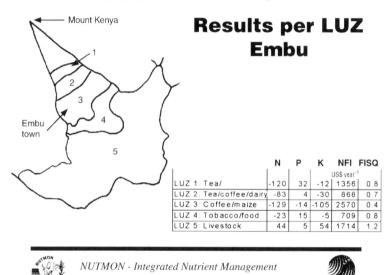

Fig. 11.1. LUZs of Embu and Mbeere.

Table 11.1. Characteristics of the LUZs.

Characteristic	LUZ 1	LUZ 2	LUZ 3	LUZ 4	LUZ 5
Altitude (m asl)	1770	1590	1280–1460	980	830
Annual mean temp. (°C)	16.8	18.2	20.2	21.4	22.6
Annual average rainfall (mm)	1750	1400	1100–1300	900	800
Main soil types	Andosol/Nitosol	Nitosol	Nitosol	Nitosol/Cambisol	Arenosol
Main enterprises/activities	Tea/dairy	Tea/coffee/dairy	Coffee/maize	Tobacco/food crops	Livestock/shifting cultivation

of the rural population is concentrated within the Tea/Dairy, Tea/Coffee/Dairy, and Coffee/Maize LUZs.

Land Use Zones

A participatory rural appraisal (PRA) was carried out (Munyi *et al.*, 1995) to qualitatively describe the various LUZs, identify major problems and lay the foundation for the actual farm selection. A pilot farm of each farm type was selected based upon willingness to participate in the monitoring programme and a number of selection criteria like cropping pattern, livestock activities, farm size, farm management practices, product marketing and off-farm activities. Table 11.2 presents the major constraints in each of the LUZs as identified during the PRA (Munyi *et al.*, 1995).

Farming Systems

In further assessing the homogeneity of the identified LUZs, it appeared that there were variations within each LUZ in terms of: topography of the farms, relative level of intensification in production, relative level of management

Table 11.2. Constraints in the LUZs.

Constraints	LUZ 1	LUZ 2	LUZ 3	LUZ 4	LUZ 5
Lack of market for milk and low price	×	×	×		
Lack of market for food crops/livestock				×	×
Limited land size for food and fodder crops	×	×			
Poor roads	×	×			×
Labour shortage during peak cropping periods	×			×	×
High input cost e.g. feeds and fertilizers	×	×	×		
Lack of credit		×	×	×	×
Low quality of inputs e.g. animal feeds, fertilizers	×	×	×		
Low soil pH	×	×			
High cost of credit		×	×		
Low soil fertility due to soil erosion/overgrazing		×	×	×	×
			×	×	
Pests and diseases (crops)				×	×
Pests and diseases (crops and livestock)			×	×	
Lack of artificial insemination				×	
Inadequate and erratic rainfall					×
Lack/poor extension services					×

at both activity and farm level, relative level of input use at both activity and farm level, relative level of resource endowment of the farmers, and distance of cultivated parcel from homestead. Three farm types were identified in each LUZ. Table 11.3 presents some characteristics of the farm types.

DATA COLLECTION AND PROCESSING

Farm Inventory and Monitoring

Before monitoring, an inventory was carried out on each farm. Information was compiled on farm size and architecture, nutrient stocks, household composition, agricultural activities and assets. During the farm inventory, the major nutrient pools at the farm and the major flows between these pools were identified (Van den Bosch et al., 1998b).

The selected farms were visited once a month for a full year during which farm management activities were monitored. A structured questionnaire was used to collect data with a monthly recall period on quantity and prices of inputs and outputs of crop and livestock activities, growth of the herd, confinement of livestock, redistribution of manure, stock of household staple crops, labour and off-farm income (De Jager et al., 1998).

Table 11.3. Characteristics of the farm types.

LUZ	Criteria	Farm type 1	Farm type 2	Farm type 3
1	Topography	Steep	Gentle	Flat
	Intensification	High	Medium	Low
	Management	High	Medium	Poor
	Input use	High	Medium	Low
	Resource endowment	High	High	Medium
2 and 3	Topography	Steep	Gentle	Steep
	Intensification	High	Medium	Low
	Management	High	Medium	Low
	Input use	High	Medium	Low
	Resource endowment	High	Medium	Medium
4	Intensification	Medium	Low	Very low
	Management	Medium	Low	Low
	Input use	Medium	Low	Low
	Resource endowment	Low	Low	Very low
5	Topography	Gentle	Steep	Gentle
	Level of manure use	Medium	Low	None
	Distance of cultivated parcel from homestead	0 km	2 km	6 km
	Resource endowment	Medium	Medium	Low

Primary data collection included soil samples, nutrient contents of products, and market prices, whereas secondary data included soil maps, agroclimate data and relevant research results. Non-traded goods and family labour opportunity costs were estimated based on average market rates.

Data Processing

To facilitate data processing, an accounting-type Farm-NUTMON model was developed (Van den Bosch *et al.*, 1998a; Fig. 11.2). Farm-NUTMON estimates nutrient balances and economic performance indicators at farm and activity level. Within the farm, three types of units are distinguished: crop activities (primary production units), livestock activities (secondary production units) and the farm households. Farm-NUTMON calculates farm-level balances for nitrogen, phosphorus and potassium from five inflows (mineral fertilizer (IN 1), organic manure (IN 2), wet and dry deposition (IN 3), biological nitrogen fixation (IN 4) and sedimentation (IN 5)), six outflows (removal of crop and animal products (OUT 1), crop residues (OUT 2), leaching (OUT 3), gaseous losses e.g. denitrification (OUT 4), water erosion (OUT 5), and human faeces (OUT 6), and six internal flows (animal feeds (FL 1), household waste (FL 2), crop residues (FL 3), grazing (FL 4), animal manure (FL 5), and farm products to household (FL 6)). Also, it performs an economic analysis at farm and activity level based on the same data set (Fig. 11.3). At activity level, cash flows (cash income minus cash receipts) and gross margins (returns minus variable costs) per unit area are calculated as measures of profitability. At farm level, net farm income (total gross margins minus fixed costs) and family earnings (net farm income plus off-farm income) are determined as well as a number of performance criteria related to land and labour returns. The economic performance indicators were analysed using basic descriptive statistical techniques.

Quantification of Nutrient Flows

Nutrient flows in Farm-NUTMON were quantified in three different ways: by asking the farmer, through the use of transfer functions and other approaches using sub-models and assumptions.

Flows directly related to farm management were, as much as possible, quantified by farmers during the monthly monitoring (chemical and organic fertilizer use, harvest of crop and animal products, redistribution of crop residues and farmyard manure, etc.). Nitrogen (N), phosphorus (P) and potassium (K) contents of the nutrient carriers were determined in the laboratory. Atmospheric deposition, gaseous losses, leaching and erosion were quantified fully on the basis of off-site knowledge using transfer functions. Some flows were estimated by means of a simple sub-model. For example,

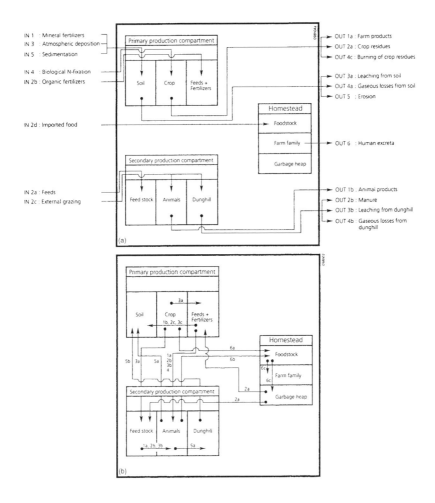

Fig. 11.2. (a) The farm concept with its compartments and subcompartments, and the nutrient flows into and out of the farm. (b) The farm concept with its compartments and subcompartments, and the internal nutrient flows.

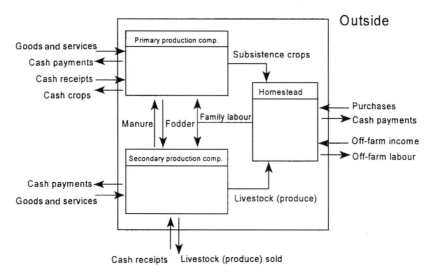

Fig. 11.3. Farm concept with its economic flows as used in ECCAL.

on a pasture, different types of feeds can be available, such as crop residues and household waste, and different animals can be grazed on this pasture with different stocking rates. The actual consumption of crop residues, household waste and grasses on the pasture by the different groups cannot usually be estimated by the farmer, neither can the quantities of manure excreted by the different groups. The quantity of nutrients taken in can be based on farmers' information on the average stocking rates, literature data on daily intake requirements, and the nutrient contents of the forages. Based on the calculated consumption rates, excretion can be estimated on the basis of feed conversion factors. Nutrient inputs due to irrigation/sedimentation and deep capture of deep rooting crops were not considered due to lack of information. Nutrient losses due to erosion (OUT 5) were quantified based on the definition of five classes for steepness of the area of the farm. Average soil loss in kg ha^{-1} year^{-1} was attached to each class based on literature studies in Kenya. Resulting soil loss data were multiplied by N, P and K contents in the soil.

RESULTS AND DISCUSSION

Land Use Zone 1

Nutrient monitoring
The nutrient budgets for nitrogen (N), phosphorus (P) and potassium (K) are presented in Tables 11.4–11.6. The results show that farms in LUZ 1 experience a net negative N balance (–119 kg ha^{-1} year^{-1}), a positive P

Table 11.4. Nitrogen balances (kg ha^{-1} year^{-1}) on yearly basis at farm level for all LUZs.

Nitrogen	LUZ 1	LUZ 2	LUZ 3	LUZ 4	LUZ 5
IN 1	75	41	21	19	0
IN 2	19	40	12	3	65
IN 3	6	5	5	5	4
IN 4	3	3	1	0	1
IN 5	0	0	0	0	0
OUT 1	−33	−14	−34	−3	−1
OUT 2	0	0	0	0	0
OUT 3	−102	−84	−54	−25	−11
OUT 4	−47	−24	−24	−10	−3
OUT 5	−29	−46	−50	−8	−10
OUT 6	−10	−5	−3	−8	−4
BALANCE	−119	−83	−126	−26	43

Table 11.5. Phosphorus balances (kg ha^{-1} year^{-1}) on yearly basis at farm level for all LUZs.

Phosphorus	LUZ 1	LUZ 2	LUZ 3	LUZ 4	LUZ 5
IN 1	40	23	19	119	0
IN 2	4	2	2	0	6
IN 3	1	1	1	1	1
IN 4	0	0	0	0	0
IN 5	0	0	0	0	0
OUT 1	−2	−2	−4	0	0
OUT 2	0	0	0	0	0
OUT 3	0	0	0	0	0
OUT 4	0	0	0	0	0
OUT 5	−11	−18	−18	−5	−5
OUT 6	0	0	0	0	0
BALANCE	32	6	−14	15	2

balance (32 kg ha^{-1} year^{-1}) and a negative K balance (−17 kg ha^{-1} year^{-1}) at farm level. Tea is the main crop enterprise in this LUZ occupying about 50% of the total farm area. Napier grass is the second most important crop occupying up to 25% of the total area. The rest of the farm area is normally planted with other minor crops such as maize, beans, Irish potatoes,

Table 11.6. Potassium balances (kg ha^{-1} year^{-1}) on yearly basis at farm level for all LUZs.

Potassium	LUZ 1	LUZ 2	LUZ 3	LUZ 4	LUZ 5
IN 1	24	11	0	18	0
IN 2	11	28	5	3	78
IN 3	3	3	3	3	3
IN 4	0	0	0	0	0
IN 5	0	0	0	0	0
OUT 1	−18	−7	−30	−3	0
OUT 2	0	0	0	0	0
OUT 3	0	0	0	0	0
OUT 4	0	0	0	0	0
OUT 5	−30	−67	−80	−20	−20
OUT 6	−7	−3	−2	−5	0
BALANCE	−17	−35	−104	−4	60

pasture, kale, and so on. The results reveal that harvested crop products, namely tea, napier, maize, beans and wood, contribute about −33 kg ha^{-1} year^{-1} of N that is depleted from the soil in this LUZ (Table 11.4). High N losses are mainly determined by leaching (−102 kg ha^{-1} year^{-1}) as well as gaseous losses (−47 kg ha^{-1} year^{-1}). Nutrient budgets for P in LUZ 1 (Table 11.5) are positive. These positive P budgets have, however, no relationship with the availability of this element to the plants. The andosols which are found in this LUZ contain high Fe and Al (hydr)oxides which make P unavailable to most plants. Potassium nutrient budgets in LUZ 1 are negative (Table 11.6). This is mainly as a result of high levels of erosion due to steeply sloping areas typical of this LUZ.

Economic performance

The economic performance of farms in the various LUZs is shown in Table 11.7. In LUZ 1, the average net farm income amounts to KSh 74,600 per farm year^{-1}. The individual crop activities tea, napier grass and maize/beans intercrop have average gross margins of KSh 278,135; 55,200 and 33,639 ha^{-1} year^{-1}, respectively. Land allocated to tea is the largest, with an average of 0.5 ha per farm, with the rest of the crops such as maize, beans and vegetables being allocated very small plots with an average of 0.05 ha. Dairy performance in this LUZ is variable with an average gross margin of around KSh 21,000 year^{-1}. Crops contribute 74% while livestock contributes the remaining 26% of the net farm income, with little or negligible off-farm income. There is a large variation in the household cash flow with an average of household net cash flow of KSh 45,100 per farm year^{-1}.

Table 11.7. Economic performance indicators.

Performance indicator	LUZ 1	LUZ 2	LUZ 3	LUZ 4	LUZ 5
Net farm income (KSh per farm year^{-1})	74,600	47,700	141,300*	39,000	94,300
Crops contribution to farm income (%)	74*	72*	71*	78	53
Livestock contribution to farm income (%)	26*	28*	29*	22	47
Off-farm income (KSh per farm year^{-1})	11,000	3,400	0	900	5,400
Farm earnings (KSh per farm year^{-1})	85,600	51,100	141,300*	39,900	99,700
Farm net cash flow (KSh per farm year^{-1})	45,100	45,900	134,900*	23,400*	7,500

Note: *indicates a large variation within the LUZ.

Land Use Zone 2

Nutrient monitoring

Nutrient monitoring results for LUZ 2 are shown in Tables 11.4–11.6. Addition of N to the soils in this LUZ occurs through use of N:P:K fertilizers 17 : 17 : 0 and 20 : 20 : 0 in coffee, and 25 : 5 : 5 + 5S in tea. In addition, manures and animal feed concentrates are also used in some farms. However, there is a net loss of N in this LUZ (Table 11.4). The losses in this LUZ are, however, lower than those found in LUZ 1. About –83 kg ha^{-1} year^{-1} of N is lost from most farms found in this LUZ. The highest level of N depletion (–84 kg ha^{-1} year^{-1}) occurs through leaching, whereas gaseous losses account for –24 kg ha^{-1} year^{-1}. Harvested crop products account for a net loss of –14 kg ha^{-1} year^{-1} of N.

Farms in LUZ 2 exhibit a positive P and negative K balance of 6 and –35 kg ha^{-1} year^{-1}, respectively (Tables 11.5 and 11.6). The main sources of P and K are fertilizers, manures and purchased napier grass in some of the farms.

Economic performance

Table 11.7 shows the performance of farms in various LUZs. Coffee and tea take an average of 0.5 ha per farm in this land use zone, with a range of 0.2–1.2 ha. The net farm income in this LUZ is KSh 47,700 per farm year^{-1}. The crop activities coffee and tea have average gross margins of KSh 48,000 and KSh 30,000 ha^{-1} year^{-1}, respectively. In this land use zone more land is allocated to food crops as compared with land use zone 1. Dairy production performs relatively well in a few farms. Crops activities contribute more to the total net farm income than livestock, with respective shares of 72% and 28%. Over 60% of the farmers in this land use zone do not have any source of off-farm income. Farm net cash flow varies within the land use zone with an average of KSh 45,900 per farm year^{-1}.

Land Use Zone 3

Nutrient monitoring

LUZ 3 consists predominantly of a coffee and maize-based cropping system. As one approaches LUZ 4 the proportion of land occupied by coffee decreases and the proportion of maize increases (Fig. 11.1). Other minor crops include bananas, beans and napier grass. The results of two of the farms monitored and analysed are presented in Tables 11.4–11.6. The results show negative balances of N (–126 kg ha^{-1} year^{-1}), P (–14 kg ha^{-1} year^{-1}) and K (–104 kg ha^{-1} year^{-1}. Highest levels of N depletion occur through leaching (–54 kg ha^{-1} year^{-1}) whereas harvested crop products and gaseous losses account for –34 and –24 kg ha^{-1} year^{-1}, respectively. The main difference in nutrient balance between this LUZ and those of LUZ 1 and 2 appears to be with regard to K which shows a high negative balance of –104 kg ha^{-1} year^{-1}. These differences in K balances may be explained by the fact that K-based fertilizer, 25 : 5 : 5 + 5S, is used in tea cultivation. It is noteworthy that the majority of the farms in LUZ 3 are located in areas which show an equilibrium or zero P and a negative K balance. These are the farms near LUZ 4. Such farms tend to be dominated by the growing of food crops, especially maize, where usage of fertilizers is not common.

Economic performance

The proportion of land allocated to coffee and maize, which are the main crops in this land use zone, varies from one farm type to another. Farm types with intensive livestock management with pure dairy breeds have more land allocated to coffee production while farm types with semi-intensive livestock management with cross breeds have most of the land (over 70%) allocated to maize production and other food crops such as beans, vegetables and bananas. Farm types with an intermediate level of livestock management have almost equal proportions of land allocated to coffee and maize with other food crops like beans, vegetables and bananas.

Economic performance of both crops and livestock show a large variation within the land use zone (Table 11.7). Dairy production contributes highly to the farm earnings from livestock in this land use zone.

On average, crops and livestock contribute around 70% and 30%, respectively, of the total farm earnings but with a considerable variation among farms. The net farm income and family earnings also show a large variation within the land use zone. A farm with highly managed zero-grazing livestock realizes a net farm income of KSh 253,886 per farm year^{-1}, against KSh 28,814 on a farm with an average managed semi-zero grazing livestock system.

Land Use Zone 4

Nutrient monitoring
In this LUZ the production system consists predominantly of tobacco/food crops with or without the presence of livestock. The results (Tables 11.4–11.6) indicate a loss of N (–26 kg ha^{-1} year^{-1}), with a positive P (15 kg ha^{-1} year^{-1}), and a near equilibrium K nutrient balance (–4 kg ha^{-1} year^{-1}). Crops grown for subsistence in this LUZ include maize, beans, bananas, sweet potatoes, cowpeas as well as sorghum. Tobacco is the main cash earner for almost all the farms in this LUZ. Like other cash crops in LUZ 1 and 2, tobacco is well fertilized using either 17 : 17 : 17 or 6 : 18 : 20 type of fertilizers. Thus, the net accumulation of K in most farms in this LUZ may therefore be attributed to the use of fertilizers rich in K for tobacco growing. However, tobacco does not contribute to any major outflow because the leaf yields are relatively low, approximately 800 kg ha^{-1} of dry leaves year^{-1}.

Economic performance
The main crop activities in this land use zone are bananas, sweet potatoes, beans, maize and tobacco. The only crop grown for market (cash crop) is tobacco. Other crops are grown for subsistence with surplus being sold to the local markets. There are significant differences in economic performance within the land use zone. The households with only poultry (without cattle and goats) have a relatively poor economic performance compared with the others. Most of the households in this land use zone do not have an off-farm income and for those which have, the amount of off-farm income is negligible.

Land Use Zone 5

Nutrient monitoring
This is a LUZ with extensive livestock grazing/subsistence crops production systems. There is no usage of any form of external inputs for all the crops grown in this LUZ. All the crop-based activities are for the production of family food. Sale of crop-based farm products in this LUZ is rare.

The main crops grown in this LUZ include maize, cowpeas, sorghum, pearl millet, green grams and pigeon peas. The main types of livestock kept in this LUZ are the indigenous Zebu cattle, the small East African goat as well as local chicken. Both cattle and goats browse in the natural forages during the day and spend the night in a common kraal found within the farmyard. The results (Tables 11.4–11.6) show a system which is at equilibrium for P (2 kg ha^{-1} year^{-1}) and positive for N (43 kg ha^{-1} year^{-1}) and K (60 kg ha^{-1} year^{-1}) budgets. There is a high build up of N and K stocks in farms in this LUZ due to browsing of livestock in the communal grazing land. Lack of land consolidation appears to be a common feature in farms in this LUZ. Therefore, manure which accumulates in the common goat and cattle kraal cannot be applied to crops grown in far-away parcels due to lack of transport. Thus, the high N balance in the farms is as a result of these huge stocks of manure which pile up in the kraals for a period of 10–20 years.

Economic performance

There is an almost exclusive subsistence production in this land use zone. The main crops grown are green grams, cowpeas, maize, pigeon peas, millet and sorghum. Most of the crop production here is mixed cropping with little or no monocropping. The proportion of land allocated to various crop activities within a household is almost the same but there are variations within the land use zone. There appear to be small variations in terms of economic performance within the land use zone. This is reflected in the gross margins that are not significantly different between households for activities that are almost similar, except for green grams. Green grams are performing relatively well in the entire land use zone.

Livestock performance in this land use zone is poor compared to the other land use zones. However, in terms of contribution to the farm earnings within the land use zone, livestock contributes more than crops (Table 11.7). Livestock types such as goats, poultry and sheep contribute substantially to the farm earnings in this LUZ, unlike in other land use zones. However, lactating cattle (milk) contribute most of the farm earnings in relation to the other livestock types, just like in the other land use zones.

Net farm incomes vary within the land use zone. However, the variation is not as large as in the other land use zones. Off-farm income in this land use zone is negligible, hence family earnings here are almost the same as the net farm income. There are low sales of farm produce in this LUZ and households here experience cash flow problems (Table 11.7).

RELATIONS BETWEEN THE NUTRIENT BALANCE AND ECONOMIC PERFORMANCE

There appears to be a negative correlation between economic performance at the activity level and the net nutrient balance (N, P, K) within a land use zone. From the analyses, most of the activities with high gross margins have low net nutrient balance and vice versa. There appears to be a relatively stronger inverse relationship between net nitrogen (N) balance and

Table 11.8. Economic performance at the activity level and the net nutrient balance for Farm type 1 in LUZ 1.

Activity	Gross margins (KSh ha^{-1} year^{-1})	Net nutrient balance (kg ha^{-1} year^{-1})		
		N	P	K
Tea	127,905	16	−144	−19.3
Maize/beans	33,639	−104	−5.9	−43

performance at the activity level (Table 11.8). Results for land use zones 2, 3, 4 and 5 follow the same trend of net nutrient balance and economic performance relationship at the activity level as LUZ 1, shown in Table 11.8.

No relationship was observed between either net farm income, family earnings or household net cash flow with net nutrient balance (N, P, K) at the farm level.

There is a relationship between the use of chemical fertilizers (IN 1), organic manure (IN 2) and household net cash flow. Farm types with a better cash flow appear to use either more of both chemical and organic fertilizers, or more chemical fertilizers alone (Table 11.9).

CONCLUSIONS

Nutrient monitoring results at farm level, from all five LUZs indicate a net negative N balance in all LUZs except LUZ 5. The highest levels of nutrient depletion are found in N. Leaching is the major source of N loss especially in high rainfall zones such as LUZ 1 and 2. For P and K the results across all LUZs show a mixed picture. Apart from LUZ 3, positive balances for P are found in all LUZs, while for K both strong negative (LUZ 3) and strong positive balances (LUZ 5) were found. In LUZ 1 and 2 farmers combine cash crops of coffee and tea with intensive zero-grazing dairy. Therefore mineral fertilizer use and feed imports are high. In LUZ 5, browsing of cattle and goats in the natural forages contribute to a high build up of N, P and K at farm level. Farm types which are dominated by the growing of food crops (maize, beans, cowpeas, sorghum, pearl millet, bananas, etc.) show a trend of strong mining of nutrients at plot and, in some cases, at farm level. This mining is attributable to low application levels of fertilizers due to high prices of fertilizers coupled with low output prices and lack of credit facilities for these subsistence crops for the purchase of manures and fertilizers. A major drop in the price of any cash crop, for example coffee in the 1980s, could also lead to a net outflow of nutrients. Crop residues and manures are

Table 11.9. Household cash flow and use of inorganic and organic fertilizers.

LUZ	1			2			3			4			5		
	N	P	K	N	P	K	N	P	K	N	P	K	N	P	K
Inorganic fertilizer (kg ha^{-1} year^{-1})	63.9	18.5	9.0	74.0	40.0	23.0	36.0	4.0	0	10.2*	10.38*	9.6*	0.0	0.0	0.0
Organic fertilizer (kg ha^{-1} year^{-1})	4.2	0.8	2.7	52.7	43.0	64.6	40.4	25.0	55.0	2.0	0.9	0.9	31.1	3.3	34.2
Net cash flow (KSh year^{-1})	45,100			45,900			134,900			23,400			7,500		

Note: *indicates that the high levels of inorganic fertilizers (provided on a credit basis) are exclusively used in tobacco production activity.

hardly transferred between farms. The highly negative balances of LUZ 1, 2 and 3 are caused by both high outputs in terms of farm products and high losses via leaching, gaseous losses and erosion. Both farm products leaving the farm and losses are much less in LUZ 4 and 5, resulting in less negative and even positive balances.

Economic performance shows large variations between farms within a LUZ and between LUZs. The market orientation and resulting net farm cash flow is relatively low in LUZ 5 compared with the other LUZs. In LUZ 1–4 the share of livestock activities ranges between 22 and 29%, but in LUZ 5 livestock plays a much more important role with 47% of the total farm income. There is an inverse relationship between gross margins and the nutrient balances at the activity level, especially for the nitrogen balance. Households with higher cash flows appear to have more positive nutrient balances due to application of chemical and organic fertilizers. The type of grazing system also influences importation of nutrients from organic fertilizers. Farms with an extensive grazing system outside the farm have better nutrient balances from organic materials compared with the farms with livestock feeding within the farm.

REFERENCES

De Jager, A., Kariuki, I., Matiri, F.M., Odendo, M. and Wanyama, J.M. (1998) Monitoring nutrient flows and economic performance in African farming systems (NUTMON). IV. Linking nutrient balances and economic performance in three districts in Kenya. *Agriculture, Ecosystems and Environment* 71, 81–92.

Jaetzold, R. and Schmidt, H. (1983) *Farm Management Handbook of Kenya. Vol. II/C. East Kenya.* Ministry of Agriculture, Nairobi, Kenya.

Munyi, V., Mills, B. and Nandwa, S. (1995) *Characterization of Soil and Water Management Systems and Farmers' Perceptions; Embu Case Study.* Kenya Agricultural Research Institute, Kenya.

Smaling, E.M.A., Stoorvogel, J.J. and Windmeijer, P.N. (1993) Calculating soil nutrient balances in Africa at different scales. II. District Scale. In: Smaling, E.M.A. An agroecological framework for integrated nutrient management with special reference to Kenya. Doctoral thesis, Agricultural University, Wageningen, The Netherlands, pp. 93–120.

Stahl, M. (1993). Land degradation in East Africa. *Ambio* 22 (8), December.

Stoorvogel, J.J. and Smaling, E.M.A. (1990) *Assessment of Soil Nutrient Depletion in Sub-Saharan Africa; 1983–2000. Vol. III; Literature Review and Description of Land Use Systems.* Report 28, The Winand Staring Centre, Wageningen, The Netherlands, 137 pp.

Van den Bosch, H., De Jager, A. and Vlaming, J. (1998a) Monitoring nutrient flows and economic performance in African farming systems (NUTMON). II. Tool development. *Agriculture, Ecosystems and Environment* 71, 49–62.

Van den Bosch, H., Gitari, J.N., Ogaro, V.N., Maobe, S. and Vlaming, J. (1998b) Monitoring nutrient flows and economic performance in African farming systems (NUTMON). III. Monitoring nutrient flows and balances in three districts in Kenya. *Agriculture, Ecosystems and Environment* 71, 63–80.

Nitrogen Flows through Fisheries in Relation to the Anthropogenic N cycle: a Study from Norway

12

M. Azzaroli Bleken

Department of Crop Sciences, Agricultural University of Norway, PO Box 5022, N-1432 Ås, Norway

ABSTRACT

A thorough study of nitrogen (N) cycling through Norwegian fisheries in 1988–1991 is presented. At most, 28% of the N in fish catch was recovered in edible fish products. Farmed fish contributed a further 4%. In addition 3% was presumably recovered in milk or meat by agricultural animals receiving fish meal in their feed. About half of the Norwegian marine catch was used by the oil and feed industry, which also received by-products from fish for human consumption. About 15% was dumped as waste, and 40% of the N was lost from fish cages (fish secretion and feed loss) and droppings by agricultural animals. Only about 19% of the N supply to farmed fish (carnivorous) was found in edible fish parts. This is a better recovery than by Norwegian beef production, but it is equal to that of pig and lower than poultry. This study shows that there is a link between the way we exploit the fish and global anthropogenic nitrogen pollution. A better use of marine catch can considerably reduce the amount of fertilizer needed to supply animal proteins for human consumption. In this respect, efforts to maximize the utilization of fish catch directly as human food are more effective than comprehensive recycling of waste products as feed.

A tentative budget of anthropogenic N contributions and removal to the North Atlantic was estimated. Removal by fish catch was only 2–3% of the total N input derived from human activity. Although a comparison between input to sea (mainly inorganic N) and removal (living animals) cannot be straightforward, clearly N depletion by sea catch is not an argument for continued N pollution of the sea.

INTRODUCTION

No other products from a natural, non-cultivated environment contribute to human nutrition as much as marine fish. At present they provide more than 10% of the animal proteins of the world food supply (FAO, 1997). The total catch of marine fish has increased from about 35 million metric tonnes live weight at the beginning of the 1960s to 80–85 million tonnes 30 years later (FAO, 1997). However, the natural fish supply is limited and many stocks are already declining due to excessive fishing. Also anthropogenic emissions of nitrogen salts are threatening fish stocks as a consequence of the eutrophication of closed basins and coastal waters (Baden et al. 1990; Cederwall and Elmgren, 1990; Rosenberg et al., 1990). On the other hand, there is an increasing interest in intensifying fish production by expanding primary production (algae) through nitrogen (N) fertilization of sea waters. This can be seen as an extension of the rapid development in aquaculture during the two last decades.

It is also argued that considerable amounts of nutrients are removed from the sea by fisheries, and therefore the anthropogenic supply of N salts to marine environments is necessary to counterbalance this removal. For some fish communities, leaching from agriculture to coastal areas could be beneficial (Hansson and Rudstam, 1990), at least within certain limits. There is, however, a strong argument against a global enrichment of reactive N (Vitousek, 1994; Galloway et al., 1995) in as much as any form of reactive N available in the biosphere is a potential source of nitrous oxide (N_2O) emission (Duxbury et al., 1993; Granli and Bøckman, 1994; Kroeze, 1994). N_2O contributes to global warming (Subak et al., 1993) and interacts with the chemistry of ozone in the stratosphere (Prinn, 1994). The concentration of N_2O in the stratosphere was constant before industrial times; since then it has been increasing at an accelerating rate (Khalil and Rasmussen, 1988; Dibb et al., 1993).

Agriculture is the main sector responsible for the global increase in reactive N, besides combustion of fossil fuels (Galloway et al., 1995). To sustain animal production, large amounts of nitrogenous fertilizers are necessary for primary production (plants). Even with optimal recycling of animal waste and by-products, less than 10% of the N input as plant nutrients is recovered in the edible animal products (Bleken and Bakken, 1997a). Since marine fish can advantageously replace other meat in the human diet, a proper use of sea catch could contribute to reducing the overall leaching of nitrogen salts from agriculture.

This study analyses the N flows through Norwegian fisheries and aquaculture to estimate the N recovery in edible food products based on marine catch. Nitrogen content in food is closely related to the amount of protein (1 g N ≈ 6.25 g protein). An attempt is made to evaluate the effect of the fisheries sector on the global nutrient balance of marine waters. This study is necessarily descriptive.

METHODS

Average annual nitrogen flows from Norwegian sea catch and imports of fish products to the final consumer have been estimated for the period 1988–1991. Amounts are given in Gg N year^{-1} (Gg = 10^9 g).

This study is based on Norwegian Official Statistics (NOS, SSB, 1992 and 1994) and FAO statistics (FAO, 1996), and on direct information from industries in the sector. Private fishing is not taken into account. Volume of trade imports and exports are based on toll declaration (Norway Statistics, Oslo, 1994, personal communication). The Institute of Aquaculture Research Ltd (E. Austreng, Ås, 1996, personal communication) has contributed the data about N-balance for aquaculture (N input in feed and N output in products and waste). The RUBIN foundation, constituted in 1991 to promote the exploitation of waste, has provided detailed accounts of the amount and utilization of offal (viscera, heads and all other parts discarded by processing). However, the different data sources were not always compatible with respect to time and items grouping as required by this study. The necessary adaptations are the author's responsibility. Data from separate sources have been extensively cross-checked.

Catch is used in the sense of nominal catch (FAO, 1996), thus as retained catch (without catch losses and dead catch). An average N concentration of 3% of fresh weight has been used for live fish and its components (fillet and offal). Nitrogen concentration in other sea animals and processed products, such as dried fish and smoked fish are taken from Blaker and Rimestad (1994), and those for fish meal were provided by the Norwegian Agricultural Inspection Service (Sundstøl et al., 1991).

As discussed below, high quality fish for consumption is largely gutted at sea; consequently, live weight was estimated using coefficients characteristic of each species (SSB, 1994).

RESULTS

Norwegian Sea Catch, 1988–1991

About 63 Gg N were taken annually from the sea (Table 12.1). This estimate includes catch from foreign ships landing in Norway as well as imports of fish-based feed and small amounts of food imports, the latter calculated as round fish (live weight). Sea catch included crustaceans, mainly deep water prawns, and molluscs (together about 4% of the total N removal), and seaweed (0.5 Gg N year^{-1}). Large amounts of seaweed are gathered along the coast, but due to the low N content their contribution to the amount of N harvested from the sea is small.

Table 12.1. Nitrogen in marine catch (round fish) by main group of sea product landed in Norway and imported, in Gg N year^{-1} (averages of 1988–1991). Imports consisted mainly of feed or of products for the fish meal industry.

Product	N (%)	Gg N year^{-1}
Seaweed	0.25	0.5
Cod fish	3.0	12.3
Herring and sprat	3.0	8.1
Mackerel, capelin	3.0	28.1
Crustaceans[a]	3.7	2.0
Other	3.0	3.4
Imported products	—	9.0
Total		63.4

Note: [a]Mainly shrimps.

Striking aspects of the sea catch are the large annual variability in total volume and species composition. For example in 1992 it was about one-third larger than the average of the 1988–1991 period.

Roughly half of the catch was used directly as a source of food for human consumption, while the other half was used by the oil and fish meal (animal feed) industry. Figure 12.1 summarizes the complex N flows from sea catch to the final products (food for human nutrition), either as fresh and processed fish, or recovered in products from agricultural animals. Recycling or dumping of offal is also shown. These flows will be presented in the following sections.

Norwegian Fish for Human Consumption

Overall 30 Gg N year^{-1} of the catch were used for human consumption (Food fish in Fig. 12.1). Including 4.3 Gg N from fish farming, the total N in round fish used as food supply was about 34 Gg N year^{-1}.

The amount of offal dumped has been estimated as the difference between total by-products (10.5 Gg N year^{-1}) and the amount known to be recycled (5.0 Gg N year^{-1} from wild fish and 0.3 Gg N year^{-1} from farmed fish). Herring and mackerel were usually landed as round fish, often close to oil-and-meal industries, and therefore waste from fillet production could conveniently be utilized for meal production. Cod and related species were gutted before landing. Also heads were usually cut (a Norwegian peculiarity). This resulted in a large amount of by-products, about one-third of the round weight, which traditionally have only partly been utilized. Cod-type species are caught either by large trawlers in pelagic waters or by small boats using lines in coastal waters. Dumping of waste was estimated to

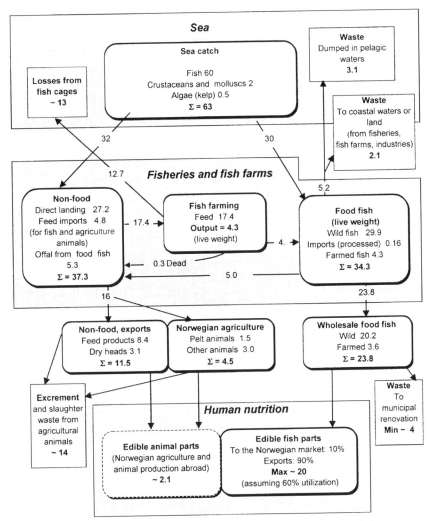

Fig. 12.1. Idealized N flows from the sea (top), through Norwegian fisheries (centre), to edible products (bottom), in Gg N year^{-1} (Gg = 10^9 g). Averages of 1988–1991. The Non-food box (top left) represents the total source of raw material available to the oil-and-meal industry, including by-products from fish processing. Fish farming (centre) receives N from the feed industry. Food fish (centre right) includes live weight of wild and farmed fish. Wholesale food fish is partly processed and cleaned, but it still includes some non-edible parts. Human nutrition is a maximum estimate of N in *edible* fish parts, as well as *edible* animal parts derived from the use of fish as feed. Small boxes with sharp corners represent losses from the food producing system. See text for full explanation.

occur in pelagic waters (3.1 Gg N year^{-1}) or close to/on land (Fig. 12.1) depending on the fishing method.

The yield of wholesale products varies greatly between fish species and processing procedure. As an example, the yield of herring-fillet is 50% of round weight, while that of pickled/salted herring is 35%, leaving 65% in by-products used by the meal-and-oil industry (RUBIN, 1993, 1995a). Mackerel is usually processed with head, either fresh, frozen or smoked, and thus little by-product occurs. The yield of cod fillet is only 33% of live weight and, as mentioned before, offal was normally not utilized by the meal industry.

Of 24 Gg N year^{-1} in gutted and partially processed fish for wholesale, 21.5 Gg N year^{-1} were sold abroad (based on the inventory of exported commodities) and 2.3 Gg N year^{-1} were available for the domestic market (average 1989–1991). The amount of N really available as food was much less. The yield of fresh fillet is estimated to vary between 30 and 60% of live weight, the latter for fat fish such as mackerel and salmon. Assuming an optimal average yield of 60%, including roe, an upper estimate of the net supply of edible parts for human consumption was about 20 Gg year^{-1} (Edible fish parts at bottom of Fig. 12.1), of which about 13% was provided by fish farming (Fig. 12.1). The actual supply was probably between 10 and 20 Gg N year^{-1}. Most (90%) of the products for human consumption was exported.

The difference between N in wholesale fish and that in edible parts was assumed to be collected as municipal waste.

Fish Used for Feed Production

The oil and meal industry received 27 Gg N year^{-1} directly from marine catch and about 5 Gg N year^{-1} as by-products from food fish. Trade import of fish-based feed and other commodities supplied a further 4.8 Gg N year^{-1}. Thus about 37 Gg N year^{-1} were available to the feed industry (Non-food in Fig. 12.1). About 30% of the product was exported. A total of 17.4 Gg N year^{-1} was used by Norwegian fish farming. A further 3 Gg N year^{-1} were added to concentrated feed for agricultural animals in Norway, and small amounts were used as silage or wet by-products for raising pelt animals (Fig. 12.1) and as bait (not shown). An unaccounted for remainder of 3 Gg N year^{-1} is considered acceptable, given the large uncertainties connected to all estimates of fish catch and fish products.

Fish Farming

Norwegian fish farming involves exclusively carnivorous fish, feeding on wild fish from sea catch, hence the N balance of fish farming is an integrated part of the N cycle of marine fish. The fish farming industry has developed rapidly during the last 15 years, both in volume and technology. The product is mainly high value fat fish of the salmon family, but other species are also becoming important. Farms are usually situated in coastal waters close to traditional landing places for wild fish, always in places with a strong current. This sector has received much attention from the public and the authorities due to local pollution problems (excrement, lost feed, dead fish etc.), and its environmental performance has improved rapidly.

On average, during the period 1988–1991, fish feed contained about 17.4 Gg N year^{-1} and the round weight production including blood was 4.3 Gg N year^{-1} (4.1 Gg N after blood loss), thus the average N-cost was 4.0 (N-cost: N in feed divided by N in products) (Table 12.2). The amount of N in fish available for wholesale after gutting and partial processing was 3.6 Gg N year^{-1}. Fillet yield is 60% of round weight, and all edible parts make up about 65%. Thus, the N in edible products from fish farming was at most $4.3 \times 0.65 \approx 2.8$ Gg N year^{-1}, and the corresponding N-cost (feed/wholly edible parts) was 17.4/2.8 = 6.2.

The amount of offal was about 1 Gg N year^{-1} (based on utilization of the product in 1991, RUBIN, 1995a), including dead fish. Only about 65% of the offal was utilized, mainly raw or as silage for agricultural animals. Due to improvements in sanitary conditions in the cages, the percentage of by-products utilized is increasing.

The lower N-cost in 1993 (Table 12.2) reflects the improvement of the feeding which is presently taking place. However, the amount of fish stock left in the cages at the end of the year varies greatly, and is difficult to

Table 12.2. N budget (Gg N year^{-1}) for Norwegian fish farming and N recovered in product for sale as live weight (round fish), gutted fish (wholesale product) and edible parts. N-costs are calculated for the process from feed to fish or to edible parts.

	1988–1991	1993
Feed	17.4	19.0
Slaughtered fish (live weight minus blood)	4.1	5.4
Dead fish in good condition (recycled to pelt animals)	0.3	?
Total nitrogen loss from cages, including blood	12.3	13.3
Live weight, including blood (c. 5% of total N)	4.3	5.7
Wholesale fish, gutted	3.6	4.6
Offal	0.7	1.1
N-cost: feed/live weight	4.0	3.4
N-cost: feed/wholesale fish	4.8	4.1
N-cost: feed/edible parts (65% of live weight)	6.2	5.2

assess. Therefore comparisons between single years should be made with caution.

The difference between input as feed and recovery in live fish represents losses from fish cages (about 13 Gg N year^{-1}, Fig. 12.1 upper left) as secreta and unutilized feed, and also includes fish escaped from cages.

Overall Contribution to Human Nutrition, and Total Losses

As mentioned before, an upper estimate of the edible fish products supplied by Norwegian catch for human consumption is around 20 Gg N, including farmed fish. Since fish meal is used in agriculture as feed concentrates, edible products from livestock should be included as well. Altogether about 14.5 Gg N were available in by-products and fish meal for feeding livestock (excluding 1.5 Gg N fed to pelt animals, see Fig. 12.1). A tentative estimate of their contribution to the food supply can be found by using the ratio N-in-feed/N-in-edible-animal-products (meat lean tissues, dairy products and eggs) for the whole animal husbandry in Norway (Table 12.6). This overall N-cost was 7 kg N per kg N. Thus, a rough estimate of the contribution by fish meal to edible food supply is: 14.5 Gg N year^{-1} : 7 ≈ 2.1 Gg N year^{-1}, or around 14% of the feed input. An eventual specific effect of fish proteins on the recovery of N by the animals is not considered, but it is unlikely to be of importance for this study.

This preliminary calculation allows us to approximately distribute the amount of N in marine catch landed (or imported) in Norway (Table 12.3). During the period analysed (1988–1991), 35% at most was recovered in edible fish products and in livestock production (from agriculture). The actual human intake was certainly much lower, due to deterioration during distribution, and due to refusal by consumers. About 43% was lost from fish cages and agricultural livestock, primarily as liquid and solid excrements,

Table 12.3. Distribution (%) of total N in Norwegian landings among edible products, secretion losses from cultivated animals, dumped waste products and others (average of 1988–1991).

Edible products, wild fish and other sea animals	28
Edible products, cultivated fish (Norwegian)	4
Edible products, terrestrial animals (or cultivated fish abroad)	3
Losses from fish cages (Norwegian)	21
Losses from agricultural animals (secretion and non-edible parts)	22
Waste dumped on land, including waste from private households	10
Waste dumped in pelagic waters (highly variable)	5
Others: algae, used as bait or recovered in pelt animals	2
Unrecovered	5

and at least 15% was dumped either in sea waters or on land, including landfills and destruction plants.

N REMOVAL BY CATCH VERSUS ANTHROPOGENIC N ENRICHMENT OF THE SEA

A Norwegian Estimate

The total Norwegian anthropogenic contribution of reactive N to the sea by atmospheric emission, agriculture, fish farming, sewage systems and runoff from industries was estimated to be around 80 Gg N year^{-1} (Table 12.4), or of about the same order of size as the removal by Norwegian landings. However, most of the fish and fish products were exported and only about 10% of the final products was consumed in Norway. Thus, the share of the catch consumed in Norway was smaller by a factor of ten (at least) than the total anthropogenic emission from Norway to the sea.

A More Global Estimate

Since Norwegian N contribution to the sea represents only a small part of the emission by all countries consuming Norwegian fish catch, a budget for the whole North Atlantic could be a more appropriate alternative. Howarth *et al.* (1996) observed that the riverine N fluxes from the main temperate regions of the North Atlantic were linearly correlated with the net anthropogenic N inputs to those regions. They found that the riverine N flux per area (y) was strongly correlated ($r^2 = 0.73$) to the anthropogenic N input per area (x) according to a linear relationship $y = a + bx$, where $b = 0.2$.

Table 12.4. Anthropogenic N contributions to the North Atlantic from Norway and N removal by Norwegian fish catch (see Table 12.1). Atmospheric deposition based on data from Barrett *et al.* (1995), other contributions based on Bleken and Bakken (1997b).

	Gg N year^{-1}
Atmospheric deposition from Norwegian emissions	20
Leachates from agriculture (only net contribution to the sea)	22
Sewage (total N load to the sewage system)	22
Losses from fish cages	13
Runoff from industries	3
Sum of anthropogenic contributions	**80**
Removal by catch	**63**

Thus a suggests a basis deliverance per unit of area, and we might assume it to represent a pristine emission not depending on the anthropogenic enrichment. Further, the relationship indicates that of 10 g of anthropogenic N, 2 g are transported to the sea. The same authors estimate the total anthropogenic addition of reactive N to the watersheds of the North Atlantic basin to be about 35 Tg year^{-1} (Tg = 10^{12} g). Therefore the delivery of anthropogenic N by riverine transport can be estimated at around 7 Tg year^{-1} (35 Tg year^{-1} × 0.2).

According to Galloway *et al.* (1996), atmospheric deposition in the North Atlantic contributes 10–11 Tg N year^{-1} (Table 12.5). Deposition in this region of the world is derived mainly from anthropogenic emissions. Hence a rough estimate of the total anthropogenic contribution to the North Atlantic is around 17 Tg N year^{-1}, while fisheries take up about 0.46 Tg N (range 0.38–0.54 Tg N year^{-1}, Nixon *et al.*, 1996), thus removal by catch is only a small percentage of the anthropogenic contribution of reactive N to this basin.

DISCUSSION

Since changes in the share of different fish species will also affect the way industries utilize the catch, the results of this study are a kind of 'instantaneous picture' of the period 1988–1991. Given the variation in quantity and composition of the catch, and the difficulty in estimating the N content of products as feed and salted fish, the moderate discrepancy (5%) in this Norwegian study between the estimates of N uptake from the sea, and those of N in wholesale commodities plus known waste, is acceptable.

This study is based on official statistics of nominal catch, and it does not consider the effect of illegal practices, such as dumping of catch exceeding

Table 12.5. Major anthropogenic N contributions to the North Atlantic and N removal by catch. See text about riverine transport. Atmospheric deposition also includes contribution by natural processes, but this is probably much lower than deposition derived from anthropogenic activities.

	Tg N year^{-1}
Riverine transport of anthropogenic N from land to the sea	7.0
Atmospheric deposition to the continental shelf[a]	1.8
Atmospheric deposition to the ocean[a]	8.7
Sum of anthropogenic contributions	**17.5**
Removal by catch[b]	**0.38–0.54**

Notes: [a]Based on Galloway *et al.* (1996).
[b]Based on Nixon *et al.* (1996).

legal quotas. Therefore I presume that fish catch is likely to be somewhat underestimated.

The N recovery by fish farming (Table 12.2) is lower than published by other authors (Hillestad *et al.*, 1996), since this study applies a more restrictive definition of recovery: it considers sale fish only, as live weight, thus including blood lost during slaughtering. Hillestad *et al.* included dead fish and escaped fish in their estimates of recovery by fish farming. Direct consumption (unregistered sale) should in principle be included, but it is unlikely to be substantial. Lacking well-documented estimates, I have neglected it. This restrictive approach conforms to the one used by Bleken and Bakken (1997a).

Compared to animal production by agriculture in Norway, farmed fish showed a lower N-cost (that is, the ratio N-in-feed/N-in-edible-parts) than beef production, but about the same as pig and higher than poultry (Table 12.6). Compared to the whole livestock production by agriculture, fish farming required about 25% less feed protein to provide an equal amount of food protein. Taking into consideration that agricultural livestock is primarily fed on vegetable proteins, while Norwegian farmed fish is carnivorous, aquaculture cannot be said to be remarkably N-efficient. World aquaculture is dominated by freshwater herbivorous fish (as Chinese carp), and therefore it is based on a more straightforward food chain than salmon production in Norway.

Table 12.6. N in edible parts as percentage of N in live weight animals, and N-cost (N-in-feed/N-in-products) in 1988–1991 for Norwegian fish farming and agricultural livestock.

	% of the N in animals found in edible parts	N-in-feed / N-in-live weight	N-in-feed / N-in-edible parts
Farmed fish, this study	65	4	5.2
Ox (beef)	52	4.7	9.0
Cattle and ox (milk and beef)	a	4.2[b]	5.9[b]
Sheep (only meat)	42	9.3	22.1
Pig	71	3.7	5.2
Poultry (including egg)	55	2.3	4.2
Overall agricultural livestock			7.0[c]

Notes: [a]52% for meat (% of live weight) and 83% for dairy products (% of milk delivered from farms).
[b]Weighted average for the whole cattle and ox production.
[c]National consumption of feed concentrates and green fodder divided by wholesale supply of edible products.
Source for agricultural livestock: Bleken and Bakken, 1997a; Vatn *et al.*, 1996.

This study shows that many fish parts are lost as waste, dumped in pelagic waters or on land, while they could certainly be used as animal feed. This situation is changing rapidly in Norway due to the requirements of the Pollution Control Authority. However, it is important to recognize that not all waste can be easily utilized. For example, appropriate plants on board for processing of offal into meal are very expensive and require large trawlers, while the quality of the product is hardly competitive, and therefore non-economic (RUBIN, 1995b).

From a global use of the resources, a more direct use of fish in human nutrition would probably be more effective than *further* efforts to recycle *all* waste products as feed. This will be exemplified hereafter. The analysis of the N-costs of alternative protein sources for human consumption provides a method for evaluating the consequences of different strategies in order to increase the food supply of protein and minimize N losses to the environment. As mentioned in the introduction, human activity mobilizes reactive N at a rate comparable with that of terrestrial natural processes, and there is a general consensus that anthropogenic sources are the main cause of the atmospheric N_2O increase, and the cause of the eutrophication of fresh water systems and marine coastal waters. The use of chemical fertilizer is the main source of anthropogenic N. Marine biological production is based on natural sources of reactive N. Thus, consuming wild fish caught from the sea does not directly alter the global amount of reactive N in the biosphere. However, since fish food can advantageously substitute for proteins from terrestrial animals in the human diet, its use can indirectly reduce the amount of N fertilizer used by agriculture.

The analysis of N-costs of protein food by Norwegian agriculture (Bleken, 1997) showed that the production of 1 kg N in edible animal products (approximately 6.25 kg proteins) requires a net input into agricultural land of 12 kg N, as chemical fertilizers, atmospheric deposition and biological fixation. The effect on the N cycling in agriculture is even larger, since the total N application to Norwegian soils, including animal manure, is 16 kg N per 1 kg N in edible animal products. Thus, using marine fish instead of meat and milk, makes a considerable amount of N fertilizer unnecessary. Supplied to animals as feed, the relative 'saving' of N-fixation would be much lower, around 2.3 kg N per kg N in edible products (1.7 kg N if animal manure is recycled, based on Bleken and Bakken, 1997a). This clearly indicates the importance of utilizing fish at the highest trophic level when considering eutrophication and N_2O emissions from the use of N fertilizer in agriculture.

Considering food supply in Norway only, if present fish consumption was replaced by meat and milk, the input of anthropogenic N should increase by 24 Gg N year^{-1}, or by about 22% of the chemical fertilizer consumption in the country. Since about 15% of the world's supply of animal proteins is provided by fish (including fresh water fish, FAO, 1997), clearly this argument is of global relevance. On the other hand, marine fish resources are limited and there are already clear signs of excessive exploitation. Thus, an increase in the use of marine fish must come from better utilization, rather than from a more intensive catch. Table 12.7 presents a number of hypothetical actions applied to the results of this study. We do not know how much of the edible product is lost during conservation, transport and due to refusal by the consumers, but losses of 20% are likely. In Table 12.7 the first line indicates a reduction of these losses, for example from 15 to 5%, of the edible fish parts (as estimated in Fig. 12.1). The second line considers a moderate divergence (20%) of fish from the oil-and-fish-meal industry (feed) to the food industry. Only round fish directly used by the feed industry is considered (thus, offal from processing is not included), and it is assumed that the net yield in edible proteins is 40% of those in the fish. The calculation also takes into account the concomitant loss in yield of farmed fish. There are different opinions on the suitability of several fish species as human food, much depending on culture and traditions. There are, however, industrial techniques able to separate the fleshy mass of less favourite fish species, and process it for direct use in human nutrition. The third line estimates a possible recycling of *all* fish waste, including waste from private households, as feed. The third action is definitely the least effective.

This exercise shows the advantages of using more fish directly for human consumption, even when the percentage of edible parts is relatively low. On a world basis, about a third of marine catch is used as fish meal (used as feed) and oil (FAO, 1997); thus, there is potential for increasing the amount of fish used for human nutrition. There are, however, strong arguments in favour of fish farming, since farmed fish has a higher feeding

Table 12.7. Hypothetical actions to increase the utilization of Norwegian fish catch (average 1988–1991) for human nutrition, and total N input to agricultural soils if the same amount of proteins was provided by agricultural animals (1 kg N = 6.25 kg proteins). All quantities in Gg N year^{-1}.

Action	Amount diverted	Increase in food supply, only edible parts	Decrease in N application to agriculture soils
Increasing the utilization of edible fish parts by 10%	2.0	2.00	32 Gg N year^{-1}
20% of the feed-fish diverted to human consumption[a]	5.4	1.8[a]	28 Gg N year^{-1}
All fish waste (15% of catch) recycled as feed	9.5	1.3	21 Gg N year^{-1}

Note: [a]Assuming a yield in edible parts of 40%. It is also assumed that the by-products are used as feed, with a N-in-feed/N-in-edible-product ratio of 7. Decrease in farmed fish production is subtracted from the yield of edible parts.

efficiency (thus a lower N-cost) than wild fish. Most sea fish species used for human consumption in Norway are carnivorous (cod and related species). Domestic fish species used by Norwegian aquaculture are also carnivorous, fed primarily on capelin, which is also predated by the cod. Wild cod requires about four times as much feed as farmed fish (Hillestad et al., 1996). Since capelin is not suitable for human consumption, fish farming seems to be a convenient and efficient management option for fish resources. However, according to this study, less than 20% of the feed is recovered in farmed fish as edible products (food). It is reasonable to assume that appropriate industrial techniques would be able to retain twice as much in edible fleshy parts for human consumption.

This study shows that the anthropogenic contribution of reactive N from Norway to the North Atlantic is much greater (more than 10 times as much) than the uptake by fish catch associated with food consumption in Norway. The budget for the whole North Atlantic showed that anthropogenic N contributions to the sea are very large compared to the removal by fish catch (removal estimated around 2–3% of the input from anthropogenic sources, Table 12.5). I have not been able to compare the budgets of the *anthropogenic* N contribution minus the removal by fisheries with the findings by other studies. Nixon et al. (1996) have compiled *total* N budgets for a number of bays and estuaries in the North Atlantic, including natural fixation, all river transport, denitrification and exchanges to other basins, as well as direct anthropogenic effects. The budget of the Baltic Sea (Nixon et al., 1996) showed an annual accumulation in the waters of 220 Gg N (period 1982–1991) in addition to 150 Gg N year^{-1} accumulated in sediments. In comparison the removal by fish catch was small, only around 70 Gg N year^{-1}, and it corresponded to only about 2% of the total N input. Most other budgets presented by Nixon et al. showed that removals by catch were only a small percentage of the total N inputs. We cannot directly compare N removal as animal proteins with transport and deposition of inorganic N. However, it is clear that N removal by catch is far from compensating for anthropogenic N enrichment of the sea, and it is not a argument for a general 'fertilization' of the sea. Furthermore, in many coastal areas excessive reactive N can lead to eutrophication, which in turn reduces the production of attractive fish species (Baden et al., 1990; Cederwall and Elmgren, 1990).

CONCLUSION

Efforts to optimize the use of marine catch should aim at utilizing as much as possible of it directly for human nutrition. The disposal of 10% of the edible parts causes a larger loss in edible proteins than what can be produced by recycling all present waste as feed stuff. Thus, to really optimize the utilization of marine resources, it is important to develop conservation techniques and distribution systems that minimize losses of the final products. If used as feed, only 14–20% of the nitrogen in the fish is recovered as edible protein, compared to 30–60% when used as food, either fresh or after processing by food industries.

Whether fish meal is used for agricultural animals or for farmed fish is relatively less important for the total recovery of food proteins. This conclusion, however, does not consider differences in the nutritive quality of the proteins and of other compounds in the final products. The ratio N-in-feed/N-in-edible-parts in Norway, was found to be 5.2 for carnivorous fish farming, only 25% less than for agriculture livestock feeding essentially on plant products.

An optimal utilization of marine fish as food can contribute to a reduction in the use of chemical fertilizer in agriculture considerably, if fish is eaten instead of meat.

N-removal from the North Atlantic basin by fisheries is trivial compared with N supply by anthropogenic activities. Thus, there is no reason to sustain N enrichment of the sea to counterbalance N removal by fish catch.

ACKNOWLEDGEMENTS

Particular thanks to Erland Austreng (Institute of Aquaculture Research Ltd, Ås), who provided most of the data about aquaculture. I also thank Øistein Bækken (RUBIN Foundation, Trondheim). Lars Bakken (Agricultural University of Norway, Ås) inspired this work. Arild Vatn (Agricultural University of Norway, Ås) helped to start this study. The Research Council of Norway provided the financial support.

REFERENCES

Baden, S.P., Loo, L.O., Pihl, L. and Rosenberg, R. (1990) Effects of eutrophication on benthic communities including fish: Swedish West Coast. *Ambio* 19 (3), 113–122.

Barrett, K., Seland, Ø., Mylona, S., Sandnes, H., Styve, H. and Tarrasón, L. (1995) *European Transboundary Acidifying Air Pollution. Ten Years Calculated Fields and Budgets to the End of the First Sulphur Protocol.* EMEP/MSC-W Report 1/1995. Oslo.

Blaker, B. and Rimestad, A.H. (1994) *Matvaretabell*, 6. rev. utg. (Food composition table by the State Food Council). Statens ernæringsråd, Oslo. (In Norwegian.)

Bleken, M.A. (1997) Food consumption and nitrogen losses from agriculture. In: Låg, J. (ed.) *Some Geomedical Consequences of Nitrogen Circulation Processes.* Det Norske Videnskaps-Akademi, pp. 19–31.

Bleken, M.A. and Bakken, L.R. (1997a) The nitrogen cost of food production: Norwegian society. *Ambio* 26, 134–142.

Bleken, M.A. and Bakken, L.R. (1997b) The anthropogenic nitrogen cycle in Norway. In: Romstad, E., Simonsen, J. and Vatn, A. (eds) *Controlling Mineral Emissions in European Agriculture*. CAB International, Wallingford, pp. 27–40.

Cederwall, H. and Elmgren, R. (1990) Biological effects of eutrophication in the Baltic Sea, particularly the coastal zone. *Ambio* 19, 109–112.

Dibb, J.E., Rasmussen, R.A., Mayewski, P.A. and Holdsworth, G. (1993) Northern hemisphere concentrations of methane and nitrous oxide since 1800 – results from the Mt Logan and 20D ice cores. *Chemosphere* 27, 2413–2423.

Duxbury, J.M., Harper, L.A. and Mosier, A.R. (1993) *Contribution of Agroecosystems to Global Climate Change. Agricultural Ecosystem Effects on Trace Gases and Global Climate Change*. Asa Special Publication no.55 Madison, pp. 1–18.

FAO (1996) Fisheries statistics. Catches and landings 1994. *FAO Fisheries Series No. 46* (78), FAO, Roma.

FAO (1997) *Global Production*. Under: Welcome to the FAO Fisheries Department. (FAO Fisheries Department home page on Internet) http://www.fao.org/waicent/faoinfo/fishery/highligh/global.htm.

Galloway, J.N., Schlesinger, W.H., Levy II, H., Michaels, A. and Schooner, J.L. (1995) Nitrogen fixation: Anthropogenic enhancement-environmental response. *Global Biogeochemical Cycles* 9 (2), 235–252.

Galloway, J.N., Howarth, R.W., Michaels, A.F., Nixon, S.W., Prospero, J.M. and Dentener, F.J. (1996) Nitrogen and phosphorus budgets of the North Atlantic Ocean and its watershed. *Biogeochemistry* 35, 3–25.

Granli, T. and Bøckman, O.C. (1994) Nitrous oxide from agriculture. *Norwegian Journal of Agricultural Sciences* (Suppl.) 12, 1–128.

Hansson, S. and Rudstam, L.G. (1990) Eutrophication and Baltic fish communities. *Ambio* 19, 123–125.

Hillestad, M., Austreng, E. and Åsgård, T. (1996) Økologisk og ressursøkonomisk perspektiv på fôring av fisk (Ecological and resource economics perspectives of fish feeding.). In: Kvenseth, P.G., Maroni, K. and Andersen, P. (eds) *Miljøhåndbok for Fiskeoppdrett*. Kystnæringen forlag, Bergen, pp. 123–133. (In Norwegian.)

Howarth, R.W., Billen, G., Swaney, D., Townsend, A., Jaworski, N., Lajtha, K., Downing, J.A., Elmgren, R., Caraco, N., Jordan, T., Berendse, F., Freney, J., Kudeyarov, V., Murdoch, P. and Zhu Zhao-Liang (1996) Regional nitrogen budgets and riverine N & P fluxes for the drainages to the North Atlantic Ocean: Natural and human influences. *Biogeochemistry* 35, 75–139.

Khalil, M.A.K. and Rasmussen, R.A. (1988) Nitrous oxide: Trends and global mass balance over the last 3000 years. *Annals of Glaciology* 10, 73–79.

Kroeze, C. (1994) Anthropogenic emissions of nitrous oxide (N_2O) from Europe. *Science of the Total Environment* 152 (3), 189–205.

Nixon, S.W., Ammerman, J.W., Atkinson, L.P., Berounsky, V.M., Billen, G., Boicout, W.C., Boynton, W.R., Church, T.M., Ditoro, D.M., Elgren, R., Garber, J.H., Giblin, A.E., Jahnke, R.A., Owens, N.J.P., Pilson, M.E.Q. and Seitzinger, S.P. (1996) The fate of nitrogen and phosphorus at the land-sea margin of the North Atlantic Ocean. *Biogeochemistry* 35, 141–180.

Prinn, R.G. (1994) The interactive atmosphere: global atmospheric–biospheric chemistry. *Ambio* 23 (1), 50–60.

Rosenberg, R., Elmgren, R., Fleischer, S., Jonsson, P., Persson, G. and Dahlin, H. (1990) Marine eutrophication case studies in Sweden. *Ambio* 19 (3), 102–108.

RUBIN (1993) *Varestrømanalyse, Biprodukter fra Fisk og Reker (videreføring)* (Commodity Flow, By-products from Fish and Shrimps). Report 003/14, RUBIN, Trondheim. (In Norwegian.)

RUBIN (1995a) *Varestrømanalyse – 1993/94* (Commodity flow – 1993/94). Report 414/4. RUBIN, Trondheim. (In Norwegian.)

RUBIN (1995b) *Kartlegging av Biprodukter i Fiskeflåten*. (Survey of By-products from Fishing). Report 003/28. RUBIN, Trondheim. In Norwegian.

SSB (1992) Fiskeristatistikk 1989–1990 (*Fishery Statistics 1989–1900*). Statistisk sentralbyrå, NOS C 4, Oslo, Norway. (In Norwegian.)

SSB (1994) NOS Fiskeristatistikk 1990–1991 (*NOS Fishery Statistics 1990–1991*). Statistisk sentralbyrå, NOS C 93, Oslo, Norway. (In Norwegian.)

Subak, S., Raskin, P. and Vonhippel, D. (1993) National Greenhouse Gas Accounts – Current Anthropogenic Sources and Sinks. *Climatic Change* 25, 15–58.

Sundstøl, F., Homb, T., Ekern, A. and Breirem, K. (1991). In: Skøien, S. (ed.) *K.K. Heie – Håndbok for Jordbruket, 1992*. Landbruksforlaget, Oslo, pp. 208–223.

Vatn, A., Bakken, L.R., Bleken, M.A., Botterveg, P., Lundeby, H., Romstad, E., Rørstad, P.K. and Vold, A. (1996) Policies for reduced nutrient losses and erosion from Norwegian agriculture. *Norwegian Journal of Agricultural Sciences* Suppl. 23, Fig. 2.4, p. 16.

Vitousek, P.M. (1994) Beyond global warming: Ecology and global change. *Ecology* 75, 1861–1876.

13 Agricultural and Ecological Performance of Cropping Systems Compared in a Long-term Field Trial

P. Mäder, T. Alföldi, A. Fließbach, L. Pfiffner and U. Niggli

Research Institute of Organic Agriculture, Ackerstrasse, CH-5070 Frick, Switzerland

ABSTRACT

This chapter presents results of a long-term field experiment in which the agronomic and ecological performance of five farming systems were compared since 1978. The five systems are biodynamic (DYN), organic (ORG), conventional with manure (CON), conventional with mineral fertilizer (MIN) and an unfertilized control (NON). The systems differ in nutrient input, fertilizer type, disease and weed control, but not in crop rotation and soil cultivation. Crop rotation is 7 years and includes potatoes, winter wheat, cabbage or beetroots, winter wheat, barley and 2 years of grass–clover. Inputs of nitrogen, phosphorus and potassium were on average 25, 40 and 50% lower, respectively, in the DYN and ORG systems than in the CON and MIN systems. The results deal with the first two crop rotations, from 1978 to 1991.

Average crop yield was around 20% higher in the CON and MIN systems than in the DYN and ORG systems. Crop and market quality, as described by the contents of nutrients, amino acids, protein and sugars in the crop, differed little between farming systems. Apparent nitrogen recovery was higher in the MIN and CON systems than in the ORG and DYN systems. Nitrogen, phosphorus and potassium budgets were on average negative for all systems, suggesting that nutrient inputs via fertilizers and manure were larger than nutrient output via crop harvests. Mean nitrogen budgets, averaged over a whole crop rotation, was around -110 kg ha^{-1} year^{-1} for the DYN, DRG and CON systems.

Soil physical parameters were not affected by the farming systems except for aggregate stability, which was highest in the DYN and ORG systems. Most parameters that were used to estimate soil biomass and activity increased in the order: NON = MIN < CON < ORG < DYN. The efficiency of energy use per crop unit was 20 to 30% higher for DYN and ORG systems than for the MIN system.

Based on the results presented, it is suggested that there is wide scope in Europe for a further transition of conventional to organic farming systems.

INTRODUCTION

Agricultural production in Western Europe has been intensified considerably over the last decades. The liberal use of chemicals, fertilizers and heavy machinery as well as the narrowing of crop rotations and the elimination of surrounding semi-natural habitats has led to soil and water pollution in many areas, and to a decrease in biodiversity. As a consequence of the application of large amounts of mineral fertilizers and the import of large amounts of animal feeds from elsewhere, nutrient budgets in high-tech farming in Western Europe are frequently unbalanced. Furthermore, intensive agriculture has contributed to the drastic decline in species diversity and depletion of the natural flora and fauna (Basedow, 1990). Natural ecosystems reinvest a major part of their productivity to maintain soil fertility and biotic stability. The removal of crops and residues in intensively managed agroecosystems limit such reinvestment. This extends further the dependence of intensively managed agroecosystems on external inputs such as fertilizers and pesticides (Altieri, 1995).

Organic farming intends to apply practices that keep the cultivated land close to natural ecosystems and is therefore more sustainable with respect to ecosystem processes and functions. The European Community has released a decree (EWG No. 2092/91) on organic farming to assure a comparable production standard. Nearly 0.5% of European agricultural land (nearly 2,000,000 ha belonging to 80,000 farms) was under organic management in 1997. Stimulated by an increased demand for organically grown food, the area is slowly increasing (Lampkin, 1998). In Switzerland, around 8% of the land is managed organically with a rapid annual increase; the greatest part is managed according to the official rules of integrated crop production.

We compared the agronomic (production potential) and ecological (soil fertility, species diversity, nutrient budgets and nutrient and energy use) performance of five farming systems in a long-term field experiment. Within a 14-year period, changes of biological and chemical soil properties occurred, depending on the production method of the farming system. The purpose of this chapter is to give an overview of the results of this long-term experiment in view of the agronomic and ecological performance of organic and conventional farming systems, with special emphasis on nutrient budgets.

SET-UP OF THE FIELD EXPERIMENT

In 1978, a field experiment was set up in the vicinity of Basle (Switzerland) by the Research Institute of Organic Agriculture, Frick, in cooperation with the Swiss Federal Research Station for Agroecology and Agriculture, Zürich-Reckenholz (Besson and Niggli, 1991). The farming systems mainly differed in fertilization strategy and plant protection management (Table 13.1). The experiment comprises five systems, namely, biodynamic (DYN), organic (ORG), two conventional (CON and MIN) and an unfertilized control system (NON). Soil tillage was performed in roughly the same way for all five systems.

The biodynamic system (DYN) was managed according to the national (DEMETER-Anbaurichtlinien, 1997) and the international standards for biodynamic farming (International DEMETER-guidelines, 1992), and the organic system (ORG) according to the Swiss national guidelines for organic farming (VSBLO, 1997). Both biological systems were fulfilling the requirements of the EU decree EWG No. 2092/91 released by the European

Table 13.1. Main differences between the farming systems (second crop rotation 1985–1991).

Treatments	NON	DYN	ORG	CON	MIN[a]
Fertilizers	–	Farmyard manure (FYM) (from 1.2 LSU ha^{-1} year^{-1})		FYM and mineral fertilizer	Mineral fertilizer only
	–	Composted manure	Rotted manure	Stacked manure	–
Plant protection					
Weed control		Mechanical		Mechanical and herbicides	
Disease control		Indirect methods		Chemical (thresholds)	
		–	Copper		
Insect control		Plant extracts, bio-control		Chemical (thresholds)	
Special treatments		Biodynamic preparations	–	CCC application in cereals	

Note: [a]The MIN system remained unfertilized during the first crop rotation (1978–1984).
Source: According to Besson and Niggli (1991).

Community. The biodynamic (DYN) and the organic (ORG) systems received organic fertilizers only. Both organic and mineral fertilizers were applied in one conventional (CON) system (Table 13.1), whereas only mineral fertilizer was applied in the second conventional system (MIN). The control system remained unfertilized (NON). The biological systems (DYN and ORG) received 45–69% of the nutrients (N, P, K) applied to the conventional systems (CON and MIN). The organic matter input by manure was almost the same for the three systems receiving manure (2000 kg ha^{-1} year^{-1}) but the manure differed in the way it was processed (Fig. 13.1). Aerobically composted farmyard manure was applied in DYN, slightly aerobically rotted farmyard manure in ORG, and anaerobically rotted, stacked farmyard manure in CON.

Plant protection was conducted according to the above-mentioned guidelines of the biodynamic and organic systems applying plant extracts and bio-control. Weed control was performed mechanically. In the conventional systems, pesticide application was related to economic thresholds (integrated plant protection). Plant protection in the control system (NON) was the same as in the biodynamic system.

The crop rotation was the same for all systems: potatoes, winter wheat, cabbage or beetroots, winter wheat, barley and two years of grass–clover. The conventional systems (CON, MIN) are according to the official federal rules for integrated plant production.

The experiment has been designed as a randomized block with four replicates. Three different crops have been grown each year on three

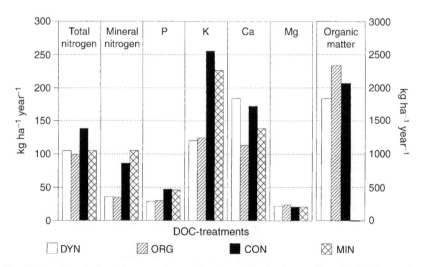

Fig. 13.1. Input of nutrients and organic matter (kg ha^{-1} year^{-1}) to the biodynamic (DYN), organic (ORG), conventional (CON) and mineral (MIN) systems. Data source: Spiess *et al.* (1993).

parallel series of plots. Single plot size is 5 m by 20 m. Execution of the experiment was performed according to agricultural farming practice. In order to ascertain the link to farming practice, advisory farmer groups were established for each system. The farmers were involved when the experiment was designed and are assembled twice a year to give advice on how to manage the respective systems properly according to the actual guidelines and practice.

The soil of the field experiment is an orthic Luvisol (sandy loam) on deep deposits of alluvial loess. The climate is rather dry and mild with a mean precipitation of 785 mm per year and an annual mean temperature of 9.5°C.

AGRONOMIC PERFORMANCE

Crop Yield

Crop yield of the two biological systems was markedly lower than in the two conventional systems CON and MIN, which is mainly attributable to less nutrient input and less plant protection. Average reduction in yield over all crops was about 17% (Table 13.2). However, it is important to note that in the second crop rotation of the two biological systems DYN and ORG, yield decreased only for potatoes, whilst it remained nearly constant for barley and grass–clover, and increased for wheat. Differences between DYN and ORG in yield were small in the first crop rotation (1978–1984) but increased during the second crop rotation (1985–1991).

Potato yield was on average 37% lower in the biological systems DYN and ORG than in the conventional systems CON and MIN. Leaf senescence of potatoes started earlier in the biological system than in the conventional system, mainly due to the low potassium supply in the biological systems (Berchtold *et al.*, 1993). Moreover, a severe incidence of leaf pathogens, such as *Phytophtora* and *Alternaria* was recorded in the biological treatments in some years, which drastically diminished the photosynthetically active leaf surface. In contrast, these pathogens could be controlled efficiently by organic and inorganic fungicides in the conventional systems. The comparatively small differences between farming systems in grass–clover yield (around 10%) may have been caused by the higher clover content, with its N-fixing *Rhizobium* symbiosis, and a higher root colonization by mycorrhizal fungi in the biological treatments.

Generally speaking, differences in crop yield between biologically and conventionally grown crops were larger for crops with a relatively short vegetation period, such as potatoes, than for crops with a relatively long vegetation period, such as cereals, grass–clover and beetroots. The latter crops build up a high root biomass, allowing for an efficient exploitation of soil nutrients. A possible delay of the initial beet growth in cold and wet

Table 13.2. Crop yield per system in Mg ha^{-1} year^{-1}; average of 3 years for each crop[a] during first (1978–1984) and second (1985–1991) crop rotation period; relative values in comparison to CON system (= 100%).

Crops	Crop rotation period	NON Mg ha^{-1}	NON %	DYN Mg ha^{-1}	DYN %	ORG Mg ha^{-1}	ORG %	CON Mg ha^{-1}	CON %	MIN Mg ha^{-1}	MIN %
Potatoes[b]	1978–84	21.7	38	34.9	61	39.5	69	57.1	100		
	1985–91	9.1	18	27.7	55	33.2	66	50.2	100	49.1	98
Cabbage[b]	1978–84	36.4	65	45.1	83	42.6	78	54.6	100		
Beetroots[b]	1985–91	33.1	42	62.0	76	60.7	75	81.4	100	64.8	80
Wheat[c]	1978–84	3.4	92	3.4	95	3.6	98	3.7	100		
	1985–91	2.7	54	3.8	74	4.1	81	5.0	100	5.2	103
Barley[c]	1978–84	3.2	87	3.6	92	3.4	87	3.9	100		
	1985–91	2.6	53	3.6	75	3.6	75	4.8	100	4.9	102
Grass–clover (1st year)[c]	1978–84	12.9	89	14.4	99	14.2	98	14.5	100		
	1985–91	11.6	75	14.5	94	14.5	94	15.5	100	14.2	92
Grass–clover (2nd year)[c]	1978–84	10.2	71	12.5	86	12.1	84	14.4	100		
	1985–91	8.2	58	11.6	82	12.2	85	14.2	100	12.7	89
Average of all crops	1978–84		74		88		88		100		
	1985–91		50		76		81		100		94

Notes: [a]For wheat average of 6 years per crop rotation period.
[b]Fresh matter.
[c]Dry matter.
Source: Adapted from Niggli et al. (1995).

spring times and in early summer, when nitrogen mineralization rate is low, can be partly compensated for during later growing stages. However, a retarded potato growth in spring, due to, for example, a cold and wet soil, has only a short period for compensation.

Crop Quality

Consumers of organically grown food expect a better product quality, not only as a result of environmentally sound production, but also with respect to a higher nutritional value. Organically and conventionally grown crops can be compared by analysing products offered on the market, products from selected organic and conventional farms or from crops cultivated within comparison trials (Vetter *et al.*, 1987). However, comparisons have rarely been done with the same variety being produced under comparable environmental conditions (Woese *et al.*, 1997).

Crop quality, as described by its content of minerals, amino acids, protein and sugars as well as the technological quality, was not significantly affected by the different systems (Fig. 13.2). The proportion of marketable potatoes from the DYN and ORG systems was 25% lower compared to potatoes from the CON and MIN systems. The proportion of marketable beetroots was similar for all systems.

Total nitrogen content was up to 25% higher in biodynamically grown potatoes than in conventionally grown potatoes. By contrast, the potassium content of potatoes and beetroots was 20% lower in the DYN and ORG systems than in the CON and MIN systems. Vitamin C, saccharose and storability of beetroots, were not significantly affected by the system in which they were produced. Differences between farming systems in nutrient content were larger for potatoes and beetroots than for the cereals barley and winter wheat. Due to a different proportion of clover versus grass in the grass–clover crop, contents of potassium, calcium and magnesium differed by up to 30% between DYN and ORG on the one hand and CON on the other hand.

Summarizing, crop quality varied somewhat between the different farming systems, but differences were relatively small.

ECOLOGICAL PERFORMANCE

Soil Fertility

Soil fertility comprises physical, chemical and biological soil characteristics (Janke and Sarrantonio, 1993; Reganold, 1995). In view of the multiple soil functions, soil fertility must be assessed by evaluating a series of structurally and functionally meaningful soil parameters.

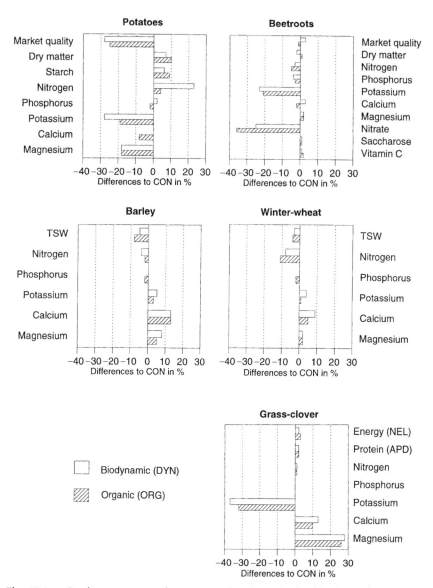

Fig. 13.2. Quality parameters for potatoes, beetroots, barley, winter wheat and grass–clover; relative comparison between crops of the biodynamic (DYN) and organic (ORG) systems on the one hand and crops of the conventional system (CON) on the other hand, in %. Average of 3 cropping years (6 years for grass–clover) during the second crop rotation period. TSW, thousand seed weight. Adapted from Spiess *et al.* (1995) and Alföldi *et al.* (1996).

Soil physical characteristics
After 14 years there were no significant differences in the volume of total or large sized pores and bulk density between systems. However, soil aggregate stability as assessed by water percolation (for method see Becher and Kainz, 1983) was enhanced markedly in the DYN and ORG systems compared with the CON and MIN systems (Table 13.3) (Siegrist, 1995). Aggregate stability was significantly correlated with earthworm biomass.

Soil chemical characteristics
Corresponding to the lower nutrient input (e.g. Fig. 13.1), amounts of extractable P and K in both biological systems were lower than in the conventional system (Table 13.4). Surprisingly, the soil in the DYN system exhibited distinctly higher P values than the soil in the ORG system, despite

Table 13.3. Selected soil physical properties in five systems after two crop rotations, after 14 years under different cultivation. Significant differences in the same row are indicated by different letters ($P = 0.05$).

Soil property	NON		DYN		ORG		CON		MIN	
[1]Aggregate stability (ml 10 min^{-1})	373	a	572	b	562	b	433	a	345	a
[2]Bulk density (g cm^{-3})	1.30	a	1.25	a	1.23	a	1.23	a	1.20	a

Source: [1]Siegrist (1995) (winter wheat 1993); [2]Alföldi *et al.* (1993) (winter wheat 1991).

Table 13.4. Soil chemical properties of five systems after two crop rotations. Significant differences are indicated by different letters in the same row ($P = 0.05$).

Soil property	NON		DYN		ORG		CON		MIN	
[1]pH (H$_2$O)	6.2	a	6.9	c	6.5	b	6.2	a	6.1	a
[1]Total organic carbon (g C$_{org}$ kg^{-1})	14.8	a	17.7	c	16.4	b	16.1	b	14.5	a
[1]Total nitrogen (g N kg^{-1})	1.34	a	1.69	c	1.50	b	1.47	ab	1.41	ab
[1]Extractable P (DL) (mg P kg^{-1})	11	a	33	c	25	b	38	d	24	b
[2]F_m (mg P min^{-1} kg^{-1})	6	a	45	b	21	a	24	a	17	a
[1]Extractable K (DL) (mg K kg^{-1})	48	a	61	b	58	b	101	d	73	c
[1]Extractable Ca (HCl/H$_2$SO$_4$) (g Ca kg^{-1})	1.81	a	2.47	b	1.96	a	1.84	a	1.78	a
[2]Extractable Mg (DL) (mg Mg kg^{-1})	68	a	101	bc	116	c	95	b	89	b

Notes: F_m, mean flux of phosphate ions between solid and liquid phase of soil as determined by ^{32}P isotopic exchange method (for method and calculation see Oberson *et al.* (1993)).
Source: [1]Mäder *et al.* (1993) (winter wheat 1991); [2]Oberson (1993) (winter wheat beetroots 1991).

the similarity in P uptake/input ratio in both systems over 14 years (Spiess *et al.*, 1993). This may be explained by: (i) an increased microbial contribution to phosphorus mobilization in the DYN system as compared with the ORG system (Oberson *et al.*, 1996); and (ii) a significantly higher P-ion exchange-kinetics in soils of the DYN system (Oberson, 1993; Oberson *et al.*, 1993).

Soil microbial characteristics

Soil microorganisms are of fundamental importance as a living component of terrestrial habitats and play key roles in ecosystem functions. Their primary role is in governing nutrient cycling processes, which are essential in the maintenance of soil fertility and plant nutrition. Furthermore, microorganisms are involved in the genesis and maintenance of soil structure (Atlas, 1984).

Most parameters that can be assessed to measure microbial pool sizes and microbial activity showed the same ranking among the systems: NON = MIN < CON < ORG < DYN (Table 13.5).

Earthworms

Earthworms are well suited as bioindicators of soil fertility. Due to their biology and behaviour, earthworm populations can indicate the structural, micro-climatic, nutritive and toxic situation in soils (Christensen, 1988). In particular, anecic species, which are large-sized, vertically burrowing

Table 13.5. Soil microbial properties of five systems after two crop rotations. Significant differences are indicated by different letters in the same row ($P = 0.05$).

Soil property	NON		DYN		ORG		CON		MIN	
Microbial biomass (mg C_{mic} kg^{-1})	361	a	603	d	528	c	443	b	359	a
C_{mic} C_{org}^{-1} ratio (g kg^{-1})	24	a	34	c	32	c	27	b	25	a
Respiration (μg CO_2-C 15 d^{-1} kg^{-1})	258	a	324	a	302	a	295	a	273	a
Dehydrogenase (mg TPF 6 h^{-1} kg^{-1})	42	a	106	d	85	c	59	b	46	a
Catalase (g H_2O_2 h^{-1} kg^{-1})	3.6	a	6.05	c	5.4	bc	4.4	ab	4.0	a
Protease (mg tyrosine equivalents 2 h^{-1} kg^{-1})	233	a	810	d	613	c	476	b	378	b
Alkaline phosphatase (mg phenol 16 h^{-1} kg^{-1})	112	a	1607	d	973	c	531	b	416	ab
Saccharase (g red. sugar 3 h^{-1} kg^{-1})	1.18	a	2.29	d	1.97	c	1.58	b	1.49	ab

Source: Adapted from Mäder *et al.* (1993) (winter wheat 1991).

earthworms, play an important role for conservation and improvement of soil structure, because they build up stable burrows (Roth and Joschko, 1991). They contribute significantly to nutrient recycling and aspects of pest and disease control, such as reduction of leaf miner pupae and scab pathogens in orchards (Kennel, 1990).

Earthworm biomass, density and the presence of anecic species were significantly higher in the biological systems than in the conventional and control systems (Fig. 13.3). Differences were most pronounced for anecic earthworms (Pfiffner *et al.*, 1993). Many factors that may influence earthworms population, such as soil type, crop rotation and environment, were similar in all five systems; soil cultivation was also similar. Thus, the intensity of plant protection and the quality and quantity of fertilization must be mainly responsible for the differences in the earthworm populations. It has to be considered also that organic matter input in the DYN, ORG and CON systems was similar. From this it can be concluded that plant protection with pesticides is a major factor responsible for the lower earthworm populations in the CON system.

Resource Utilization

Nutrients

Table 13.6 shows inputs of nutrients via fertilizers and manure, nutrient output via nutrient uptake by the crop, and nutrient budgets, calculated as

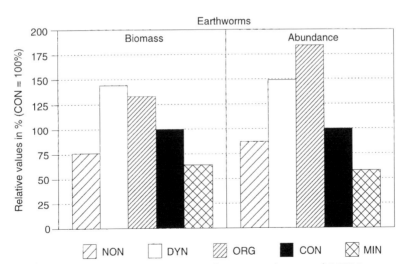

Fig. 13.3. Biomass and abundance of earthworms in the control (NON), biodynamic (DYN), organic (ORG) and mineral (MIN) systems, relative to the conventional system (CON = 100%). Average of 3 years (1990 and 1991 after beetroots; 1992 after potatoes). Data: Pfiffner *et al.* (1993).

Table 13.6. Fertilization (input), nutrient uptake by plants and nutrient budget in the first and second crop rotation (CRP) (all values in kg ha⁻¹ year⁻¹).

Crop rotation	NON			DYN			ORG			CON			MIN		
	Input	Uptake	Budget	Input	Uptake	Budget	Input	Uptake	Budget	Input	Uptake	Budget	Input	Uptake	Budget
1st CRP (1978–1984)															
N	0	177	−177	109	201	−92	106	196	−90	131	227	−95	0	179	−179
P	0	27	−27	27	32	−4	32	32	0	49	36	+13	0	27	−27
K	0	115	−115	128	176	−48	125	177	−53	264	246	+18	0	113	−113
Ca	0	62	−62	162	62	+100	115	60	+55	167	66	+101	0	61	−61
Mg	0	15	−15	21	16	+5	26	17	+9	17	16	0	0	15	−15
2nd CRP (1985–1991)															
N	0	171	−171	100	233	−133	93	235	−142	144	277	−133	105	249	−144
P	0	21	−21	30	33	−3	27	34	−6	44	41	+3	46	36	+10
K	0	75	−75	112	163	−51	123	181	−58	247	294	−47	226	225	+1
Ca	0	60	−60	207	73	+134	110	71	+40	176	74	+102	138	66	+72
Mg	0	15	−15	22	20	+2	23	20	+2	25	20	+5	21	20	+1
1st and 2nd CRP (1978–1991)															
N	0	174	−174	105	217	−112	99	216	−116	138	252	−114	53	214	−161
P	0	24	−24	29	32	−4	30	33	−3	47	38	+8	23	31	−8
K	0	95	−95	120	170	−50	124	179	−56	255	270	−15	113	169	−56
Ca	0	61	−61	184	67	+117	113	65	+47	172	70	+102	69	63	+6
Mg	0	15	−15	22	18	+4	24	18	+6	21	18	+3	11	17	−7

Source: Spiess *et al.* (1993).

the difference between the first two items, during the first and the second crop rotation. Differences between systems in nitrogen and phosphorus uptake agree well with the differences in crop yield. Potassium uptake increased more than proportionally to yield, due to luxury consumption.

Nutrient budgets were often negative, as the uptake exceeded the amount of nutrients applied. Budgets of potassium and to a lesser extent for phosphorus were more negative in the DYN and ORG systems than in the CON system, because of differences in input (Spiess *et al.*, 1993). Note that the nutrient budget presented in Table 13.6 is an estimate of the difference between inputs via manure and fertilizers on the one hand, and outputs via harvested products on the other hand. It does not include nutrient inputs via atmospheric deposition and biological nitrogen-fixation. It also does not include nutrient losses via volatilization, leaching and runoff.

Table 13.7 presents the apparent nitrogen recovery (ANR) of four farming systems, averaged over a crop rotation. We have assumed that the unfertilized crop of the control system (NON) can serve as a valid control for nutrient uptake from the soil. We have also assumed that the apparent nitrogen recovery (ANR) can be calculated from the difference in nitrogen uptake between the fertilized crops in the DYN, ORG, CON and MIN systems and the unfertilized crops of the control system (NON), divided by

Table 13.7. Nitrogen input via fertilizers and manures (total amount and amount of mineral N) of four farming systems during the first and the second crop rotation (CRP). Apparent nitrogen recovery (ANR, %) was calculated according to the following equation: ANR (%) = [(N uptake in crops of system x) – (N uptake of crops of the unfertilized system)]/ total N content in fertilizers and manures in fertilized systems.

		N fertilization		
Crop rotation	Systems	Total amount (kg N ha^{-1} year^{-1})	In mineral form[a] (kg N ha^{-1} year^{-1})	ANR (%)
1st CRP	DYN	109	46	22
1978–1984	ORG	106	39	18
	CON	131	83	38
	MIN[b]	—	—	—
2nd CRP	DYN	100	26	62
1985–1991	ORG	93	31	69
	CON	144	89	74
	MIN	105	105	74

[a]As NH_4–N and/or NO_3–N.
[b]This system remained unfertilized during the first CRP.
Source: Adapted from Besson and Spiess (1995).

the total amount of nitrogen applied. In the first crop rotation nearly 40% of the nitrogen applied was recovered in the crop of the conventional system (CON) whereas in the biological systems (DYN, ORG) apparent nitrogen recovery (ANR) was much lower. This can be explained by the higher proportion of mineral nitrogen applied in the conventional system and by a higher input of pesticides. In the second crop rotation, ANR was distinctly higher in all farming systems (Table 13.7), because the crops took advantage of better conditions (higher-yielding varieties, more accurate fertilization and plant protection) (Besson *et al.*, 1995). In the biological systems, a higher protease activity and enhanced respiration was measured, increasing the potential to mineralize organically bound nitrogen in these systems. This has recently been confirmed for the biodynamic system in an incubation experiment (Meyre, 1997).

Energy and manpower

The energy use efficiency was calculated for the different systems considering direct and indirect energy for each system, with reference to the crop yield. Direct energy is defined as the energy used in the crop production process and indirect energy is the energy used for the production of machinery, fertilizers and pesticides. Energy consumption on an area basis was between 30 and 50% higher in the conventional farming system (CON) than in both biological systems (DYN and ORG), but depending on the crop, the yield achieved was only 10–35% higher. The energy use per crop unit was 15–30% lower in the biological system than in the conventional systems, except for potatoes (Table 13.8). This difference was mainly due to indirect energy for fertilizer and pesticide production. Manpower per unit surface area was similar in the organic and the conventional systems, except for beetroots. Owing to lower yields, manpower per crop unit in the biological plots was higher than in the conventional plots. The preparation of

Table 13.8. Energy input to produce 1 Mg of crop for biodynamic (DYN), organic (ORG) and conventional (CON) production systems. Average of two crop rotation periods. Comparison with CON in italics.

Crops	DYN		ORG		CON	
	GJ Mg^{-1}	rel.(%)	GJ Mg^{-1}	rel.(%)	GJ Mg^{-1}	rel.(%)
Potatoes	0.76	*106*	0.76	*106*	0.71	*100*
Beetroots	0.26	*75*	0.25	*71*	0.40	*100*
Wheat	3.13	*74*	2.84	*67*	4.21	*100*
Barley	2.51	*68*	2.85	*78*	3.67	*100*
Grass–clover	0.33	*67*	0.40	*82*	0.49	*100*

Source: According to Alföldi *et al.* (1995).

biodynamic sprays in the DYN system considerably increased manpower per area as well as per crop unit.

DISCUSSION AND CONCLUSIONS

Crop yields are a function of inputs of nutrients and energy. In western Europe, conventional farming systems generally use large amounts of nutrients and pesticides. By contrast, inputs of nutrients and pesticides in organic systems are small. In organic systems, nutrient inputs occur via farmyard manure, derived from within the farm-system. Conversely, the bulk of nutrients applied in conventional systems has been derived from elsewhere, and has been produced at high energy costs. Consequently, the efficiency of resource utilization is higher in organic than in conventional systems. As can be observed from the results of the field experiment presented here, yield and also soil quality parameters continue to diverge between farming systems, even 14 years after conversion. This clearly emphasizes the need for long-term experiments, to be able to evaluate both agronomic and ecological performance of farming systems.

The budgets for nitrogen, phosphorus and potassium, as defined in this study, were negative for all farming systems, indicating that none of the treatments received excessive amounts of nutrients. Inputs of the conventional systems are also considered to be moderate in view of application of fertilizers and manures.

Results of the labelling techniques measurements (Table 13.4) suggest that the crops did not suffer from phosphorus deficiency in the biological systems, despite the fact that amounts of extractable soil phosphorus were low. It was hypothesized that soil microbial processes are governing flux-rates of phosphorus between solid and soluble phases. This was most pronounced for the biodynamic system, which revealed highest microbial activity and highest P fluxes.

Organic systems can be characterized by low energy and nutrient inputs, and by healthier soils, at the costs of low crop yields. From an ecological point of view, organic farming should be performed on a large scale, on landscapes and whole regions, to evaluate its potential at regional level. In areas where the environmental indicators point at severe pollution of the environment, or at a decrease in biodiversity and soil fertility, biological farming or other less intensive farming systems might help to solve some of the problems.

From the financial point of view, the farmers' income depends on crop yield and costs of external inputs. Fertilizers, pesticides and energy are still relatively cheap. Consequently, farmers will make use of these production factors. However, the considerable societal costs of intensive agriculture, for example by groundwater pollution and decreases in biodiversity, should be taken into consideration too.

Nowadays, various countries subsidize organic farming by direct payment to compensate for lower yield and the farmers' role in environmental protection. Sweden, Austria, Switzerland, Finland, Denmark and Germany in particular (see Internet http://www.aber.ac.uk/~wirwww/organic) show a remarkable increase in the number of farms in conversion to organic farming (2–9% of the agricultural area). Moreover, products of organic farming systems have higher prices than products of conventional farming systems, as consumers are partly willing to pay higher prices for biologically grown food. Other possible incentives and policy measures to reduce intensive agricultural practices and/or to promote organic farming, such as taxes on fertilizers and pesticides, have not been implemented in practice (yet).

REFERENCES

Alföldi, T., Stauffer, W., Mäder, P., Niggli, U. and Besson, J.-M. (1993) DOK-Versuch: vergleichende Langzeit-Untersuchungen in den drei Anbausystemen biologisch-Dynamisch, Organisch-biologisch und Konventionell. III. Boden: Physikalische Untersuchungen, 1. und 2. Fruchtfolgeperiode. *Schweizerische Landwirtschaftliche Forschung* 32 (4), 465–477.

Alföldi, T., Spiess, E., Niggli, U. and Besson, J.-M. (1995) DOK-Versuch: vergleichende Langzeituntersuchungen in den drei Anbausystemen biologisch-Dynamisch, Organisch-biologisch und Konventionell. IV. Aufwand und Ertrag: Energiebilanzen, 1. und 2. Fruchtfolgeperiode. *Schweizerische Landwirtschaftliche Forschung*, Sonderheft DOK, 2, 1–16.

Alföldi, T., Mäder, P., Niggli, U., Spiess, E., Dubois, D. and Besson, J.-M. (1996) Quality investigations in the long-term DOC-trial. In: Raupp, J. (ed.) *Fertilization Systems in Organic Farming. Quality of Plant Products Grown with Manure Fertilization*. Proceedings of the fourth meeting in Juva (Finland), 6–9 July 1996. Publication of the Institute for Biodynamic Research, Darmstadt, pp. 34–43.

Altieri, M.A. (1995) *Agroecology: The Science of Sustainable Agriculture*, 2nd edn. Westview Press, Boulder, Colorado, 433 pp.

Atlas, R.M. (1984) Diversity of microbial communities. *Advances in Microbial Ecology* 7, 1–47.

Basedow, T. (1990) Effects of insecticides on Carabidae and the significance of these effects for agriculture and species number. In: Stork, N.E. (ed.) *The Role of Ground Beetles in Ecological and Environmental Studies*. Intercept, Andover, pp. 115–125.

Becher, H.H. and Kainz, M. (1983) Auswirkungen einer langjährigen Stallmistdüngung auf das Bodengefüge im Lössgebiet bei Straubing. *Z. für Acker und Pflanzenbau* 152, 152–158.

Berchtold, A., Besson, J.-M. and Feller, U. (1993) Effects of fertilization levels in two farming systems on senescence and nutrient contents of potato leaves. *Plant and Soil* 154, 81–88.

Besson, J.M. and Niggli, U. (1991) DOK-Versuch: vergleichende Langzeit-Untersuchungen in den drei Anbausystemen biologisch-Dynamisch, Organisch-biologisch und Konventionell. I. Konzeption des DOK-Versuches: 1. und 2. Fruchtfolgeperiode. *Schweizerische Landwirtschaftliche Forschung* 31, 79–109.

Besson, J.-M. and Spiess, E. (1995) Processus chimiques et microbiologiques dans le sol cultivé selon les systèmes cultureaux biologiques et conventionnels (Essai DOC). In: Mäder, P. and Raupp, J. (eds) *Fertilization Systems in Organic Farming*. Proceedings of the second meeting in Oberwil (Switzerland), 15–16 September 1995. Publication of the Research Institute of Organic Agriculture, Oberwil, and the Institute for Biodynamic Research, Darmstadt, pp. 3–12.

Besson, J.-M., Spiess, E. and Niggli, U. (1995) N uptake in relation to N application during two crop rotations in the DOC field trial. *Biological Agriculture and Horticulture* 11, 69–75.

Christensen, O. (1988) Lumbricid earthworms as bio-indicators relative to soil factors in different agroecosystems. In: Veeresh, G.K., Rajagopal, D. and Viraktamath, C.A. (eds) *Advances in Management and Conservation of Soil Fauna*. Oxford and IBH Publishing, New Dehli, pp. 839–849.

DEMETER-Anbaurichtlinien (1997) Verein für Biologisch-dynamische Landwirtschaft, Münchenstein, Switzerland.

International DEMETER-Guidelines, Agriculture, Horticulture and Fruit Growing (1992) Conference for International DEMETER-Affairs, Darmstadt.

Janke, R.R. and Sarrantonio, M. (1993) *Regenerating Soil Quality in the Post-modern World*. Sustainable Soil Management Symposium, California, 22 April, 1993. University of California, Davis, pp. 1–14.

Kennel, W. (1990) The role of the earthworm Lumbricus terrestris in integrated fruit production. *Acta Horticulturae* 285, 149–156.

Lampkin. N. (1998) Ökologischer Landbau in Westeuropa. *Ökologie und Landbau* 26, 2. See also http://www.aber.ac.uk/~wirwww/organic.

Mäder, P., Pfiffner, L., Jäggi, W., Wiemken, A., Niggli, U. and Besson, J.-M. (1993) DOK-Versuch: Vergleichende Langzeit-Untersuchungen in den drei Anbausystemen biologisch-Dynamisch, Organisch-biologisch und Konventionell. III. Boden: Mikrobiologische Untersuchungen. *Schweizerische Landwirtschaftliche Forschung* 32 (4), 509–545.

Meyre, S. (1997) Einfluss der Bewirtschaftungsverfahren des DOK-Versuchs auf die Mineralisierung und Verfügbarkeit von Stickstoff im Boden. Dissertation Nr. 12405, Institut für terrestrische Ökologie, Eidgenössische Technische Hochschule, 139 pp.

Niggli, U., Alföldi, T., Mäder, P., Pfiffner, L., Spiess, E. and Besson, J.-M. (1995) DOK-Versuch: vergleichende Langzeit-Untersuchungen in den drei Anbausystemen biologisch-Dynamisch, Organisch-biologisch und Konventionell. VI. Synthese, 1. und 2. Fruchtfolgeperiode. *Schweizerische Landwirtschaftliche Forschung*, Sonderheft DOK 4, 1–34.

Oberson, A. (1993) Phosphordynamik in biologisch und konventionell bewirtschafteten Böden des DOK-Versuchs. Diss. ETH Zürich, Nr. 10119, 171 pp.

Oberson, A., Fardeau, J.-C., Besson, J.-M. and Sticher, H. (1993) Soil phosphorus dynamics in cropping systems managed according to conventional and biological agricultural methods. *Biology and Fertility of Soils* 16, 111–117.

Oberson, A., Besson, J.-M., Maire, N. and Sticher, H. (1996) Microbiological processes in soil organic phosphorus transformations in conventional and biological cropping systems. *Biology and Fertility of Soils* 21,138–148.

Pfiffner, L., Mäder, P., Besson, J.-M. and Niggli, U. (1993) DOK-Versuch: vergleichende Langzeit-Untersuchungen in den drei Anbausystemen biologisch-Dynamisch, Organisch-biologisch und Konventionell. III. Boden: Untersuchungen über die Regenwurmpopulationen. *Schweizerische Landwirtschaftliche Forschung* 32, 547–564.

Reganold, J.P. (1995) Soil quality and profitability of biodynamic and conventional farming systems: A review. *American Journal of Alternative Agriculture* 10 (1), 36–45.

Roth, C.H. and Joschko, M. (1991) A note on the reduction of runoff from crusted soils by earthworm burrows and artificial channels. *Journal of Soil and Plant Nutrition*, 154, 101–105.

Siegrist, S. (1995) Experimentelle Untersuchungen über die Verminderung der Bodenerosion durch biologischen Landbau in einem NW-schweizerischen Lössgebiet. *Die Erde* 126, 93–106.

Spiess, E., Stauffer, W., Niggli, U. and Besson, J.-M. (1993) DOK-Versuch: vergleichende Langzeit-Untersuchungen in den drei Anbausystemen biologisch-Dynamisch, Organisch-biologisch und Konventionell. IV. Aufwand und Ertrag: Nährstoffbilanzen, 1. und 2. Fruchtfolgeperiode. *Schweizerische Landwirtschaftliche Forschung* 32 (4), 565–579.

Spiess, E., Daniel, R., Stauffer, W., Niggli, U. and Besson, J.-M. (1995) DOK-Versuch: vergleichende Langzeituntersuchungen in den drei Anbausystemen biologisch-Dynamisch, Organisch-biologisch und Konventionell. V. Qualität der Ernteprodukte: Stickstoff- und Mineralstoffgehalte, 1. und 2. Fruchtfolgeperiode. *Schweizerische Landwirtschaftliche Forschung*, Sonderheft DOK 3.

Vetter, H., von Abercron, M., Bischoff, R., Kampe, W., Klasink, A. and Ranfft, K. (1987) Qualität pflanzlicher Nahrungsmittel – 'alternativ' und 'modern' im Vergleich, Teil III. AID-Schriftenreihe Verbraucherdienst, Nr. 3100.

VSBLO (1997) Richtlinien für die Erzeugung, Verarbeitung und den Handel von Produkten aus biologischem (ökologischem) Anbau. Vereinigung Schweizerischer Biologischer Landbauorganisationen (VSBLO BIO SUISSE), Basel.

Woese, K., Lange, D., Boess, C. and Bögl, K.W. (1997) A comparison of organically and conventionally grown foods – Results of a review of the relevant literature. *Journal of Science of Food Agriculture* 74, 281–293.

Comparative Nutrient Budgets of Temperate Grazed Pastoral Systems

K.M. Goh[1] and P.H. Williams[2]

[1]Soil and Physical Sciences Group, Soil, Plant and Ecological Sciences Division, PO Box 84, Lincoln University, Canterbury, New Zealand
[2]New Zealand Institute of Crop and Food Research, Canterbury Agricultural and Science Centre, Private Bag 4704, Christchurch, New Zealand

ABSTRACT

Budgets of inputs and outputs of nitrogen (N), phosphorus (P), potassium (K) and sulphur (S) obtained in recent case studies of nutrient cycling in major temperate grazed pastoral systems are presented and compared. The systems include extensive and intensive dairy, beef cattle, sheep and ley farming (or mixed cropping) systems. Considerable differences existed and these varied according to the farming system studied, the intensity of farming, kind of nutrient, the method of measurement and site. In general, N input from fertilizers, N-fixation and supplement feeds are greater in dairy farms than in other farming systems and this varied from 0 to 330 kg N ha^{-1} year^{-1}. Biological N-fixation provided almost all the N in dairy farms in New Zealand but not in Europe. Extensive upland sheep farms received no N fertilizers but are N accumulating systems due mainly to N-fixation (8–135 kg N ha^{-1} year^{-1}). Nitrogen outputs in intensive dairy farms were higher than those in beef production systems (531 vs. 309 kg N ha^{-1} year^{-1}). In dairy farms, P outputs are 17 kg P ha^{-1} year^{-1}. Outputs of K in animal products from dairy farms varied from 10 to 17 kg K ha^{-1} year^{-1} but its output in sheep farms has not been measured. A net loss of up to 100 kg K ha^{-1} year^{-1} has been estimated to occur in New Zealand dairy farms. Estimated S leaching losses in sheep farms varied from 16 to 22 kg S ha^{-1} year^{-1}. On the whole, inputs and outputs of nutrients are better balanced in ley farming systems than in other farming systems largely because ley farming incorporates a rotation of cropping which depletes nurients and a pasture phase which replenishes nutrient losses, especially for N.

Implications of the data presented for improving the management of the different farming systems are discussed.

INTRODUCTION

The productivity and sustainability of managed grasslands depends on nutrient cycling and transformations within the soil–plant–animal grazed systems, and the balance between nutrient supply from natural and fertilizer sources and nutrient outputs from the systems (Snaydon, 1987). Quantitative consideration of the balance between nutrient inputs and outputs together with nutrient transfers between and within the soil–plant–animal pool provides a useful approach for understanding nutrient transformation processes in these systems (Ryden, 1984; Nguyen and Goh, 1994a; Whitehead, 1995).

Many case studies of nutrient budgets have been conducted in managed grasslands. However, widely different experimental or farm practices and methods of measurements are used in different parts of the world. For example, methods used for input measurements varied from simple measurements of dry matter production by harvesting the herbage and analysing its nutrient content to measurements of nutrient yields in the wool of sheep (Nguyen and Goh, 1992b). For output measurements such as leaching, methods used varied from inserted suction cups in the soil profile, to stand-alone lysimeters and field tension lysimeters (Goh et al., 1979; Cuttle et al., 1992). In addition, few studies measure all the inputs and outputs of a nutrient in a system. Thus, results from case studies are not only regional and site specific, but also case specific. For this reason, this review partitions case studies into extensive and intensive dairy, beef cattle, sheep and ley farming (or mixed cropping) systems.

The selection of case studies or sites for presentation in this review is very limited as only a few studies have been conducted and most of these are from Australasia and Europe. Furthermore, as most managed grasslands are intensively managed, the information on extensive or relatively unimproved grasslands is even more limited. Because of the increasing interest in alternative farming systems, nutrient budgets of conventional and alternative farming systems are also compared and presented. Major nutrients considered are nitrogen (N), phosphorus (P), potassium (K) and sulphur (S) because N, P, K and S balances are closely linked in grazed pastures, especially in legume-based pastures where regular applications of superphosphate fertilizer to sustain or increase pasture production simultaneously supply P and S, and result in increases in N supply due to N-fixation (Nguyen and Goh, 1992b, Metherell et al., 1993). The consumption of pasture by grazing animals, especially dairy cows, is known to deplete soil K continuously and thus affect the sustainability of productive dairy pastures (Williams et al., 1990a).

DAIRY PRODUCTION SYSTEMS

Dairy production represents an intensive form of farming in many parts of the world. This system has become strongly dependent on inputs of fertilizer and purchased feed, both of which add nutrients (N, P and K) to the farm system. Other inputs of nutrients occur through symbiotic and non-symbiotic fixation of N_2 and deposition from the atmosphere (Fig. 14.1). Nutrients are removed from the farm in milk (4000–6500 l per cow), surplus animals, and any manure and home-grown pasture or crops that are sold.

Within the farm, the major influence on nutrient flows is the dairy cow. Only 5–30% of the nutrients ingested by dairy cows are retained to meet animal needs (Haynes and Williams, 1993). The remainder is excreted in dung and urine. Excreta deposited indoors when animals are being milked or housed is collected and spread back on to the land as slurry or manure. Excreta deposited on to pasture or crops during periods of grazing are unevenly distributed across the field. Furthermore, nutrients in dung and

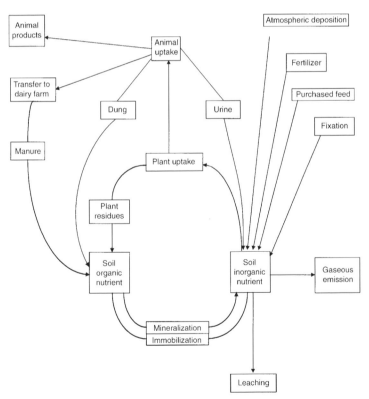

Fig. 14.1. Flow diagram showing the major nutrient cycling pathways in a dairy farm (adapted from Scholefield *et al.*, 1991).

urine patches are highly concentrated at rates of 300–1000 kg N ha^{-1}, 200–300 kg P ha^{-1} and 300–1000 kg K ha^{-1}. Excessive inputs of nutrients in fertilizer and manure and the uneven distribution of nutrients in dung and urine can lead to losses of N via volatilization, denitrification and leaching.

Nitrogen Budgets

Some examples of N budgets for dairy farming systems in New Zealand, England and the Netherlands are shown in Table 14.1. On the New Zealand dairy farms, cows graze outside on grass–clover pastures all year round. Animals are yarded for approximately 4 h each day for milking during the milking season (spring to autumn). In the winter when the cows are not being milked, they are retained on pasture and fed silage conserved on the farm the previous spring. On a typical dairy farm in England and The Netherlands, cows graze on grass pastures during the summer. Over winter they are housed inside and fed silage, concentrates and fodder crops.

Table 14.1. Nitrogen budgets for dairy production systems (Aarts *et al.*, 1992; Jarvis *et al.*, 1996; Ledgard *et al.*, 1996c) (kg N ha^{-1} year^{-1}).

Budget	New Zealand[a] (1993–1994)[a]	New Zealand[b] (1993–1994)	England[c] (1989–1991)	The Netherlands[c] (1983–1986)
Inputs				
Fertilizer		225	250	330
Feed			52	178
Deposition			25	46
N-fixation	212	165	10	
Other				7
Total	212	390	337	561
Outputs				
Milk	76	89	67	68
Meat				16
Volatilization	15	45	49	105
Denitrification	7	11	55	286
Leaching	13	18	56	
Other				56
Total	111	163	227	531
Budget				
Inputs − Outputs	101	227	110	30

Notes: [a]Years when data were collected.
[b]Data from experimental plots.
[c]Data obtained or estimated for whole farms.

Fertilizer nitrogen inputs on the New Zealand farm tend to be lower than those of the other farming systems due to a reliance on N-fixation by the pasture clovers. There are wide variations in the reported amounts of N fixed in legume-based pastures (Cookson *et al.*, 1990; Peoples *et al.*, 1995; Whitehead, 1995), depending on locality, methods of measurements (Goh *et al.*, 1978; Chalk, 1985), legume production and persistence, soil N status, fertilizer applications and competition with associated grasses (Ledgard and Steele, 1992; Ledgard *et al.*, 1996a). However, the usual range in grass–clover pastures in New Zealand is between 100 and 350 kg N ha^{-1} (Ledgard *et al.*, 1990), but less than 100 kg N ha^{-1} in unimproved pastures. In the United Kingdom, the amount of N fixed in clover-based pastures varied from nil to 400 kg N ha^{-1} year^{-1} (Whitehead *et al.*, 1986; Whitehead, 1995) while 270–370 kg N ha^{-1} were reported as being fixed in Switzerland (Boller and Nosberger, 1987). The rate at which the fixed N is mineralized may not be sufficient to meet the pasture needs. Nitrogen deficiencies can occur particularly during the early spring when soil temperatures are low, limiting mineralization. To overcome this problem there is a trend towards using moderate rates of fertilizer N (50–200 kg N ha^{-1}). This can reduce the amount of N$_2$ fixed by the clovers, but the reduction is more than compensated for by the fertilizer N.

On the more intensively managed farms in England and The Netherlands, inputs of fertilizer N of up to 400 kg N ha^{-1} are used to maintain the productivity of grass swards. Considerable amounts of N are also brought on to the farm in purchased feed (50–200 kg N ha^{-1}). Atmospheric deposition of N also supplies a significant quantity of N in the European farming system.

In general, the conversion of N inputs into animal products is very low on dairy farms with only 20–40% of the N inputs leaving the farm in milk and meat (Table 14.1; Klausner, 1995). This surplus of N leads to very large losses of N via leaching, denitrification and volatilization. Volatilization losses of ammonia from urine occur due to the high N content and pH of the urine generated by the hydrolysis of urea (main form of N in urine) to ammonium (Sherlock and Goh, 1984). Commonly 15–25% of the urine N can be lost in this manner (Haynes and Williams, 1993). Dung patches can also be a source of ammonia volatilization. However, since most of the dung N is present in organic forms, only a small amount is usually lost per annum (e.g. 1–5%, Whitehead, 1990; Scholefield *et al.*, 1991). Ammonia can also be volatilized from slurry with 25–35% of the slurry N lost following application on to pasture (van der Meer *et al.*, 1987). This loss can be minimized by injecting the slurry beneath the soil surface. Other losses of ammonia can occur during storage of the slurry.

Emission of N via denitrification in grassland is small (e.g. 2–5 g N ha^{-1} day^{-1}; Ryden, 1984; Luo *et al.*, 1994). However, this is significantly increased by fertilizer application (Jarvis *et al.*, 1995). Grazing further enhances emissions as urine and dung patches are hot spots for denitrification (De Klein and Logtestijn, 1994; Jarvis *et al.*, 1995; Allen *et al.*, 1996; Williams *et al.*, 1997, unpublished data). Leaching losses from cut swards receiving up to 400 kg N ha^{-1} are generally low (e.g. 8–20 kg N ha^{-1} year^{-1}; Dowdell and Webster, 1980). However, losses are much higher in grazed pastures due to an accumulation of nitrate in the soil profile from dung and urine patches. Garwood and Ryden (1986) found that the mean soil nitrate concentration was six to tenfold higher under grazed swards compared with cut swards. Consequently leaching losses were six to ninefold higher from the grazed sward.

The size of the leaching, denitrification and volatilization losses varies between the systems and is very dependent on soil and climatic conditions. The leaching and denitrification losses measured on the New Zealand farm are very low due to the dry winter the year that the measurements were taken (Ledgard *et al.*, 1996b,c). The following year was considerably wetter with three times more drainage and the leaching losses were 74 kg N ha^{-1} for the no fertilizer farm and 101 kg N ha^{-1} for the fertilized farm. This suggests that losses for the New Zealand farm (Table 14.1) may be underestimated. On the English farm, a large amount of N was thought to be lost by denitrification in addition to the 55 kg N ha^{-1} already measured (Jarvis, 1993). Regardless of whether N losses are underestimated or not, recent data (Table 14.1) provide evidence that excessive inputs of N are being used on dairy farms and the excess N may be transmitted to waterways and the atmosphere. Not all of the excess N is lost to the environment. A proportion will remain on the farm where it is immobilized in the soil and accumulates in the soil organic N pool. Typically, 30–100 kg N ha^{-1} are thought to accumulate per annum (Aarts *et al.*, 1992; Berentsen and Giesen, 1995). While these amounts are small in relation to the total amount of soil organic N present in the topsoil of pastoral soils, accumulation over a long period may lead to increased soil N and consequently increased leaching losses. Due to the magnitude of these losses, a number of strategies are being devised in Europe and North America to reduce the N losses, such as restricted grazing and N intake of animals, limiting N application rates of fertilizer and manure and optimizing pasture and crop requirements with manure N application rates (Aarts *et al.*, 1992; Berentsen and Giesen, 1995; Klausner, 1995; Jarvis *et al.*, 1996).

Phosphorus Budgets

Phosphorus budgets on dairy farms show that P is accumulating on dairy farms at rates of 14–92 kg P ha^{-1} year^{-1} (de Willigen and Ehlert, 1996; Haygarth *et al.*, 1998). This accumulation occurs due to the large quantities of P imported on to the farm relative to that exported in animal products. While the net balance is similar between various farming systems around the world, the components of the balance vary. The P budgets for Dutch,

New Zealand and English dairy systems from Aarts *et al.* (1992), Williams (1995) and Haygarth *et al.* (1998) are compared in Table 14.2. In the English and Dutch systems, most of the P inputs are derived from feed but on the New Zealand farm, fertilizer is the main source. The high P fertilizer application is required to maintain the productivity and N_2 fixing activity of the clover component of the New Zealand grass–clover pasture (Haynes, 1981).

The net gain in P on dairy farms will result in the accumulation of P in the soil. The P surplus may contribute to environmental pollution through runoff and soil erosion which can enhance eutrophication of waterways. Steps are being taken to reduce the environmental impact of dairy farming by regulating the application rate of manure, preventing application of manure during times of high loss (winter) and reducing the P content of concentrates (de Willigen and Ehlert, 1996).

Potassium Budgets

As for N and P, high inputs of K can occur on dairy farms in feed and fertilizer. For example, 108 kg K ha^{-1} year^{-1} was imported on to a typical Dutch farm in purchased feed and 30 kg K ha^{-1} year^{-1} in fertilizer (Aarts *et al.*, 1992). There was also a small annual input from the atmosphere of

Table 14.2. Phosphorus budgets for dairy production systems (Aarts *et al.*, 1992; Williams, 1995; Haygarth *et al.*, 1998) (kg P ha^{-1} year^{-1}).

Budget	New Zealand[b] (1994)[a]	England[c] (1994)	The Netherlands[d] (1983–1986)
Inputs			
Fertilizer	56	16.0	15
Feed	3	27.0	32
Deposition		0.2	
Other			1
Total	59	43.2	48
Outputs			
Milk	10	16.0	12
Meat		0.5	4
Other			
Total	10	16.5	16
Budget			
Inputs – Outputs	49	26.7	32

Notes: [a]Years when data were collected.
[b]Data from experimental plots.
[c]Data from lysimeter.
[d]Data from whole farms.

4 kg K ha^{-1}. Only 10–30% of this K was removed from the farm in animal products (Aarts et al., 1992; Klausner, 1995) and consequently there is a large loss of K from the farm via leaching, runoff and erosion. While this may not cause environmental concern, it represents an inefficient utilization of K on dairy farms.

In contrast to the Dutch farm, K inputs on New Zealand dairy farms are relatively small, typically ranging from 0 to 50 kg K ha^{-1} year^{-1} applied as fertilizer (Williams et al., 1990a). Outputs in animal products can range from 10 to 17 kg K ha^{-1} year^{-1} which are similar to the Dutch farm. However, since the dairy cows graze outside on pasture all year round on the New Zealand farm, losses associated with the recycling of excretal K within the pasture will be quite large in comparison with systems where cows are housed inside for part of the year. Recycling of K in urine patches (which accounts for 90% of the excreted K) is particularly inefficient. Some of the urine K moves down through the soil profile immediately following a urination event by macropore flow of the urine. Measurements have shown that cattle urine can flow 40–50 cm down the soil profile following urination (Williams and Haynes, 1994). Potassium carried to this depth can be considered to be lost from the soil–plant system at least in the short term (Williams et al., 1990b) and 20–50 kg K ha^{-1} may be lost by this process per annum (Williams et al., 1990a). As a result of this inefficient recycling of K, losses of K are high relative to inputs and there is a net loss of up to 100 kg K ha^{-1} year^{-1} from most New Zealand dairy farms. However, the total soil K content of soils used for dairying is very high and K deficient symptoms are rarely seen. Nevertheless, continuous depletion of soil K under dairying is not sustainable in the long term and it is not known how long the soil K reserves will continue to supply K at a rate fast enough to maintain highly productive dairy pastures.

Internal Cycling of Nutrients

There are differences between the farming systems described in Tables 14.1 and 14.2 in the internal cycling of nutrients (N, P and K). In systems where the dairy cows graze outdoors all year round only a small amount of manure (and nutrient) is collected from the dairy cows. However, when cows are housed for part of the year, large amounts can accumulate which provides potential problems for environmental pollution. In situations where guidelines are given for the rate of application of manure on to pasture or crops, the guidelines are often based on the N content of the manure. However, the P:N ratio of manure is often higher than the P:N requirements of the plant and thus P accumulates in the soil. If allowances are made for N losses from the manure via leaching, volatilization and denitrification then even greater amounts of P accumulate in the soil. Increased soil P will also lead to increased P content of the herbage eaten and so the manure produced by the cows will become enriched too. Thus the problem of nutrient surplus and disposal of P enriched manure becomes compounded.

BEEF PRODUCTION SYSTEMS

Beef cattle farming systems vary from extensively managed rangelands to intensive pastoral farming and the extremely intensive feedlots. Nutrient flows are similar to those on dairy farming systems. Nutrient inputs occur through fertilizer application, purchased feed, atmospheric deposition and N-fixation. Nutrients are removed in animals sold off the farm. Nitrogen can be lost via leaching, volatilization and denitrification.

Extensive Rangeland

In the extensive rangeland system, cattle graze on undeveloped, unfertilized swards. As little as 10–20% of the above-ground herbage may be consumed in any one year by the cattle due to the low stock density and the poor feed quality (Coleman *et al.*, 1977). Annual inputs of nutrients into this system are mainly through wet and dry atmospheric deposition (Table 14.3) which are very low (e.g. 2–10 kg N ha^{-1}; Woodmansee, 1979). Output of nutrients is also low with 3 kg N ha^{-1} removed in animal products (Table 14.3). All of the manure excreted in rangelands is deposited on to the soil. A high proportion of the excreta N is thought to be lost via volatilization (e.g. 80% of the urine N and 20% of the dung N; Woodmansee, 1979). Volatilization losses are affected by the presence or absence of dung beetles. These insects mix the dung into the soil thereby reducing the potential for volatilization loss. Although losses of N due to volatilization in rangeland are high relative to other losses, the total amount lost is still very small compared with the intensively managed beef production systems. Furthermore, the rangeland system appears balanced. The main environmental problem associated with rangeland farming is overgrazing which leads to soil erosion and surface water contamination rather than to an imbalance of nutrient inputs and outputs.

Improved Grassland

Beef production from improved grassland may be balanced in situations where inputs are low (e.g. no fertilizer is applied; Table 14.3). However, animal production from such systems is very low too. As beef production from grassland intensifies the inputs increase either through fertilizer use or N-fixation by clover plants (Table 14.3). These inputs will result in greater

Table 14.3. Nitrogen budgets for beef production systems on rangeland in South Dakota, USA, and pastures in England from experimental field sites (Woodmansee, 1979; Ryden, 1984; Scholefield et al., 1991) (kg N ha^{-1} year^{-1}).

Budget	Rangeland (1978)[a]	Unfertilized ryegrass pasture (1983–1984)	Ryegrass/ white clover (1986–1987)	Fertilized ryegrass pasture (1986–1987)
Inputs				
Fertilizer				420
Deposition	9	25	15	15
N-fixation	<0.5		160	
Non-symbiotic fixation	<0.5			
Total	9	25	175	435
Outputs				
Meat	3	11	23	29
Volatilization	7	8	10	80
Denitrification			4	40
Leaching			23	160
Other				
Total	10	19	60	309
Budget				
Inputs – Outputs	−1	6	115	126

Notes: [a] Years when data were collected.

pasture dry matter production. Utilization of this pasture is high (up to 80% utilization of the above-ground herbage) in intensively managed pastures and consequently the amount of N cycling through the grazing cattle is also high (> 400 kg N ha^{-1} year^{-1}). This in turn results in large amounts of N being deposited on to the pasture in dung and urine patches. As described in the dairy section, losses of N via volatilization, denitrification and leaching increase as the amount of N deposited in excreta on to the pasture increases. So overall, as N inputs increase there are concomitant increases in the amount of N consumed by the grazing cattle and the losses of N from the farm (Table 14.3; Watson et al., 1992).

Feedlots

Production of beef in feedlots is an extreme form of intensification. Although only a small proportion of the global beef production comes from feedlots (e.g. in the USA 33% of the beef is produced on feedlots; Eghball and Power, 1994), the problems relating to manure management are more intense than on rangeland and grassland. Most feedlots are open and unpaved. The animals are fed concentrates and retain about 10% of the ingested N (Whitehead, 1990). The rest of the nutrients are excreted in dung and urine. The manure is scraped off the surface one to four times per year and stockpiled. It is spread on to crop land in the autumn or spring.

The main N losses in this system occur through ammonia volatilization and denitrification from the manure. Runoff losses are also likely from the feedlot during periods of heavy rain. Approximately 50% of the excreted N may be lost by these processes (Eghball and Power, 1994). There is very little leaching from the feedlot as the surface has zero infiltration due to stock trampling and dispersion of the soil particles from the sodium deposited in the animal excreta. A further 25% of the excreted N is lost when the manure is removed from the feedlot, stored and handled. Further losses of N may occur after application of the manure on to cropland via surface runoff, leaching, volatilization and denitrification. However, providing the rate and timing of application is appropriate for the crop requirements, these losses can be minimal (Chang and Janzen, 1996).

Major N losses from the feedlot operation are associated with the production and handling of the animal manure. Eghball and Power (1994) estimated that 10 million animals in feedlots in the USA produce 475.9×10^6 kg N per annum in manure. If 75% of this is lost then substantial amounts of N are being transferred to the atmosphere and waterways per annum. Similarly 157×10^6 kg P are produced per annum in animal excreta in feedlots. The main mechanism for P loss is runoff from the feedlot, so substantial amounts of P may also be leaking into the environment from this husbandry system. In recognition of these problems, management practices aimed at reducing nutrient losses are being introduced on some feedlots. These include more frequent removal of the manure, using bedding material like straw and mixing denitrification inhibitors and stabilizers into the manure (Eghball and Power, 1994).

SHEEP PRODUCTION SYSTEMS

Compared with dairy production systems, sheep production systems are less intensive and N fertilizers are infrequently used. However, P and S fertilizers, especially superphosphate, are commonly used in sheep grazed pastures in New Zealand (Nguyen and Goh, 1990) to enhance and maintain pasture legume production as biological N-fixation is relied upon to provide most of the N needs.

The range of sheep stocking rates varies widely from less than one sheep to 20–25 or more sheep ha^{-1}. Thus, the grazing intensity has considerable effects on the transfer and cycling of nutrients between the soil–plant–animal system (Ryden, 1984; Goh and Nguyen, 1997). For example, O'Connor and Harris (1992) pointed out that many of the sheep-excreted nutrients in New Zealand extensive hill country grazed pastures are transferred to situations where they may only sparingly contribute to further recycling within the grazed areas.

Nitrogen Budgets

Some of the typical reported N budgets in intensive and extensive sheep production systems in New Zealand and the United Kingdom are shown in Table 14.4. It is evident from the data presented that similar magnitudes of N inputs and outputs occur in extensive upland unimproved pastures in the United Kingdom and New Zealand. Considerable improvement in N inputs

Table 14.4. Nitrogen budgets (kg N ha^{-1} year^{-1}) for an intensive (Quin, 1982) and two extensive (Lambert et al., 1982) sheep production systems in New Zealand (NZ) and a typical extensive upland pasture in the United Kingdom (UK) (Batey, 1982).

Budget	Lowland[a] improved pasture (NZ) (1982)[c]	Hill country[b] improved pasture (NZ) (1976–1977)	Hill country[b] unimproved pasture (NZ) (1976–1977)	Upland extensive pasture (UK) (1977–1980)
Inputs				
Fertilizer	0	0	0	0
Symbiotic N-fixation	130	135	17	8
Non-symbiotic N-fixation		13	13	
Atmosphere		3	3	
Other[d]	10			14
Total inputs	140	151	33	22
Outputs				
Ammonia volatilization	30	11[e]	4[e]	3
Denitrification	10			1
Stock retention	10	9	4	2
Stock transfers	20			
Leaching	70	14	10	5
Gaseous loss during burning				5
Total outputs	140	34	18	16
Budget				
Inputs – Outputs	0	117	15	6

Notes: [a]Data from experimental plots.
[b]Data estimated from various farms.
[c]Years when data were collected.
[d]'Other' includes N from non-symbiotic N-fixation and atmosphere.
[e]Denitrification is included in the ammonia volatilization.

occurred solely from symbiotic N-fixation in New Zealand when these pastures were fertilized with superphosphate, lime and molybdenum fertilizers to enhance clover growth (Lambert *et al.*, 1982).

The intensity of animal grazing in a grass–clover pasture is known to affect the size and balance of various components of both C and N cycling between plants, animals and soil (Ryden, 1984). This occurs largely through the effects of grazing animals on the proportion of clover and grass in the pasture, thus affecting amounts of N fixed and assimilated, and also the N content of the pasture herbage. For example, in contrast to intensive sheep grazed systems (Quin, 1982), extensive sheep grazed systems (Batey, 1982; Lambert *et al.*, 1982) are N accumulating systems. The extent of N input is largely determined by the magnitude of biological N-fixation (Table 14.4). Due to higher stocking rate of sheep in the lowland intensive (14 ewes ha^{-1}) than in the upland extensive (0.25–1.5 ewes ha^{-1}) systems, higher gaseous and leaching losses of N occurred in the lowland intensive system.

In comparing sheep grazed grass–clover and fertilized grass pastures at three grazing intensities in the United Kingdom, Parsons *et al.* (1991) reported that low total N inputs were associated with low N outputs and losses. This is contrary to the data obtained for improved lowland intensive and improved upland extensive sheep pastures (Table 14.4), which, although showing similar amounts of N-fixation, also showed that the lowland improved intensive pastures were associated with higher N outputs.

Grass–clover pastures have been reported to be more efficient than fertilized grass pastures in the utilization and recycling of N within the sheep grazed systems (Parsons *et al.*, 1991; Ruz-Jerez *et al.*, 1995). However, Cuttle *et al.* (1992) found higher nitrate leaching in grazed grass–clover (8–33 kg N ha^{-1} per winter) than fertilized (2–7 kg N ha^{-1} per winter) grazed pastures in two of the three winters studied in the United Kingdom. These differences were attributed to a decline in the clover content of pastures from 35 to 4% and differences in stocking rates rather than the efficiency of N use and cycling. In New Zealand, Ruz-Jerez *et al.* (1995) found that nitrate leaching losses in grass–clover, herbal-ley and fertilized grass pastures (6.9, 7.5 and 10.7 kg N ha^{-1}, respectively) were similar in the first year of the 2 years studied. In the second year, however, nitrate leaching in the fertilized grass pasture was four times greater than that of the other two pastures studied. This substantial leaching of nitrate (23 kg N ha^{-1}) occurred during the winter when a high nitrate concentration in the soil solution coincided with a relatively large drainage (99 mm). Ruz-Jerez *et al.* (1994) also found lower denitrification losses under clover-based pastures. Thus, factors other than pasture type such as nitrate concentration in the soil solution and amount of water available for drainage are important in influencing the comparative leaching losses of nitrate from legume and non-legume-based pastures.

In improved sheep grazed pastures in Australasia, N is predominantly supplied by biological N-fixation (Table 14.4). Inputs of N from rainfall and non-symbiotic N-fixation are generally regarded as insignificant (Goh and Nguyen, 1992). Regular applications of superphosphate fertilizer to New Zealand pastures to maintain pasture production have led to the increase in clover growth and N-fixation as demonstrated in long-term sheep grazing experiments (Nguyen *et al.*, 1989; Nguyen and Goh, 1990). This has resulted in the accumulation of soil N.

Phosphorus Budgets

As mentioned above, regular application of P (and S) as superphosphate is common in improved sheep grazed pastures in Australasia (Blair, 1983; During, 1984; Lewis *et al.*, 1987; Nguyen and Goh, 1990). The efficient use of the applied P depends on the efficient cycling of P through the various P transformation pathways.

Large amounts of P (40–60 kg P ha^{-1}) are usually taken up by a good quality pasture (During, 1984). About 33 kg P ha^{-1} is ingested by sheep from a good quality pasture with an annual dry matter production of 12 t ha^{-1}. Depending on the stocking rates, the average pasture utilization of 75–90% indicates that 10–25% of P and other nutrients taken up by pasture plants is returned to the soil as plant residues. The remainder is subjected to various pathways of P loss processes such as the removal of animal products, animal excreta transfers and leaching. Phosphorus removal in animal products is relatively minor (Table 14.5).

Data from sheep grazed pastures (Table 14.5) receiving long-term superphosphate fertilizer applications for over 35 years in New Zealand show that most of the applied P was recovered (83–94%) in the soil–plant–animal systems (Nguyen and Goh, 1992a). More than half of the applied P (52%) was recovered in the soil within the major plant rooting zone (0–300 mm soil depth). The P losses from leaching were insignificant and were unlikely to occur beyond the major plant rooting zone. Less than 6% of the applied P was estimated to be lost through sheep transfer of excreta to stock camps and P transport from the top to the bottom of the irrigated border strip via the applied irrigation water.

Currently, there is considerable interest in P runoff from agricultural land into freshwater catchments (Macklon *et al.*, 1994; He *et al.*, 1995; Haygarth *et al.*, 1995) because P levels as low as 30 µg l^{-1} can cause eutrophication (Sharpley and Smith, 1989; Turner and Rabalais, 1994). Haygarth and Jarvis (1996) estimated annual P runoff of about 3 kg ha^{-1} from grazed grassland lysimeters in the United Kingdom, while a range of values were reported by other workers in Europe (e.g. 0.08–6.26 kg P ha^{-1} year^{-1} by Ryden *et al.*, 1973; 0.5–4.9 kg P ha^{-1} year^{-1} by Kofoed, 1984; and 2.7 kg total P ha^{-1} year^{-1} by Probst, 1985).

About 12% of the sheep excretal P was estimated to be transferred from the main grazing area to stock camps (Table 14.5). In less intensively grazed

Table 14.5. Phosphorus inputs, outputs and pools (kg P ha^{-1}) in the soil–plant–animal system in sheep grazed pastures treated with three rates of annual superphosphate applications over 35 years in New Zealand from long-term field trial plots (1952–1987)[a] (Nguyen and Goh, 1992a).

Phosphorus pools and budgets	Superphosphate application (kg P ha^{-1} year^{-1})		
	0	188	376
Weathering	4	4	4
Superphosphate fertilizer			
annual basis	0	17.5	35
over 35 years	0	612.5	1225
Soil P pool	2205	2533	2869
Change in soil P due to pasture development and superphosphate applications			
over 35 years	85	413	749
annual basis	2.4	11.8	21.4
Change in soil P due to superphosphate applications over 35 years	—	328	664
Recovery (%) of applied P in soil	—	53.6	54.2
Herbage P pool	8.1	29.1	44.6
Animal P removal	0.6	1.2	1.4
Excretal P output	5.9	22.1	34.3
Excretal transfer			
stock camps	1.3	3.1	4.7
off-farm and off-trial sites	0.2	0.6	0.9
total transfer losses	1.5	3.7 (12.6)[b]	5.6 (11.7)[b]
Plant P residue	1.6	5.8	8.9
Root P residue	5.6	7.9	7.6
P in soil–plant–animal pools	10.2	26.7	39.3
Recovery (%) of applied P in soil–plant–animal pools	—	94.3	83.1

Notes: [a]Years when data were collected.
[b]Expressed as % of applied P after accounting for the amount of excretal P transfer in the control treatment.

hill pastures in New Zealand, 22–50% of the excretal P was transferred to stock camps (Gillingham, 1980). Although most of the sheep-ingested P was returned to camp and non-camp areas in the form of faeces (Nguyen and Goh, 1992a), not all of this P was retained in the soil for subsequent utilization by pasture plants. The returned faecal P may be physically decomposed very rapidly under moist, temperate climatic conditions in New Zealand (Rowarth *et al.*, 1985) and some of it may be lost through leaching or surface runoff.

Applied P from annual superphosphate applications resulted in the accumulation of both inorganic and organic P fractions in soils to a depth of 225 mm (Nguyen and Goh, 1992a). The accumulation of soil inorganic P was most pronounced when superphosphate was applied annually in excess of pasture P requirements.

Potassium Budgets

Complete budgets of K in sheep grazed pastures have not been reported. However, estimates of some inputs and outputs of K in New Zealand pastures have been presented (Goh and Nguyen, 1992). About 13.8 kg K was estimated to be ingested by one stock unit. Only 0.02% (0.03 kg K per stock unit) of this was removed in animal products. Nearly all of the ingested K (99.98%) was returned to the pastures, mainly in the form of urine (Haynes and Williams, 1993). About half to 86% of the ingested K was recycled in the excreta and a considerable proportion was lost from the soil (Goh and Nguyen, 1992).

Sulphur Budgets

The need for assessing S requirements in intensively managed grassland ecosystems is becoming increasingly important because of the worldwide emphasis on sustaining non-renewable fertilizer resources and the reduction of atmospheric S emission. Sulphur has been established for more than two decades as an essential element for animal production and the maintenance of clover-based pastures in Australasia (During, 1984). Sulphur deficiency has become more widespread in Europe (Syers *et al.*, 1987; Zhao *et al.*, 1996) because of the continued reduction in the inputs of S from atmospheric deposition in many areas of western Europe during the last two decades (Whelpdale, 1992; Department of Environment, 1995) and the shift to the use of high-analysis compound fertilizers containing little or no S.

Mass balance S models based on S inputs and outputs are widely used for routine S fertilizer recommendations in New Zealand (Nguyen and Goh, 1993, 1994a,b; Goh and Nguyen, 1997). Results from a long-term field trial in New Zealand show that major inputs of S are from applied fertilizers, especially superphosphate, rainfall and irrigation water, while major S outputs are S leaching losses and sheep excreta transfers to camp sites (Table 14.6).

Rainfall represents the major atmospheric S inputs (7.7–15 kg ha^{-1} year^{-1}) to areas close to the sea, industrial and urban centres (Ledgard and Upsdell, 1991), but all rainfall S is not likely to be available to plants as the sulphate in rainwater may be subjected to leaching (Goh and Nguyen,

Table 14.6. Sulphur inputs to and outputs from (kg S ha^{-1} year^{-1}) the soil–plant–animal system in sheep grazed pastures receiving three rates of annual superphosphate application over 35 years in New Zealand from long-term field trial plots (1952–1987)[a] (Nguyen and Goh, 1992b).

Budget	Superphosphate applications (kg ha^{-1} year^{-1})		
	0	188	376
Fertilizer	0	21	42
Irrigation	6	6	6
Rainfall	3	3	3
Animal S removal	1.4	2.9	3.4
Plant S uptake	6.4	27.8	37.4
Plant S residue	1.3	5.6	7.5
Root S residue	6.3	7.9	7.0
Excretal S output	3.7	19.3	26.5
Excretal S transfers			
stock camps	0.8	2.7	3.6
off-farm and off-trial sites	0.1	0.5	0.7
total transfer losses	0.9	3.2 (11.0)[b]	4.3 (8.1)[b]
Recycled excretal S	2.8	16.1	22.2
Estimated leaching losses	6.7	23.9	43.2
Recovery in soil–plant–animal system	—	35.2	21.2
Efficiency (%) of S utilization from applied and recycled S	—	57.2	46.2

Notes: [a]Years when data were collected.
[b]Expressed as % of applied S after accounting for the amount of excretal S transfer in the control treatment.

1997). Sulphur may also be deposited by dry deposition due to the adsorption of sulphur dioxide (SO_2) by plants and soil. The rate of dry S deposition is proportional to atmospheric SO_2 concentration. Atmospheric SO_2 has been shown to be an important source of S in the United Kingdom and as much as 50 kg S ha^{-1} year^{-1} has been estimated to originate from wet and dry deposition (Syers *et al.*, 1987).

Sulphur losses due to volatilization and soil erosion are generally insignificant. Runoff losses depend on soil drainage, land and vegetation characteristics (Close and Woods, 1986; Goh and Nguyen, 1992). Sulphur removal in animal products is also relatively small (Table 14.6). The major S output is leaching losses, which are substantial (11–43 kg S ha^{-1} year^{-1}) and increase with increasing fertilizer rates applied (Table 14.6). These losses may originate from recycled sheep excreta, especially urinary S and background sources (e.g. fertilizer, irrigation, rainfall). About 60% of excretal S is urinary S (Nguyen and Goh, 1994b) and about 57–75% of ^{35}S-labelled urine applied to a sheep grazed irrigated pasture in New Zealand was not recovered and was estimated to be leached. In addition, soil organic S was found at 300 mm soil depth (Nguyen and Goh, 1992b) suggesting that leaching involved not only sulphate-S but also organic S. Significant amounts of applied S were not recovered; the recovery was only between 21 and 35% (Table 14.6). These leaching S losses may limit the ability of soil S reserves to sustain adequate S for pasture production when S fertilizer application is withheld.

Significant proportions (13–50%) of excreta S are not returned to the main grazing area but are transferred to sheep camp sites, particularly (22–50% of excreta S output) in pastures with low (< 75%) pasture utilization (Goh and Nguyen, 1992; Nguyen and Goh, 1994a). About 8–11% of excreta transfer losses were found in long-term trials (Table 14.6).

LEY FARMING SYSTEMS

In contrast to monocultures of fieldcrops, crop rotations in ley farming (or mixed cropping) systems are generally considered as crucial components of soil and resource conserving agricultural systems (Keeney, 1990; Karlen *et al.*, 1994). Ley farming implies a temporal alternation of food crop and pasture grass or legume utilized to raise livestock and restore soil fertility (Thompson, 1953; Reeves and Ewing, 1993; Pannell, 1995).

In New Zealand, arable crops are grown on the Canterbury Plains for 2–5 years followed by 2–5 years of grass–clover sheep grazed pastures (Haynes and Francis, 1990). In southern Australia, the traditional ley farming system is based on self-regenerating annual pasture legumes such as subterranean clover (*Trifolium subterraneum* L.) and cereal cropping. In other countries, such as Sweden, there has been a large decline in the use of ley crops and an increase in continuous cereals (Andersson, 1986). These changes impact significantly on soil nutrient status and budgets.

Nitrogen Budgets

Most of the reported case studies on N budgets in ley farming systems examined one or two phases of the rotation. Complete measured N budgets for all phases of the rotation in a given ley farming system have not been recorded.

Paustian *et al.* (1990) compared four cropping systems in Sweden over a 2-year period. Their results (Table 14.7) showed a net loss of total N of 10–40 kg N ha^{-1} year^{-1} for both barley treatments with or without N fertilizer applications and for the fertilized grass (*Festuca pratensis* Huds.) ley. On the other hand, the lucerne (*Medicago sativa* L.) ley showed a net

Table 14.7. Nitrogen inputs, outputs and budgets (kg ha^{-1} year^{-1}) in four cropping systems for the topsoil (0–270 mm) from field experimental plots (1982 to 1983)[a] in Sweden (Paustian et al., 1990).

Budget	Cropping system[b]			
	B0	B120	GL	LL
Inputs				
Atmosphere	5.0	5.0	5.0	5.0
Seed N	5.0	5.0	negl.	negl.
Fertilizer	0	120.0	200.0	0
N-fixation	nd	nd	nd	380.0
Total inputs	10.0	130.0	210.0	390.0
Outputs				
Harvest	36.0	127.0	241.0	246.0
Gaseous losses	4.0	5.0	10.0	20.0
Leaching	5.0	10.0	1.0	10.0
Total outputs	45.0	140.0	250.0	270.0
Budget				
Inputs – Outputs	–35.0	–10.0	–40.0	+120.0

Notes: [a]Years when data were collected.
[b]BO, barley receiving no N fertilizer; B120, barley receiving 120 kg N ha^{-1}; GL, grass ley receiving 200 kg N ha^{-1}; LL, lucerne ley; negl., negligible; nd, not determined.

gain of 120 kg N ha^{-1}, comprising 90 kg N ha^{-1} year^{-1} in the plant standing crop and 30 kg N ha^{-1} year^{-1} in the soil. Nitrogen outputs and inputs were about balanced in the unfertilized barley. Net N loss in fertilized barley was 40 kg N ha^{-1} year^{-1}. Fertilizer additions and harvest export were the main flows of N. Nitrogen exported at harvest exceeded fertilizer and seed N input by about 3 and 4 kg N ha^{-1} year^{-1} in unfertilized barley and grass ley, respectively. Non-symbiotic N-fixation was not measured but was regarded to be less than 1 kg N ha^{-1} year^{-1}. Total gaseous and leaching losses of N varied between 1 and 2 kg N ha^{-1} year^{-1}. Denitrification rates were lowest in barley but both fertilized and unfertilized barley leached more N than the grass leys.

Where legume leys are used in the ley farming system, a common practice in Australasia, biological N-fixation by *Rhizobium* in association with legumes is the major N input to the arable systems (Haynes et al., 1993; Nguyen et al., 1995). With forage legumes (e.g. grass–clover pasture), the amount of N fixed is estimated to be in the range of 106–145 kg N ha^{-1} year^{-1} under dryland and 152–226 kg ha^{-1} year^{-1} under irrigation in New Zealand (Crush, 1979). Short-term Australian studies using ^{15}N have shown that the N input depends directly on the legume biomass produced. With subterranean clover, the proportion of N from biological N-fixation was 70–97% (Vallis *et al.*, 1977; Bergersen and Turner, 1983) although this was only 20% in a medic pasture grown under high inorganic N availability (Butler, 1988; Evans *et al.*, 1989).

Agronomic studies in Canterbury in New Zealand have shown that in the first year of cropping after a grass–clover pasture, fertilizer N is not required to achieve optimum arable crop yields (Stephen, 1982). By the second year, about 50 kg N ha^{-1} of fertilizer N is required and for third and following years 100–150 kg N ha^{-1} is needed.

Phosphorus Budgets

The fate of applied P and P budgets have not been studied extensively. Selles *et al.* (1995) reported from a long-term (24 years) crop rotation experiment in Canada that most of the applied P is removed through grain harvest because P leaching was assumed to be low in the semi-arid environment of the trials. Grain harvest accounted for about half (48–65%) of the applied P. The difference between fertilizer P input and P removed in grain was used to indicate whether the systems were accumulating or depleting soil P. It was found that P fertilized systems had a net gain in P, suggesting an accumulation of available P with time while the fallow–wheat–wheat rotation with N applied system had a net loss of P, suggesting a depletion of available P with time.

In another long-term (34 years) crop rotation study in Canada, Campbell *et al.* (1993) showed that P-Olsen was leached beyond the root zone of wheat (> 1.5 m) in fertilized fallow-containing cropping systems on Chernozenic soils. These workers also provided evidence that deep-rooted crops such as legumes may solubilize and use P at depth.

Soil Organic Matter Budgets

Mixed cropping systems have the potential to affect soil organic matter level as, under pastures, soil organic matter generally increases with time (Jackman, 1964; Nguyen and Goh, 1990) until an apparent steady stage is reached (15–26 years) while, under cropping or cultivation, soil organic matter decline occurs over the entire experimental duration of 50 years (Russell, 1980). Haynes *et al.* (1991) found that the total organic matter content of soils in Canterbury, New Zealand, remained relatively constant throughout the cropping rotation and attributed this to the equal length of time (2–5 years) under cropping and pasture. However, Juma *et al.* (1993) concluded after 50 years of research on Gray Luvisolic soils in Canada that soil organic matter content was 20% higher where a 5-year rotation has been used than where a 2-year, wheat and fallow rotation was followed.

Long-term studies on the Canadian Prairies have shown that crop rotations, when properly used, can maintain or enhance soil organic matter level and improve organic matter quality (Biederbeck *et al.*, 1984; Campbell *et al.*, 1991, 1992; Campbell and Zentner, 1993; Kumar and Goh, 1999). Most of these studies show that, with adequate fertilizers, cropping frequently and including legume green manure or hay crops in cereal rotation reduce the loss of soil organic matter and increase its N-supplying power. Changes in soil organic matter were related to the amount of crop residues produced by the rotation system and the ability of the residues in protecting the soil from erosion (Campbell and Zenter, 1993). Soil organic matter remained constant in the well-fertilized fallow winter cereal–wheat rotation because of its shorter fallow rotation period compared with that of the spring-seeded crops, thus reducing soil erosion. Furthermore, winter cereal–wheat rotation is able to produce adequate amounts of crop residues in drought years. The shorter winter fallow period was also found to utilize N more efficiently as less nitrate was leached. Campbell and Zenter (1993) found that the level of soil organic matter was related inversely to the apparent N deficit (i.e. N exported in grain minus N applied as fertilizer).

Nutrient Budgets in Conventional and Alternative Agricultural Systems

In spite of increasing interest in agricultural sustainability (Keeney, 1990), there is little information on the comparative sustainability of conventional and alternative systems of agriculture such as organic or biodynamic (Koepf, 1981) systems. Budgets for N, P and S in three pairs of adjacent conventional and alternative mixed cropping systems in New Zealand were compared recently (Nguyen *et al.*, 1995). Results obtained show that the N and S levels in conventional farms were generally unchanged or gaining (Table 14.8). Thus the supply of these nutrients is not likely to be limiting production. However, P supply is limiting. Nitrogen budgets were positive at all three alternative farms studied but net removal of P and in one case S occurred. The alternative farm with positive P and S budgets received additions of compost, phosphate rock and elemental S.

Major N inputs in alternative farms originated predominantly or entirely from biological N-fixation while N fertilizers supply most of the N in conventional farms (Nguyen *et al.*, 1995). Substantial amounts of P and S and particularly N were removed in both conventional and alternative farms in harvested grain and animal products. Overall, less N, P and S were removed from alternative than conventional farms as alternative farms produced less grain yields and also lower N, P and S in harvested grain (Nguyen *et al.*, 1995).

Table 14.8. Nitrogen, phosphorus and sulphur total inputs, outputs and budgets (kg ha^{-1} year^{-1}) for three pairs of adjacent conventional and alternative mixed cropping farms in Canterbury, New Zealand (1990–1991)[a] (Nguyen et al., 1995).

Farming system and site	N			P			S		
	Inputs[b]	Outputs[c]	Budget[d]	Inputs	Outputs	Budget	Inputs	Outputs	Budget
Conventional									
Kowai	655	505	150	60	78	–18	253	44	209
Temuka	690	695	–5	63	93	–30	57	49	8
Templeton	455	431	24	69	72	–3	83	43	40
Alternative									
Kowai (biodynamic)	825	398	427	< 0.01	54	–54	< 0.001	36	–36
Temuka (organic)	778	398	380	79	51	28	220	30	190
Templeton (organic)	529	292	237	21	54	–33	33	31	2

Notes: [a]Years when data were collected.
[b]N-fixation + fertilizer application.
[c]Crop + animal product removals.
[d]Inputs – outputs.

CONCLUSIONS

Results of case studies as presented provide information on the relative importance of each major input or output of nutrients in temperate grazed pastoral systems and also their state of equilibrium or disequilibrium. Important inputs or outputs which require attention are identified.

This review reveals that considerable differences existed in the quantitative data reported for nutrient budgets, inputs and outputs. The results varied according to the type of farm, the intensity of farming, type of nutrient, method of measurement and site. Thus, it is extremely difficult to generalize on the information reported. However, in general, N is more dynamic in cycling than P, K or S. Fertilizer N inputs varied from none in Australasian pastures to more than 400 kg N ha^{-1} year^{-1} in European pastures. The amount of N-fixation is mostly insignificant in improved pastures in Europe but it constitutes a major N input in upland unimproved sheep pastures in Europe and New Zealand (8–135 kg N ha^{-1} year^{-1}) and in lowland dairy farms in New Zealand (160–200 kg N ha^{-1} year^{-1}). Intensive dairy farms removed more N from the farms than beef production systems (531 vs. 309 kg N ha^{-1} year^{-1}). The P outputs are considerably less than N removals in dairy farms. Potassium outputs in animal products of 10–17 kg ha^{-1} year^{-1} are about similar in dairy farms in Europe and Australasia but a net loss of up to 100 kg K ha^{-1} year^{-1} has been estimated to occur in New

Zealand dairy farms. Estimated S leaching losses in sheep farms in New Zealand varied from 16 to 22 kg S ha^{-1} year^{-1} but these have not been determined in other farming systems.

It is obvious that the reported data are site and case specific. Further research using standardized procedures and methods needs to be conducted on a wider range of conditions. However, nutrient budget information can only be a useful means for controlling and optimizing nutrient cycling processes if the information is used in conjunction with quantitative data on the magnitude and rates of important nutrient cycling processes and their interactions such as mineralization, immobilization, leaching and gaseous loss processes.

To be truly sustainable, a grazed pastoral system needs to be productive and have a balanced nutrient budget with no net gains or losses in the system. In a balanced nutrient budget system, the reliance on outside sources of non-renewable nutrient inputs is not necessary or is minimal. The ley farming system with appropriate lengths of cropping and restorative pasture phases was found to be more likely to achieve a balance of nutrient inputs and outputs especially when the supply of N is based on biological N-fixation. However, a farming system with a balanced nutrient budget must also be productive and yield a return to the farmer and at the same time be socially acceptable to the community. Thus, an understanding of the whole farm system and the quantification of its nutrient budgets, their processes and interactions is necessary if these objectives are to be achieved.

ACKNOWLEDGEMENTS

We thank several researchers, in particular Drs A.C. Edwards, P.M. Haygarth, S.C. Jarvis, E.A. Stockdale and C.A. Watson of the United Kingdom and Associate Professor G.J. Blair of Australia for providing information and reprints of their published work.

REFERENCES

Aarts, H.F.M., Biewinga, E.E. and Van Keulen, H. (1992) Dairying farming systems based on efficient nutrient management. *Netherlands Journal of Agricultural Science* 40, 285–299.

Allen, A.G., Jarvis, S.C. and Headon, D.M. (1996) Nitrous oxide emissions from soils due to inputs of nitrogen from excreta return by livestock on grazed grassland in the U.K. *Soil Biology and Biochemistry* 28, 597–607.

Andersson, R. (1986) Losses of nitrogen and phosphorus from arable land in Sweden. PhD thesis, Swedish University of Agricultural Sciences, Uppsala, Sweden.

Batey, T. (1982) Nitrogen cycling in upland pastures of the U.K. *Philosophical Transactions of the Royal Society, London* B296, 551–556.

Berentsen, P.B.M. and Giesen, G.W.J. (1995) An environmental–economic model at farm level to analyse institutional and technical change in dairy farming. *Agricultural Systems* 49, 153–175.

Bergersen, F.J. and Turner, G.L. (1983) An evaluation of ^{15}N methods for estimating nitrogen fixation in a subterranean clover perennial ryegrass sward. *Australian Journal of Agricultural Research* 34, 391–401.

Biederbeck, V.O., Campbell, C.A. and Zenter, R.P. (1984) Effect of crop rotation and fertilization on some biological properties of a loam in Southern Saskatchewan. *Canadian Journal of Soil Science* 64, 355–367.

Blair, G.J. (1983) The phosphorus cycle in Australian agriculture. In: Costin, A.B. and Williams, C.H. (eds) *Phosphorus in Australia*. Australian Centre for Resource and Environmental Studies, Australian National University, Canberra, Australia, pp. 96–111.

Boller, B.C. and Nosberger, J. (1987) Symbiotically fixed nitrogen from field-grown white and red clover mixed with ryegrasses at low level of ^{15}N fertilization. *Plant and Soil* 104, 219–226.

Butler, J.H.A. (1988) Growth and N_2 fixation by field grown *Medicago littoralis* in response to added nitrate and competition from *Lolium multiflorum*. *Soil Biology and Biochemistry* 20, 863–868.

Campbell, C.A. and Zentner, R.P. (1993) Soil organic matter as influenced by crop rotations and fertilization. *Soil Science Society of America Journal* 57, 1034–1040.

Campbell, C.A., Bowren, K.E., Schnitzer, M., Zentner, R.P. and Townley-Smith, L. (1991) Effect of crop rotations and fertilization on soil organic matter and some biochemical properties of a thick Black Chernozem. *Canadian Journal of Soil Science* 71, 377–387.

Campbell, C.A., Brandt, S.A., Biederbeck, V.O., Zentner, R.P. and Schnitzer, M. (1992) Effect of crop rotations and rotation phase on characteristics of soil organic matter in a Dark Brown Chernozemic soil. *Canadian Journal of Soil Science* 72, 403–416.

Campbell, C.A., Lafond, G.P., Biederbeck, V.O. and Winkleman, G.E. (1993) Influence of legumes and fertilization on deep distribution of available phosphorus (Olsen P) in a thin Black Chernozemic soil. *Canadian Journal of Soil Science* 73, 555–565.

Chalk, P.M. (1985) Estimation of N_2 fixation by isotopic dilution: An appraisal of techniques involving ^{15}N enrichment and their application. *Soil Biology and Biochemistry* 17, 389–410.

Chang, C. and Janzen, H.H. (1996) Long-term fate of nitrogen from annual feedlot manure applications. *Journal of Environmental Quality* 25, 785–790.

Close, M.E. and Woods, P.H. (1986) Leaching losses from irrigated pasture: Waiau Irrigation Scheme, North Canterbury. *New Zealand Journal of Agricultural Research* 29, 339–349.

Coleman, D.C., Andrews, R., Ellis, J.E. and Singh, J.S. (1977) Energy flow and partitioning in selected man-managed and natural ecosystems. *Agro-Ecosystems* 3, 45–54.

Cookson, P., Goh, K.M., Swift, R.S. and Steele, K.W. (1990) Effects of frequency of application of ^{15}N tracer and rate of fertilizer nitrogen applied on estimation of nitrogen fixation by an established mixed grass + clover pasture using ^{15}N-enriched technique. *New Zealand Journal of Agricultural Research* 33, 271–276.

Crush, J.R. (1979) Nitrogen fixation in pasture. IX. Canterbury Plains, Kirwee, New Zealand. *New Zealand Journal of Experimental Agriculture* 7, 35–38.

Cuttle, S.P., Hallard, M., Daniel, G. and Scurlock, R.V. (1992) Nitrate leaching from sheep-grazed grass/clover and fertilized grass pastures. *Journal of Agricultural Science, Cambridge* 119, 335–343.

De Klein, C.A.M. and Logtestijn, R.S.P. (1994) Denitrification and N_2O emission from urine-affected grassland soil. *Plant and Soil* 163, 235–242.

Department of Environment (1995) *Digest of Environmental Statistics*, Report No. 17. HMSO, London.

De Willigen, P. and Ehlert, P.A.I. (1996) Nitrogen and phosphorus balances and losses from the soil in The Netherlands. In: Currie, L.D. and Loganathan, P. (eds) *Proceedings of the Workshop on Recent Developments in Understanding Chemical Movement in Soils*. Massey University, Palmerston North, New Zealand, pp. 7–27.

Dowdell, R.J. and Webster, C.P. (1980) A lysimeter study using nitrogen-15 on the uptake of fertilizer nitrogen by perennial ryegrass swards and losses by leaching. *Journal of Soil Science* 31, 65–75.

During, C. (1984) *Fertilizers and Soils in New Zealand Farming*, 3rd edn. New Zealand Government Printer, Wellington, 355 pp.

Eghball, B. and Power, J.F. (1994) Beef cattle feedlot manure management. *Journal of Soil and Water Conservation* 49, 113–122.

Evans, J., O'Connor, G.E., Turner, G.L., Coventry, D.R., Fettell, N., Mahoney, J., Armstrong, E.L. and Walscott, N.D. (1989) N_2 fixation and its value to soil N increase in lupin, field pea and other legumes in South-eastern Australia. *Australian Journal of Agricultural Research* 40, 791–805.

Garwood, E.A. and Ryden, J.C. (1986) Nitrate loss through leaching and surface runoff from grassland: effects of water supply, soil type and management. In: van der Meer, H.G., Ryden, J.C. and Ennik, G.C. (eds) *Nitrogen Fluxes in Intensive Grassland Systems*. Martinus Nijhoff Publishers, Dordrecht, pp. 99–113.

Gillingham, A. (1980) Hill country topdressing. Application variability and fertility transfer. *New Zealand Journal of Agriculture* 141, 39–43.

Goh, K.M. and Nguyen, M.L. (1992) Fertilizer needs for sustainable agriculture in New Zealand. In: Henriques, P. (ed.) *The Proceedings of the International Conference on Sustainable Land Management*. Hawke's Bay Regional Council, Napier, Hawke's Bay, New Zealand, pp. 119–133.

Goh, K.M. and Nguyen, M.L. (1997) Estimating net annual soil sulfur mineralisation in New Zealand grazed pastures using mass balance models. *Australian Journal of Agricultural Research* 48, 477–484.

Goh, K.M., Edmeades, D.C. and Robinson, D.W. (1978) Field measurements of symbiotic nitrogen fixation in an established pasture using acetylene reduction and a ^{15}N method. *Soil Biology and Biochemistry* 10, 13–20.

Goh, K.M., Edmeades, D.C. and Hart, P.B.S. (1979) Direct field measurements of leaching losses of nitrogen in pasture and cropping soils using tension lysimeters. *New Zealand Journal of Agricultural Research* 22, 133–142.

Haygarth, P.M. and Jarvis, S.C. (1996) Soil derived phosphorus in surface runoff from grazed grassland lysimeters. *Water Research* 31, 140–148.

Haygarth, P.M., Ashby, C.D. and Jarvis, S.C. (1995) Short-term changes in the molybdate reactive phosphorus of stored soil waters. *Journal of Environmental Quality* 24, 1133–1140.

Haygarth, P.M., Chapman, P.J., Jarvis, S.C. and Smith, R.V. (1998) Phosphorus budgets for two contrasting grassland farming systems in the UK. *Soil Use and Management* 14, 160–167.

Haynes, R.J. (1981) Competitive aspects of the grass-legume association. *Advances in Agronomy* 33, 227–261.

Haynes, R.J. and Francis, G.S. (1990) Effects of mixed cropping farming systems on changes in soil properties on the Canterbury Plains. *New Zealand Journal of Ecology* 14, 73–81.

Haynes, R.J. and Williams, P.H. (1993) Nutrient cycling and soil fertility in the grazed pasture ecosystem. *Advances in Agronomy* 49, 119–199.

Haynes, R.J., Swift, R.S. and Stephen, R.C. (1991) Influence of mixed cropping rotations (pasture-arable) on organic matter content, water stable aggregate and clod porosity in a group of soils. *Soil Tillage Research* 19, 77–87.

Haynes, R.J., Martin, R.J. and Goh, K.M. (1993) Nitrogen fixation, accumulation of soil nitrogen and nitrogen balance for some field-grown legume crops. *Field Crop Research* 35, 85–92.

He, Z.L., Wilson, M.J., Campbell, C.O., Edwards, A.C. and Chapman, S.J. (1995) Distribution of phosphorus in soil aggregate fractions and its significance with regard to phosphorus transport in agricultural runoff. *Water, Air and Soil Pollution* 83, 69–84.

Jackman, R.H. (1964) Accumulation of organic matter in some New Zealand soils under permanent pasture. I. Patterns of change of organic carbon, nitrogen, sulphur and phosphorus. *New Zealand Journal of Agricultural Research* 7, 445–471.

Jarvis, S.C. (1993) Nitrogen cycling and losses from dairy farms. *Soil Use and Management* 9, 99–105.

Jarvis, S.C., Scholefield, D. and Pain, B. (1995) Nitrogen cycling in grazing systems. In: Bacon, E. (ed.) *Nitrogen Fertilization in the Environment*. Marcel Dekker, New York, pp. 381–419.

Jarvis, S.C., Wilkins, R.J. and Pain, B.F. (1996) Opportunities for reducing the environmental impact of dairy farming managements: a systems approach. *Grass and Forage Science* 51, 21–31.

Juma, N.G., Izaurralde, R.C., Robertson, J.A. and McGill, W.B. (1993) Crop yield and soil organic matter trends over 60 years in a Typic Cryoboraff at Breton, Alberta. In: *The Brenton Plots*. Department of Soil Science, University of Alberta, Edmonton, Canada, pp. 31–46.

Karlen, D.L., Varvel, G.E., Bullock, D.G. and Cruse, R.M. (1994) Crop rotations for the 21st century. *Advances in Agronomy* 53, 1–45.

Keeney, D. (1990) Sustainable agriculture: Definition and concepts. *Journal of Production Agriculture* 3, 281–285.

Klausner, S. (1995) Nutrient management planning. In: Steele, K. (ed.) *Animal Waste and the Land-Water Interface*. Lewis Publishers, New York, pp. 383–392.

Koepf, H.H. (1981) The principles and practice of biodynamic agriculture. In: Stonehouse, B. (ed.) *Biological Husbandry*. Butterworths, London, pp. 237–250.

Kofoed, D.A. (1984) Pathways of nitrate and phosphate to ground and surface waters. In: Winteringham F.P.W. (ed.) *Environment and Chemicals in Agriculture*. Proceedings of a symposium held in Dublin, 15–17 October, 1984. Elsevier Applied Science Publishers, London, pp. 27–69.

Kumar, K. and Goh, K.M. (1999) Crop residues and management practices: effects on soil quality, soil nitrogen dynamics, crop yield and nitrogen recovery. *Advances in Agronomy* (in press).

Lambert, M.G., Renton, S.W. and Grant, D.A. (1982) Nitrogen balance studies in some North Island hill pastures. In: Gander, P.W. (ed.) *Nitrogen Balances in New Zealand Ecosystems*. Plant Physiology Division, Department of Scientific and Industrial Research, Palmerston North, New Zealand, pp. 35–39.

Ledgard, S.F. and Steele, K.W. (1992) Biological nitrogen fixation in mixed legume/grass pastures. *Plant and Soil* 141, 137–153.

Ledgard, S.F. and Upsdell, M.P. (1991) Sulphur inputs from rainfall throughout New Zealand. *New Zealand Journal of Agricultural Research* 34, 105–111.

Ledgard, S.F., Brier, G.J. and Upsdell, M.P. (1990) Effect of clover cultivar on production and nitrogen fixation in grass-clover swards under dairy cow grazing. *New Zealand Journal of Agricultural Research* 33, 243–249.

Ledgard, S.F., Sprossen, M.S. and Steele, K.W. (1996a) Nitrogen fixation by nine white clover cultivars in grazed pasture, as affected by nitrogen fertilization. *Plant and Soil* 178, 193–203.

Ledgard, S.F., Selvarajah, N., Jenkinson, D. and Sprosen, M.S. (1996b) Groundwater nitrate levels under grazed dairy pastures receiving different rates of nitrogen fertiliser. In: Currie, L.D. and Loganathan, P. (eds) *Proceedings of the Workshop on Recent Developments in Understanding Chemical Movement in Soils*. Massey University, Palmerston North, New Zealand, pp. 229–236.

Ledgard, S.F., Sprosen, M.S., Brier, G.J., Nemaia, E.K.K. and Clark, D.A. (1996c) Nitrogen inputs and losses from New Zealand dairy farmlets, as affected by nitrogen fertilizer application: year one. *Plant and Soil* 181, 65–69.

Lewis, D.C., Clarke, A.L. and Hall, W.B. (1987) Accumulation of plant nutrients and changes in soil properties of sandy soils under fertilised pasture in South-Eastern South Australia. I. Phosphorus. *Australian Journal of Soil Research* 25, 193–202.

Luo, J., Tillman, R.W. and Ball, P.R. (1994) Nitrogen loss by denitrification from a pasture. In: Currie, L.D. and Loganathan, P. (eds) *Proceedings of the Workshop on the Efficient Use of Fertilisers in a Changing Environment*. Massey University, Palmerston North, New Zealand, pp. 139–151.

Macklon, A.E.S., Mackie-Dawson, L.A., Sim, A., Shand, C.A. and Lilly, A. (1994) Soil P resources, plant growth and rooting characteristics in nutrient poor upland grasslands. *Plant and Soil* 163, 257–266.

Metherell, A.K., Cole, C.V. and Parton, W.J. (1993) Dynamics and interactions of carbon, nitrogen, phosphorus and sulphur cycling in grazed pastures. *Proceedings of the XVII International Grassland Congress 1993*, pp. 1420–1421.

Nguyen, M.L. and Goh, K.M. (1990) Accumulation of soil sulphur fractions in grazed pastures receiving long-term superphosphate applications. *New Zealand Journal of Agricultural Research* 33, 111–128.

Nguyen, M.L. and Goh, K.M. (1992a) Nutrient cycling and losses based on a mass-balance model in grazed pastures receiving long-term superphosphate applications in New Zealand. 1. Phosphorus. *Journal of Agricultural Science, Cambridge* 119, 89–106.

Nguyen, M.L. and Goh, K.M. (1992b) Nutrient cycling and losses based on a mass-balance model in grazed pastures receiving long-term superphosphate applications in New Zealand. 2. Sulphur. *Journal of Agricultural Science, Cambridge* 119, 107–122.

Nguyen, M.L. and Goh, K.M. (1993) An overview of the New Zealand sulphur cycling model for recommending sulphur requirements in grazed pastures. *New Zealand Journal of Agricultural Research* 36, 475–491.

Nguyen, M.L. and Goh, K.M. (1994a) Sulphur cycling and its implications on sulphur fertilizer requirements in grazed grassland ecosystems. *Agriculture, Ecosystems and Environment* 49, 173–176.

Nguyen, M.L. and Goh, K.M. (1994b) Distribution, transformation and recovery of urinary sulphur and sources of plant-available sulphur in irrigated pasture soil-plant systems treated with ^{35}sulphur-labelled urine. *Journal of Agricultural Science, Cambridge* 122, 91–105.

Nguyen, M.L., Rickard, D.S. and McBride, S.D. (1989) Pasture production and changes in phosphorus and sulphur status in irrigated pastures receiving long-term applications of superphosphate fertilizer. *New Zealand Journal of Agricultural Research* 32, 245–262.

Nguyen, M.L., Haynes, R.J. and Goh, K.M. (1995) Nutrient budgets and status in three pairs of conventional and alternative mixed cropping farms in Canterbury, New Zealand. *Agriculture, Ecosystems and Environment* 52, 149–162.

O'Connor, K.F. and Harris, P.S. (1992) Biophysical and cultural factors affecting the sustainability of High Country pastoral land uses. In: Hendriques, P. (ed.) *The Proceedings of the International Conference on Sustainable Land Management.* Hawke's Bay Regional Council, Napier, Hawke's Bay, New Zealand, pp. 304–313.

Pannell, D.J. (1995) Economic aspects of legume management and legume research in dryland farming systems of Southern Australia. *Agricultural Systems* 49, 217–236.

Parsons, A.J., Orr, R.J., Penning, P.D. and Lockyer, D.R. (1991) Uptake, cycling and fate of nitrogen in grass-clover swards continuously grazed by sheep. *Journal of Agricultural Science, Cambridge* 116, 47–61.

Paustian, K., Andren, O., Clarholm, M., Hanson, A.C., Johansson, G., Lagerlof, J., Lindberg, T., Petersson, R. and Sohlenius, B. (1990) Carbon and nitrogen budgets of four agroecosystems with annual and perennial crops, with and without N fertilization. *Journal of Applied Ecology* 27, 60–84.

Peoples, M.B., Herridge, D.F. and Ladha, J.K. (1995) Biological nitrogen fixation: An efficient source of nitrogen for sustainable agricultural production. *Plant and Soil* 174, 3–28.

Probst, J.L. (1985) Nitrogen and phosphorus exportation in the Garonne Basin (France). *Journal of Hydrology* 76, 281–305.

Quin, B.F. (1982) The influence of grazing animals on nitrogen balances. In: Gander, P.W. (ed.) *Nitrogen Balances in New Zealand Ecosystems.* Plant Physiology Division, Department of Scientific and Industrial Research, Palmerston North, New Zealand, pp. 95–102.

Reeves, T.G. and Ewing, M.A. (1993) Is ley farming in mediterranean zones just a passing phase? *Proceedings of the XVII International Grassland Congress*

1993, New Zealand Grassland Association, Palmerston North, New Zealand. pp. 2169–2177.

Rowarth, J.S., Gillingham, A.G., Tillman, R.W. and Syers, J.K. (1985) Release of phosphorus from sheep faeces on grazed, hill country pastures. *New Zealand Journal of Agricultural Research* 28, 497–504.

Russell, J.S. (1980) Crop sequences, crop pasture rotations and soil fertility. *Proceedings of the Australian Agronomy Conference*, Australian Society of Agronomy, Melbourne, Australia. pp. 15–29.

Ruz-Jerez, B.E., White, R.E. and Ball, P.R. (1994) Long-term measurement of denitrification in three contrasting pastures grazed by sheep. *Soil Biology and Biochemistry* 26, 29–39.

Ruz-Jerez, B.E., White, R.E. and Ball, P.R. (1995) A comparison of nitrate-leaching under clover-based pastures and nitrogen-fertilized grass grazed by sheep. *Journal of Agricultural Science, Cambridge* 125, 361–369.

Ryden, J.C. (1984) The flow of nitrogen in grassland. *Proceedings of the Fertiliser Society* 229, 1–43.

Ryden, J.C., Syers, J.K. and Harris, R.F. (1973) Phosphorus in runoff and streams. *Advances in Agronomy* 25, 1–45.

Scholefield, D., Lockyer, D.R., Whitehead, D.C. and Tyson, K.C. (1991) A model to predict transformations and losses of nitrogen in UK pastures grazed by beef cattle. *Plant and Soil* 132, 165–177.

Selles, F., Campbell, C.A. and Zentner, R.P. (1995) Effect of cropping and fertilization on plant and soil phosphorus. *Soil Science Society of America Journal* 59, 140–144.

Sharpley, A.N. and Smith, S.J. (1989) Prediction of soluble phosphorus transport in agricultural runoff. *Journal of Environmental Quality* 18, 313–316.

Sherlock, R.R. and Goh, K.M. (1984) Dynamics of ammonia volatilization from simulated urine patches and aqueous urea applied to pasture. I. Field experiments. *Fertiliser Research* 5, 181–195.

Snaydon, R.W. (1987) General introduction. In: Snaydon, R.W. (ed.) *Ecosystems of the World. Managed Grasslands Analytical Studies 17B*. Elsevier, Amsterdam, pp. 3–5.

Stephen, R.C. (1982) Nitrogen fertilisers in cereal production. In: Lynch, P.D. (ed.) *Nitrogen Fertilisers in New Zealand Agriculture*. New Zealand Institute of Agricultural Science, Wellington, pp. 77–93.

Syers, J.K., Skinner, R.J. and Curtin, D. (1987) Soil and fertiliser sulphur in U.K. agriculture. *Proceedings of the Fertiliser Society* 264, 5–43.

Thompson, C.K. (1953) Improved pastures: Their establishment and use (with particular reference to higher rainfall areas of southern Rhodesia). *The Rhodesian Agricultural Journal* 50, 25–39.

Turner, R.E. and Rabalais, N.N. (1994) Coastal eutrophication near the Mississippi river delta. *Nature (London)* 368, 619–621.

Vallis, I., Henzell, E.F., Evans, T.R. (1977) Uptake of soil nitrogen by legumes in mixed swards. *Australian Journal of Agricultural Research* 28, 413–425.

Van der Meer, H.G., Thompson, R.B., Snijders, P.J.M. and Geurink, J.H. (1987) Utilization of nitrogen from injected and surface-spread cattle slurry applied to grassland. In: van der Meer, H.G., Unwin, R.J., van Dijk, T.A. and Ennik, G.C. (eds) *Animal Manure on Grassland and Fodder Crops*. Kluwer Academic Publishers, Dordrecht, pp. 47–71.

Watson, C.J., Jordan, C., Taggart, P.J., Laidlaw, A.S., Garrett, M.K. and Steen, R.W.J. (1992) The leaky N-cycle on grazed grassland. *Aspects of Applied Biology* 30, 215–222.

Whelpdale, D.M. (1992) An overview of the atmospheric sulphur cycle. In: Howarth, R.W., Stewart, J.W.B. and Ivanov, M.V. (eds) *Sulphur Cycling on the Continents*. John Wiley & Sons, Chichester, pp. 5–26.

Whitehead, D.C. (1990) Atmospheric ammonia in relation to grassland agriculture and livestock production. *Soil Use and Management* 6, 63–65.

Whitehead, D.C. (1995) *Grassland Nitrogen*. CAB International, Wallingford, 397 pp.

Whitehead, D.C., Pain, B.F. and Ryden, J.C. (1986) Nitrogen in UK grassland agriculture. *Journal of the Royal Agricultural Society of England* 147, 190–201.

Williams, P.H. (1995) Contaminant budgets for farming systems on the Taierei Plains. *Crop and Food Research Report* 56, 1–25.

Williams, P.H. and Haynes, R.J. (1994) Comparison of initial wetting pattern, nutrient concentrations in soil solution and the fate of ^{15}N-labelled urine in sheep and cattle urine patch areas of pasture soil. *Plant and Soil* 162, 49–59.

Williams, P.H., Gregg, P.E.H. and Hedley, M.J. (1990a) Mass balance modelling of potassium losses from grazed dairy pasture. *New Zealand Journal of Agricultural Research* 33, 661–668.

Williams, P.H., Gregg, P.E.H. and Hedley, M.J. (1990b) Use of potassium bromide solutions to simulate dairy cow urine flow and retention in pasture soils. *New Zealand Journal of Agricultural Research* 33, 489–495.

Woodmansee, R.G. (1979) Factors influencing input and output of nitrogen in grasslands. In: French, N.R. (ed.) *Perspectives in Grassland Ecology*. Springer-Verlag, New York, pp. 117–134.

Zhao, F.J., Hawkesford, M.J., Warrilow, A.G.S., McGrath, S.P. and Clarkson, D.T. (1996) Responses of two wheat varieties to sulphur addition and diagnosis of sulphur deficiency. *Plant and Soil* 181, 317–327.

Epilogue 15

E.M.A. Smaling,[1] O. Oenema[2] and L.O. Fresco[3]*

[1]*Wageningen University and Research Centre, Laboratory of Soil Science and Geology, PO Box 37, 6700 AA Wageningen, The Netherlands;* [2]*Wageningen University and Research Centre, Research Institute for Agrobiology and Soil Fertility (AB-DLO), PO Box 14, 6700 AA Wageningen, The Netherlands;* [3]*Research, Extension and Training Division, Sustainable Development Department, UN Food and Agriculture Organisation (FAO), Viale delle Terme di Caracalla, 00100 Rome, Italy*

INTRODUCTION

In the first chapter of this book, Meine van Noordwijk hinted at the fact that nutrient budgets and nutrient cycling have been discussed since the middle of the 19th century. 'This book', he says, 'is thus based on a long research tradition – and on wheels being reinvented'. He also described the movements of the scientific pendulum as regards soil fertility research. This includes a situation where theories and experiments were largely at odds (from Aristotle up to Von Liebig's days), via a century of research where empiricism dominated ('survival of the fittest'), to a mechanistic era where simulation models have become integrated with experiments.

We feel this book provides more than old wine in new bottles. Given the marked man-induced increases of carbon and nutrient flows worldwide, it is timely to rethink the underlying causes, consequences and possible solutions, in both temperate and tropical regions, in both terrestrial and aquatic ecosystems, and at different spatial and temporal scales. The focus of this book has largely been on managed ecosystems, and therefore excludes chapters on nutrient stocks in ore deposits, ocean floor sinks, and sequestration of atmospheric C. Hence, 'stocks' are limited to nutrients in the soils of arable land, and the biomass found on top of it (man, animals, crops, trees), and to productive aquatic systems. 'Flows' relate to inputs and outputs of carbon and the major nutrients (nitrogen (N), phosphorus (P), potassium (K), sulphur (S), calcium (Ca) and magnesium (Mg)).

---

* This text has been written in a personal capacity and does not represent the views of the Food and Agriculture Organisation of the United Nations or its Member States.

The contributions to this book have made it clear that nutrient cycles and budgets always refer to a particular system or object, and thus have clear spatial and temporal dimensions (Fresco and Kroonenberg, 1994). At global level and at non-geological time-scales, we assume all carbon and nutrient cycles are closed. At all levels below, however, mankind has, over the past century, been responsible for a steep increase in the transport of carbon and nutrients. As a result, sources and sinks of carbon and nutrients have become spatially conspicuous, with the nutrient-aggrading livestock sector in The Netherlands and the declining soil fertility in, say, central Kenya, as two opposite cases in point. The transport of carbon and nutrients will continue to increase now that trade liberalization will further enhance the number of agricultural commodities shipped from place to place.

Depletion of sources and excessive filling-up of sinks put agricultural and environmental sustainability seriously at risk. Hence, the topic of cycles and budgets is no longer merely of scientific interest. There is a clear sense of societal urgency at both researcher, political and farmer level about nutrient flows becoming excessive. In 1997, The Netherlands swine industry was strongly curtailed as a result of both environmental and animal welfare motives. Legislation is developing, ranging from farmer mineral bookkeeping to the banning of bush fires. Until now, the effectiveness of legislation has been greater in high-income than in low-income countries.

Nutrient surpluses, imbalances between nitrogen and other nutrients (Jin Jiyun *et al.*, Chapter 8, this volume), and deficiencies due to mining (Gitari *et al.*, Chapter 11, this volume) are all widespread. The intensity, scale and consequences of these surpluses, imbalances and deficiencies have strongly increased during the 20th century. To counteract the negative effects, nutrient flows have to be managed. This emphasizes the need for integrated nutrient management (INM) at different spatial and temporal scales. INM is defined here as the 'judicious' manipulation of nutrient stocks and flows, in order to arrive at a 'satisfactory' and 'sustainable' level of agricultural production, at minimum environmental cost. Our use of inverted commas already indicates that there are no clearcut objective values for nutrient management, at least not beyond certain depletion or toxicity thresholds. In fact, there are basically two ways of looking into INM: through the hard science approach, attempting to quantify or estimate what is meant by judicious, satisfactory and sustainable, or through the soft science approaches, leaving room for subjectivity. The first approach uses and generates fundamental, scientific knowledge on nutrient flows and INM technologies, whereas the latter uses a combination of scientific, experiential and religious–cultural knowledge, and is tailored to a particular agroecological and socioeconomic situation. INM in technical terms, as used in this book, should lead to managed ecosystems that have nutrient stocks close to a 'target soil fertility', where the balance (outputs minus inputs) of all major nutrients is close to zero for a given time step, and nutrient use efficiency is maximized (Janssen, Chapter 2 this volume).

In this final chapter, we will try to formulate answers to the key questions: what are the causes, what are the consequences, and how do we fight the consequences of nutrient imbalances? The previous chapters as well as other recent work are used for reference. Finally, we will formulate ten commandments which are, in our view, key to achieving managed ecosystems that are based on INM.

NUTRIENT DEPLETION VERSUS NUTRIENT ACCUMULATION

Global Change

The issue of nutrient balances is of profound importance to any understanding of global change. Over the last century, human activities have led to a doubling of the amount of N fixed, which has hence entered the global N cycle. Anthropogenic sources, fertilizer, legume planting and fossil fuel use now contribute as much as natural sources. This has immediate consequences for the concentration of the greenhouse gas nitrous oxide (N_2O) in the atmosphere, acidification of soils and surface waters and subsequent N loading of groundwater and estuaries (Lubchenko, 1998).

With respect to Africa, major data gaps preclude the drawing of definitive conclusions on the effects of global change on African nutrient balances, and vice versa, on the effect of changes in African agriculture on global change. In particular, deforestation, biomass burning (of forest and savanna), intensification of agricultural production, and increased runoff are bound to have considerable impact both locally and regionally. Their contributions are, however, difficult to quantify (Africa and Global Change, IGBP report no 29, Stockholm 1994). Another dimension is the effect of increased carbon dioxide (CO_2) on soil fertility. On a gradual rise in CO_2, for example, a potential increase in fertility may be expected through additional litter production, with higher temperatures leading to faster decomposition rates and mineralization. In particular, if increased CO_2 leads to better root systems, nutrient (and water) availability to plants may increase, especially from deeper layers that would otherwise be unavailable. However, plant physiological responses are subject to very complex feedback mechanisms, resulting in great levels of uncertainties (Bazzaz and Sombroek, 1996). Furthermore, increases in atmospheric CO_2 may be accompanied by changes in climate, that is, changes in temperature and rainfall regimes, which, in turn, may have considerable effects on soil fertility and nutrient flows as well. We conclude here that further attention is needed on the interactions between changes in atmospheric CO_2, regional climate, soil fertility, nutrient flows and crop productivity in Africa.

Nutrient Depletion in Africa

A recent Special Issue of *Agriculture, Ecosystems and Environment* (Smaling, 1998) brings together a number of case studies on nutrient flows and balances in sub-Saharan African managed ecosystems. Some authors attempted to calculate all nutrient flows, whereas others just looked into those flows that can be easily measured or those that are meaningful to farmers. This does not mean, however, that the less visible flows are less important. The major reason for not including nutrient flows such as leaching, denitrification and erosion, is sheer lack of facilities and expertise to measure them on a routine basis. Therefore, many researchers use proxies such as transfer functions, expressing the value of these flows as a function of land properties that are easy to obtain (Elias *et al.*, 1998; Van den Bosch *et al.*, 1998). Almost all case studies concentrate on the two macronutrients N and P. Some also included K and secondary nutrients, but only a few included the carbon cycle.

On the scale of a landscape or village, all studies show that soils are being depleted in nutrients in sub-Saharan African smallholder farming systems. At the same time, however, many farmers appeared to cherish certain plots and niches at the expense of others, the banana homegardens in Uganda and Tanzania being a case in point (Baiju-kya and De Steenhuijsen Piters, 1998; Wortmann and Kaizzi, 1998). A very common and important feature is that communal rangelands and grasslands are exploited everywhere; these provide free lunches in terms of nutrients, entering the farmland with the animals. Hence, the decreasing size of communal lands in many semi-arid parts of Africa constitutes a serious threat to the sustainability of the mixed farming systems. Cases from densely populated areas nonetheless show that farmers become increasingly creative in recycling organic materials (Harris, 1998; Wortmann and Kaizzi, 1998). Hence, a sweeping statement, saying that the situation in Africa is alarming, is not untrue, but there is considerable variation between individual cases at household and village level (Scoones and Toulmin, 1998).

To link household objectives and wealth to nutrient management and mining, De Jager *et al.* (1998) followed a budget approach and found a strong correlation between market orientation of farm households and the nutrient balance. In spite of higher input use, outputs were so high that the balance turned out to be more negative than in subsistence farming. Details on Embu District in central Kenya are worked out in this book (Gitari *et al.*, Chapter 11). Shepherd and Soule (1998) built a dynamic model linking farm performance to nutrient management in west Kenya. They found that poor farmers just do not have the means to adopt sustainable soil management practices, and have to obtain up to 90% of their income from outside the farm.

Also focused on Africa is the recent publication by Buresh *et al.* (1997a) on the replenishment of soil fertility in Africa. Problems related to shortages of N and P are discussed separately, but the replenishment itself needs to be worked out in an integrated manner. Sanchez *et al.* (1997) indicate that P replenishment strategies are mainly mineral-fertilizer based, with biological supplementation. N replenishment strategies are mainly biologically based, with mineral-fertilizer supplementation. Both strategies are further worked out by Buresh *et al.* (1997b) and Giller *et al.* (1997), respectively. Africa has ample phosphate rock deposits. Decomposing organic inputs may facilitate the dissolution of rock phosphates in P-depleted soils. Leguminous tree fallows and herbaceous cover crops grown *in situ* play a major role in N capture and internal cycling in ways compatible with farmer constraints. Soil fertility replenishment was found to be profitable in three case studies, but smallholder farmers lack the capital and access to credit, to make the initial investment. A cost-shared initial capital investment to purchase P fertilizer and germplasm for growth of organic inputs combined with effective microcredit for recurring costs such as fertilizers and hybrid seed is seen as the way forward (Sanchez *et al.*, 1997). The international NGO Sasakawa-Global 2000 believes that only increases in the use of mineral fertilizers can help Africa become food secure (Quinones *et al.*, 1997), whereas Palm *et al.* (1997) indicate that the combined, timely application of organic and mineral fertilizers should be most effective in increasing crop yields at high nutrient use efficiency levels. There are, however, few quantitative data to provide guidelines in this respect.

In the 1993 UN Food and Agriculture Organization (FAO) conference on Integrated Plant Nutrition Systems (IPNS) (Dudal and Roy, 1995), the following conclusions, generally applicable to tropical countries, were arrived at:

- An increased and more efficient use of mineral fertilizers in developing countries is required in the medium term.
- The use of organic sources cannot replace the use of mineral fertilizers.
- Advice on IPNS should refer to nutrient balance sheets and nutrient cycling models at different spatial scales.
- There is a need to build a computerized framework to record and process site characteristics, targeted outputs, estimated losses, inputs of both organic and mineral nature, labour and cash requirements, and anticipated benefits.
- There is a need for a quantitative assessment of the decline of food production resulting from the depletion of plant nutrients.
- There should be active farmers' participation in IPNS, conditioned by economic incentives, improved land tenure and sound agricultural policies.
- Spreading IPNS should be through site-specific diagnosis of constraints, needs and available resources (at field, farm and village level), followed by the design and testing of innovative plant nutrition systems on reference farms and on pilot farms. Positive results should subsequently be

extended through the farming community with the help of farmers' organizations and extension services, and support at decision-making level.

In short, FAO's plea concerns nutrient budgets and cycles as a point of departure for monitoring of productivity changes, defining and implementing conducive policies, and stimulating local initiative. The most remarkable conclusion is that it will not be possible to revive the agricultural sector in large parts of the tropics without the use of mineral fertilizers. Current structural adjustment thinking, however, renders this option null and void for poor farmers.

Nutrient Accumulation in Europe and North America

Following the rapid intensification of agricultural production in many industrialized countries, notably after World War II, surpluses of nitrogen and phosphorus per unit area of agricultural land rapidly increased. Availability of cheap fertilizers stimulated a liberal use of fertilizers and contributed to a boost in crop yield. Moreover, import of cheap animal feed and concentrates from other, mainly developing countries boosted animal production. Over 70% of the nutrients contained in the imported animal feed and concentrates end up in animal manures, which are applied to agricultural land. An extreme example that may well illustrate the case is agriculture in The Netherlands. Some basic data are provided in Table 15.1. There is some uncertainty in the budgets of 1950, 1960 and 1970, because of limited availability of data. Yet, the trend is clear. The surface area of agricultural land slightly decreased, but the number of livestock and the surpluses of nitrogen strongly increased, especially between 1950 and 1985. Surpluses decreased from 1985 onwards, following the introduction of milk quota and governmental policies and measures to reduce nutrient losses (Oenema and Roest, 1998; Oenema et al., 1999). The major part of the phosphorus surplus has accumulated in the soil, the major part of the nitrogen surplus has been lost to the wider environment via ammonia volatilization, nitrate leaching and denitrification. The amount of phosphorus accumulated in agricultural soil after 1950 ranges from 1000 to 5000 kg P ha^{-1}. Rather similar data have been presented for Germany by Weissbach and Ernst (1994, and references therein). The cumulative mean surplus in German agricultural soils has been estimated at 2800 kg N, 1100 kg P and 2400 kg K ha^{-1} for the period 1950–1990.

Data for the mean nitrogen surplus ha^{-1} of agricultural land for 12 member states of the European Union have been presented by Brouwer and Hellegers (1997; Table 15.2). There are close correlations between mean nitrogen input via fertilizers, mean nitrogen input via animal manures

Table 15.1. Changes in agriculture of The Netherlands in the period 1950 to 1995: surface area of agricultural land (in millions ha), livestock numbers (in millions), inputs of nitrogen (N) and phosphorus (P) via imported fertilizers and animal feeds, outputs of N and P via crop production and animal production, and surpluses of nitrogen and phosphorus (in million of kg of N and P) for the whole agricultural sector. 'Surplus' has been defined here as the difference between inputs via imported fertilizers and animal feeds, and outputs via crop production and animal production. (CBS, 1992; 1997; Oenema, unpublished data).

	1950	1960	1970	1980	1990	1995
Surface area	2.3	2.3	2.2	2.0	2.0	2.0
Number of milking cows	1.4	1.6	1.9	2.4	1.9	1.7
Number of pigs	2	2	6	10	14	15
Number of chicken	41	45	55	81	93	84
N-input via fertilizer	156	224	396	485	412	406
N-input via animal feeds	55	113	215	352	424	473
N-output via crop products	20	28	34	39	36	50
N-output via animal products	40	60	93	134	183	220
N-surplus	151	249	484	664	617	609
P-input via fertilizer	52	49	48	36	33	27
P-input via animal feeds	25	40	60	81	80	91
P-output via crop products	8	9	10	9	9	10
P-output via animal products	9	12	16	22	32	40
P-surplus	60	68	85	92	79	69

and mean nitrogen surplus. This illustrates the important role of the livestock production sector in nitrogen surpluses.

There are large differences between farm types within the EU and also within member states. Livestock farms have the largest share of the nitrogen surplus. Also, within a category of farms, nitrogen surpluses vary widely. For example, 25% of the farms growing cereals have a negative nitrogen surplus (deficit), whilst 10% of the farms have a surplus of more than 100 kg ha^{-1}. Frequency distributions of nitrogen surplus of livestock farms are highly skewed; a small percentage of farms has an extremely large nitrogen surplus (Brouwer and Hellegers, 1997). These results are very much in line with the data presented by Goh and Williams (Chapter 14, this volume).

The situation in some Central European countries before 1990 was rather similar to that in Western European countries. Data for Poland indicate that mean phosphorus surpluses decreased rapidly from a mean of about 15 to 20 kg P ha^{-1} year^{-1} in the 1970s and 1980s to about 5 kg P ha^{-1} year^{-1} in the 1990s following the changes in the political and economic system and the concomitant abolition of subsidies on fertilizers (Sapek, 1997). Mean nitrogen surpluses decreased from nearly 100 kg ha^{-1} year^{-1} in

Table 15.2. Estimated mean nitrogen inputs via fertilizers and animal manures, and estimated mean nitrogen surplus ha^{-1} of agricultural land in member states of the European Union in 1990/91 (modified after Brouwer and Hellegers, 1997).

Country	Fertilizers	Animal manures	Surplus
Belgium	163	196	170
Denmark	142	109	114
Germany	128	98	121
Greece	46	64	46
Spain	38	40	19
France	98	62	73
Ireland	60	93	63
Italy	46	55	18
Luxembourg	128	128	121
Netherlands	218	343	321
Portugal	32	40	6
United Kingdom	92	68	59
EU (12 countries)	86	73	71

1985 to less than 70 kg ha^{-1} year^{-1} in 1991 (Sapek, 1996). However, there are large regional differences. Before 1960, mean P surplus of agricultural land was still negative.

Areas with intensive agriculture in the USA, notably areas with animal husbandry, also have large surpluses of nitrogen and phosphorus. The same holds for areas with intensive agricultural production in Asia, as shown for China by Jin Jiyun *et al.* (Chapter 8, this volume). Clearly, the scale and magnitude of nitrogen and phosphorus surpluses of agricultural land have steadily increased this century. At the same time, there has been a shift in scale of the contamination effects. Surpluses of nitrogen and phosphorus contribute to pollution of groundwaters with nitrate, to eutrophication of surface waters and pristine areas, and to acidification of forest soils. Nitrogen surpluses are also related to emissions of nitrous oxide and contribute to global change.

Summarizing, the intensification of agricultural production in many countries in Europe, northern America and Asia has been accompanied by increases in nitrogen and phosphorus surpluses, and by increased losses of nitrogen and phosphorus from agricultural land to the wider environment. These surpluses and losses have in many places become unacceptably large, both in size and scale. The challenge is to decrease surpluses and losses to an acceptable level within a decade or so, whilst maintaining productivity. Scientists and policy-makers must take the lead and define acceptable levels of surpluses and losses.

Case Studies in this Book

In this book, a deliberate choice was made not to focus on a particular hemisphere, nor on a particular agricultural subsector. Cases also range between larger and smaller spatial scales. Macro-level cases include marine fisheries in Norway, plantation forestry in South Africa, fertilizer use in annual cropping in China, and livestock systems in Australasia. At field and farm level, cases include mixed farming systems in Mozambique and Kenya, agriculture–aquaculture in the Philippines, and ecological agriculture in Switzerland. The findings can be summarized as follows:

1. Integrated agriculture–aquaculture systems with low use of mineral fertilizers, and also including trees and livestock, have a much higher nutrient use efficiency than high-input rice monocultures, and also have gross margins which are more than twice as high. The net N balance of both systems is near neutral, but emissions in the monoculture are much higher.

2. A historic overview of fertilizer use in China shows a huge dependence on organic fertilizers for thousands of years, before mineral fertilizers gradually took over in the 1950s. The problem in Chinese agriculture is one of unbalanced nutrition rather than of negative nutrient balances. After periods of huge N application, P fertilizer was introduced as soils ran out of native supply; the same now holds for K which is increasingly utilized. As a consequence, nutrient balances are presently positive for N and P, but negative for K. Experiments have shown that combined NPK application increases N use efficiency. Meanwhile, however, nitrogen and phosphorus use efficiency over the past 20–30 years decreased in rice, and remained constant in maize and wheat.

3. Nutrient depletion was observed in plantation forestry in South Africa. Soils that are already poor lose high amounts of K and Ca when wood is harvested. Acid deposition is a problem in the area, and may mask some of the nutrient depletion effects.

4. Cases in mixed cash and subsistence farming in Kenya and Mozambique showed considerable differences in the nutrient balance, depending on crop choice and input levels. The overall conclusion, however, was that nutrients are being mined in smallholder agriculture. In the Kenya case, farm household economic performance was linked to nutrient management, showing that over 30% of farmers' income is derived from native soil fertility.

5. A case study on nutrient flows in Norway's fisheries sector shows that the proportion of N available in edible fish products was at most 33% of the catch. Efforts to maximize the utilization of fish catch as human food are more effective than comprehensive recycling of waste products as feed.

6. A review of temperate grazed pastoral systems shows that pastures in New Zealand largely depend on biological N-fixation, as opposed to those in Europe. Out of extensive and intensive dairy, beef cattle, sheep and ley farming systems, the last came closest to a balanced and sustainable system.

These studies show that the concept of nutrient budgeting is applicable to a wide variety of ecosystems and at different scales. The studies also show that differences exist between studies in budgeting approach (see also Oenema and Heinen, Chapter 4, this volume). Most important, however, is the conclusion that nutrient budgets can be used to manage and optimize nutrient flows in all these different systems.

CONCEPTUAL HIGHLIGHTS

Based on the contributions to this book and other recent work (Dudal and Roy, 1995; Buresh *et al.*, 1997a; Smaling, 1998), we believe that the following issues are essential to understanding and managing the nutrient balance.

Nutrient Stocks and Flows at Particular Spatial and Temporal Scales

Nutrient stocks and flows can be compared to a bank account, subject to withdrawals and payments. For nitrogen, for example, mineral N (NH_4-N and NO_3-N) is vulnerable capital: 'cash in the pocket' (Giller *et al.*, 1997). Nitrogen in soil organic matter corresponds to monthly or annual savings, whereas the recalcitrant part of soil organic matter that contributes relatively little to N supply within 5–10 years can be labelled 'gold in the bank'.

Nutrient stocks differ spatially. Soil map polygons, representing different soil characteristics such as topsoil depth, organic C content and pH, occur at each possible spatial scale, that is, from the FAO 1:5,000,000 soil map of the world to the map that farmer Njoroge in subhumid Murang'a District in Kenya draws himself when asked to indicate soil productivity differences. Farming often reflects spatial differences, which is most obvious in upland–slope–valley systems, but many other niches can be distinguished too, according to stoniness or texture for example. Temporal scales of nutrient stocks range from geological processes taking millions of years (erosion cycles bringing marketable deposits within reach) to seconds and minutes (replenishment of soil solution upon nutrient uptake by plants).

Until the appearance of agriculture, 10,000 years ago, nutrient flows were largely the result of biogeochemical processes, including photosynthesis, mineralization, weathering, erosion, redeposition and volcanism. Nowadays, the population of over 5 billion people uses both terrestrial and aquatic ecosystems intensively for productive purposes. Bringing 'new' nutrients to production systems is possible through different pathways, but the costs of capturing and applying them will be increasing in the near future. Fertilizer production is a high-energy process, and structural adjustment policies have rendered fertilizer use already uneconomical in many developing countries. At the same time, nutrients end up outside our immediate reach, such as in deep pit latrines, lake, sea and ocean floors, and of course the atmosphere.

Without clear system boundaries, figures and calculations lose a lot of value as they lack context. As said earlier, at the global level, cycles may be considered as closed. Below this level, there are subsystems which are more or less in equilibrium, such as undisturbed tropical rain forests and tundras and taigas. A more accurate definition perhaps is that nutrient cycles are 'closed' in situations where demand for nutrients by growing biomass exceeds current supply (van Noordwijk, Chapter 1, this volume). In natural ecosystems, this is the case, but not in early and late successional stages. In traditional slash and burn systems, this is also the case. In the vast majority of managed ecosystems, however, population pressure and the rapid development of market-driven agriculture and forestry have caused nutrients to move in vast amounts, from A to B, from one sphere to the other, from crop subsystems to livestock subsystems, and from solid to liquid to gaseous phase and back.

A considerable loss of information occurs during upscaling. In general, results for certain scales are based on the individual items in the balance sheets derived for the lower scales, including spatial and temporal variability. Van der Hoek and Bouwman (Chapter 3, this volume) determine nutrient flows in 'animal species' at lower system levels, Dutch and British farms at the intermediate level, and 'animal production' and 'the livestock sector' at the country level.

Nutrient Use Efficiency

The term 'efficiency' generally indicates an output/input ratio of transfers (flows) into and out of pools. Thus, efficiency depends on the boundaries where inputs and outputs are measured (van Noordwijk, Chapter 1, this volume). Nutrient use efficiency (NUE) should be disentangled into application efficiency, uptake efficiency, utilization efficiency, and harvest efficiency. Farm-level NUE can be understood from the NUE of the various components of the farm, but taking account of inputs which are based on outputs of other components. National and farm NUE can be greatly enhanced by recycling, including the urban and industrial sectors, even if field-level efficiency of using recycled wastes is lower than that of using new external inputs. However, agricultural development is increasingly based on farm specialization, increased distance between production sites and markets and a reduction of recycling.

Van Noordwijk (Chapter 1) recognizes five constraints to obtaining high NUEs. From lower to higher spatial scales, these are:

- chemical occlusion and microbial immobilization in soils;
- spatial heterogeneity of fields and farms;
- leaching and gaseous losses;
- export of harvest beyond system boundary;
- economic conditions preventing the (sound) use of fertilizers.

Overall system efficiencies can be studied on the basis of regional balances. The paradox is, however, that improvements in the efficiency can only be studied using knowledge on individual animal categories and farm scale, as outlined in the previous section. Van der Hoek and Bouwman (Chapter 3), for example, show that Dutch dairy farms realized increases in milk production from 8000 to 17,700 kg ha^{-1} over a period of time, and also managed to increase N efficiency from 11.7 to 16%. The nitrogen surplus, however, went up simultaneously by 190 kg ha^{-1} because of the use of imported feedstuffs.

In systems that include aquaculture, unused nutrients from the crop or livestock production system can be recycled efficiently. Animal manure and other organics that would otherwise be lost or used inefficiently are retained within the productive nutrient cycle (Dalsgaard and Prein, Chapter 7; Bleken Azzaroli, Chapter 12, this volume).

Towards the Perfect Managed Ecosystem

It is often stated that sustainability is reached when nutrient inputs equal nutrient outputs. This notion, we realize now, is naive and demands critical examination (Janssen, Chapter 2, this volume).

In the first place, it is not only the balance, but also the absolute values and relative sizes of individual budget items that need attention. This calls for a subdivision in nutrients that are directly available to plants and those that are not. Each nutrient flow may be split into a fraction (f) of available nutrients and a fraction ($1-f$) of nutrients that are not available within the time step considered. A major flow in managed ecosystems is nutrients removed in harvested products. Its value is a reflection of the biomass removed from a system. The nutrients withdrawn must have been 'available' during the growing period considered. Not all nutrients are equally available in time. A poor and degraded soil cannot sustain high crop yields unless measures are taken (erosion control, fertilizer and manure application, etc.). A rich soil, however, will release more nutrients during the growth cycle of a plant than needed. Particularly when N, P and K are unbalanced, losses of the nutrient that is amply available will be inevitable. In other words, every agroecosystem has a 'target fertility', where uptake efficiency is optimized and losses minimized.

Janssen (Chapter 2) suggests that the 'perfect agroecosystem' without emissions is a steady-state system at intermediate time scales, in which the amounts of nutrients added are sufficient to maintain soil fertility at the particular level at which crop growth is not limited by nutrient deficiency. Crops can then just absorb all available nutrients (Janssen, Chapter 2). This implies that the maximum yield obtainable is bound to the given cultivar, prevailing soil/climate, and a situation where nutrient inputs equal nutrient outputs. In this case, there are no other outputs than the removal of harvested products.

Bias and Error

Different nutrient budgeting approaches exist, including farm gate balance, surface balance and system balance approaches. Moreover, different data acquisition strategies exist, depending on the aim of the study, the complexity of the system under study, and available equipment and resources. As indicated by Oenema and Heinen (Chapter 4, this volume), care must be taken when comparing nutrient budgets that have been obtained by different approaches and different data acquisition strategies.

Flaws in nutrient budgets are mainly the result of incomplete understanding of nutrient cycling, which may lead to biases, and of variability in nutrient stocks and flows, which may lead to errors. Five possible sources of biases have been distinguished, i.e. personal, sampling, measurement, data manipulation and fraud, which is consciously introducing bias, and two sources of errors, sampling and measurement errors. Both biases and errors contribute to uncertainty in the nutrient budget. It is clearly indicated that some items can be estimated accurately and at low costs, such as nutrient inputs via fertilizer and nutrient outputs via harvested products, whilst the accuracy of estimated nutrient losses via leaching, volatilization and erosion is in general very low. Spatial and temporal variability of nutrient flows is a major contributor to low accuracy and low precision of the estimated nutrient flows and hence, budget.

Intriguingly, biases and errors occur in both simple one-dimensional systems, involving one trophic level, and in complex three-dimensional ecosystems, involving different trophic levels. Moreover, biases and errors in small-scale and large-scale studies can be equally large, when expressed percentage-wise.

Importantly, uncertainty in nutrient budgets is tractable and uncertainty can be quantified. Application of different budgeting approaches at compartment and system level, in combination with monitoring of nutrient stocks in soil over a number of years, appears to be most useful to check for consistency in the nutrient budgets and to trace biases and errors. Biases and errors also point at the fact that nutrient cycling is complex, and understanding of nutrient cycling processes and losses is still incomplete. There is also a lack of appropriate equipment to quantify all nutrient losses in the field.

THE ROAD TO BETTER NUTRIENT MANAGEMENT

Technologies

INM is an open-ended strategy geared at the optimal use of nutrients within a system, and the addition of new nutrients to supplement those that are taken away. Basically, INM strategies and technologies can be categorized as those that:

- *save* nutrients from being lost from the system, such as erosion control, restitution of residues, recycling of household waste and animal manure, and integration of trees or aquaculture which provide a safety net function;
- *add* nutrients to the system, such as the application of mineral fertilizers and amendments, concentrates for livestock, organic inputs from outside the farm, and N-fixation in wetland rice and by leguminous species;
- improve nutrient use *efficiency*, including improvement of application efficiency, uptake efficiency, utilization efficiency and harvest efficiency, by improved application techniques, use of higher yielding crop varieties, improved pest, weed and irrigation practices, etc.

Technologies can also be grouped on the basis of their beneficial impact on the different budget items of the nutrient balance (Stoorvogel, Chapter 5, this volume; Smaling and Braun, 1996). Next, it is important to realize how manageable the different flows are (Janssen, Chapter 2, this volume; Stoorvogel, Chapter 5, this volume). Each possible nutrient management strategy is judged implicitly by farmers in terms of risk. Risk of crop failure and thereby loss of fertilizer money, and risk of yield reduction are important drivers for adjustment of fertilizer input. The perception of risk also differs widely among farmers. Unfortunately, little effort has been made so far to induce quantitative risk analysis in nutrient management planning. Given the fact that nutrient losses occur in spatially and temporally discrete events and also that nutrient contamination of the wider environment becomes more and more an economic issue, risk analyses should receive more attention in nutrient management planning.

Nutrient Policies

Without proper policies, nutrient transports and redistribution via fertilizers and agricultural produce will continue unabated. This could create a situation where part of the agricultural land loses productivity to the extent that farmers have no option but to migrate. The other side of the spectrum shows that some agricultural systems with very high nutrient inputs as compared to outputs can no longer be maintained because of adverse environmental effects and consumer concern. Macroeconomic policy reform, however, is no guarantee for sustainable land use due to persistent poverty and structural imperfections. Positive and negative environmental effects are often not noted, and are mostly less important than growth and distribution issues (Kuyvenhoven *et al.*, Chapter 6, this volume).

Given the current structural adjustment framework, policy conclusions are:

- Although there are fewer price and market interventions, the effect of output price adjustments on the soil nutrient balances is far from conclusive.
- More emphasis is needed on public expenditure for infrastructure, support services and institutional development.
- A shift is needed from commodity to factor taxation.
- Appropriate and targeted interventions should be allowed in the case of genuine market failures (environmental externalities, degradation of common resources, national goals like equity or food security not satisfied by markets).

Kuyvenhoven *et al.* (Chapter 6), for example, show that in Senegal subsidy removal reduced fertilizer use. Soil quality decreased and farmers reacted by increasing groundnut seeding densities, expanding cultivated area. The devaluation of the West African Franc CFA insufficiently restored the price ratio between groundnut and fertilizer. In Costa Rica, on the other hand, decreasing fertilizer prices changed technologies and improved nutrient balances and farm household income. Here, subsidies on fertilizers turned out to be an appropriate instrument to induce the desired modification in land use in Costa Rica. The effects of subsidies on nutrient use efficiency need further examination.

Facilitating Social Learning

Although this book does not provide accounts from the social and communication sciences, farmers' own readiness to invest in INM is of paramount importance. So far, we have defined INM mainly in technical terms. However, it would be a mistake to ignore its crucial social learning dimension. INM can only be carried out by farmers who are experts at managing their complex soils and who are able to capture the opportunities provided in the local situation to improve soil fertility. It is especially with respect to its learning dimension that INM can benefit from integrated pest management (IPM). In fact, it is the FAO-assisted IPM Training Programme in Indonesia that has introduced a new approach to agricultural innovation and tested it under field conditions, scaling it up to a point where more than 400,000 small-scale rice growers have now been trained (Röling and Van de Fliert, 1994). It is obvious that, just as IPM is not primarily about bugs, INM is not primarily about soils. In fact, nutrients are just another entry point for this approach. The insight dawns upon us that agriculture cannot be developed without banking on the intelligence, creativity and competence of farmers. Instead of on *adoption*, the emphasis now is on continuous *learning*. Farmers become experts, not by adopting science-based technologies, but by becoming better learners. A framework to achieving this is provided by Deugd *et al.* (1998), and a practical example for southern Mali is worked out by Defoer *et al.* (1998).

OUR TEN COMMANDMENTS OF INTEGRATED NUTRIENT MANAGEMENT

We conclude with what we consider to be the ten most important issues surrounding INM:

1. The ever increasing scale and intensity of the nutrient imbalance of managed ecosystems during the last decades is of global, continental and regional concern; we should launch a supra-national effort, geared at the promotion of INM at political and farm household levels to reverse the trend and to minimize the consequences.

2. We should adopt nutrient budgets of agroecosystems as the appropriate indicators for evaluating nutrient management plans and policies at different spatial and temporal scales, but only in combination with nutrient use efficiency and soil fertility levels.

3. We think that both direct and indirect subsidies and levies on fertilizers, animal feeds and nutrient surpluses are very important tools for managing nutrient budgets.

4. We believe that agroecosystems can and have to recycle the nutrient-containing wastes from processing industries and human populations; investment in nutrient-containing agricultural produce would greatly facilitate this process at different spatial scales.

5. We should invest research funds in novel nutrient management strategies, such as nitrogen fixation by non-leguminous species, the possible disposal of organic waste in those parts of the ocean where marine life is not very developed, in integrated agriculture–aquaculture systems, and in ways to capture nutrients that are bound to be lost 'forever' in pit latrines and on lake, sea and ocean floors.

6. We must define threshold values, and calculate and monitor how systems may become 'perfect agroecosystems', and how the concept can be translated into something meaningful for those involved in achieving that.

7. We must find ways to link endogenous drivers (better INM practices as part of livelihood strategies) with exogenous drivers (conducive policies) as a joint investment in INM.

8. We also must internalize soil fertility depletion and environmental pollution in environmental economics, ending a period of the natural resource 'soil' being regarded as a good without a price.

9. We should establish, in different parts of the world, catchments where changes in land quality can be monitored on a long-term basis so as to continuously validate and modify our assumptions on spatial and temporal nutrient dynamics, as a function of land use and management.

10. It is essential to bank on farmers' capacities for continuous learning and innovation to manage nutrients in an integrated and time and space-sensitive manner.

REFERENCES

Baijukya, F.P. and De Steenhuijsen Piters, B. (1998) Nutrient balances and their consequences in the banana-based land use systems of Bukoba District, North-west Tanzania. *Agriculture, Ecosystems and Environment* 71, 147–158.

Bazzaz, F. and Sombroek, W.G. (1996) Global climatic change and agricultural production: an assessment of current knowledge and critical gaps. In: Bazzaz, F. and Sombroek, W. (eds) *Global Climatic Change and Agricultural Production*. Wiley/FAO, pp. 319–330.

Brouwer, F. and Hellegers, P. (1997) Nitrogen flows at farm level across European Union agriculture. In: Romstad, E., Simonsen, J. and Vatn, A. (eds) *Controlling Mineral Emission in European Agriculture*. CAB International, Wallingford, pp. 11–26.

Buresh, R.J., Sanchez, P.A. and Calhoun, F. (1997a) *Replenishing Soil Fertility in Africa*. SSSA Special Publication 51, Soil Science Society of America, Madison, Wisconsin, and ICRAF, Nairobi, 251 pp.

Buresh, R.J., Smithson, P.C. and Hellums, D.T. (1997b) Building soil phosphorus capital in Africa. In: Buresh, R.J., Sanchez, P.A. and Calhoun, F. (eds) *Replenishing Soil Fertility in Africa*. SSSA Special Publication 51, Soil Science Society of America, Madison, Wisconsin, and ICRAF, Nairobi, pp. 111–150.

CBS (1992) Mineralen in de landbouw, 1970–1990: fosfor, stikstof, kalium. Milieustatistieken, SDU, CBS, Den Haag, 47 pp. (In Dutch.)

CBS (1997) Mineralen in de landbouw, 1995. Kwartaalbericht Milieustatistieken. CBS-97/4, Den Haag, pp. 18–23. (In Dutch.)

Defoer, T., De Groote, H., Hilhorst, T., Kanté, S. and Budelman, A. (1998) Participatory action research and quantitative analysis for nutrient management in Southern Mali: a fruitful marriage? *Agriculture, Ecosystems and Environment* 71, 215–228.

De Jager, A., Kariuki, I., Matiri, F.M., Odendo, M. and Wanyama, J.M. (1998) Monitoring nutrient flows and economic performance in African farming systems (NUTMON). IV. Linking farm economic performance and nutrient balances in three districts in Kenya. *Agriculture, Ecosystems and Environment* 71, 81–92.

Deugd, M., Röling, N. and Smaling, E.M.A. (1998) A new praxeology for integrated nutrient management: facilitating innovation with and by farmers. *Agriculture, Ecosystems and Environment* 71, 269–284.

Dudal, R. and Roy, R.N. (eds) (1995) Integrated plant nutrition systems. *FAO Fertilizer and Plant Nutrition Bulletin* 12, FAO, Rome.

Elias, E., Morse, S. and Belshaw, D.G.R. (1998) Nitrogen and phosphorus balances of Kindo Koisha farms in southern Ethiopia. *Agriculture, Ecosystems and Environment* 71, 93–114.

Fresco, L.O. and Kroonenberg, S.B. (1992) Time and spatial scales in ecological sustainability. *Land Use Policy* 9, 155–168.

Giller, K.E., Cadish, G., Ehaliotis, C., Adams, E., Sakala, W.D. and Mafongoya, P.L. (1997) Building soil nitrogen capital in Africa. In: Buresh, R.J., Sanchez, P.A. and Calhoun, F. (eds) *Replenishing Soil Fertility in Africa*. SSSA Special Publication 51, Soil Science Society of America, Madison, WI, and ICRAF, Nairobi, pp. 151–192.

Harris, F. (1998) Farm-level assessment of the nutrient balance in Northern Nigeria. *Agriculture, Ecosystems and Environment* 71, 201–214.

Lubchenko, J. (1998) Entering the century of the environment: a new social contract for science. *Science* 279, 491–497.

Oenema, O. and Roest, C.W.J. (1998) Nitrogen and phosphorus losses from agriculture into surface waters: the effects of policies and measures in The Netherlands. *Water Science and Technology* 37, 19–30.

Oenema, O., Boers, P.C.M., van Eerdt, M.M., Fraters, B., van der Meer, H.G., Roest, C.W.J., Schröder, J.J. and Willems, W.J. (1999) Leaching of nitrate from agriculture to groundwater: the effects of policies and measures in The Netherlands. *Journal of Environmental Pollution* (in press).

Palm, C.A., Myers, R.J.K. and Nandwa, S.M. (1997) Combined use of organic and inorganic nutrient sources for soil fertility maintenance and replenishment. In: Buresh, R.J., Sanchez, P.A. and Calhoun, F. (eds) *Replenishing Soil Fertility in Africa*. SSSA Special Publication 51, Soil Science Society of America, Madison, Wisconsin, and ICRAF, Nairobi, pp. 193–218.

Quinones, M.A., Borlaug, N.E. and Dowswell, C.R. (1997) A fertilizer-based Green Revolution for Africa. In: Buresh, R.J., Sanchez, P.A. and Calhoun, F. (eds) *Replenishing Soil Fertility in Africa*. SSSA Special Publication 51, Soil Science Society of America, Madison, Wisconsin, and ICRAF, Nairobi, pp. 81–96.

Röling, N. and Van de Fliert, E. (1994) Transforming extension for sustainable agriculture: the case of Intergrated Pest Management in rice in Indonesia. *Agriculture and Human Values*, II (2+3), Spring and Summer, pp. 96–108.

Sanchez, P.A., Shepherd, K.D., Soule, M.J., Place, F.M., Buresh, R.J., Izac, A-M.N., Mokwunye, A.U., Kwesiga, F.R., Ndiritu, C.G. and Woomer, P.L. (1997) Soil fertility replenishment in Africa: an investment in natural resource capital. In: Buresh, R.J., Sanchez, P.A. and Calhoun, F. (eds) *Replenishing Soil Fertility in Africa*. SSSA Special Publication 51, Soil Science Society of America, Madison, Wisconsin and ICRAF, Nairobi, pp. 1–46.

Sapek, A. (1996) Nitrogen balance on national, regional and farm level in Poland. In: Van Cellemput *et al.* (eds) *Progress in Nitrogen Cycling Studies*. Kluwer Academic Publishers, Dordrecht, pp. 371–375.

Sapek, A. (1997) Phosphorus cycle in Polish agriculture. In: Sapek, A. (ed.) *Phosphorus in Agriculture and Water Quality Protection*, Falenty IMUZ Publisher, Poland, pp. 8–18.

Scoones, I. and Toulmin, C. (1998) Soil nutrient balances: what use for policy? *Agriculture, Ecosystems and Environment* 71, 255–268.

Shepherd, K.D. and Soule, M.J. (1998) Soil fertility management in West Kenya: dynamic simulation of productivity, profitability and sustainability at different resource endowment levels. *Agriculture, Ecosystems and Environment* 71, 131–146.

Smaling, E.M.A. (ed.) (1998) Nutrient balances as indicators of productivity and sustainability in sub-Saharan African agriculture. *Agriculture, Ecosystems and Environment* 71/1,2,3 (Spec.Issue), 283 pp.

Smaling, E.M.A. and Braun, A.R. (1996) Soil fertility research in sub-Saharan Africa: new dimensions, new challenges. *Communications in Soil Science and Plant Analysis* 27, 365–386.

Van den Bosch, H., Gitari, J.N., Ogaro, V.N., Maobe, S. and Vlaming, J. (1998) Monitoring nutrient flows and economic performance in African farming systems (NUTMON). III. Monitoring nutrient flows and balances in three districts in Kenya. *Agriculture, Ecosystems and Environment* 71, 63–80.

Weisbach, F. and Ernst, P. (1994) Nutrient budgets and farm management to reduce nutrient emissions. In: 't Mannetje, L. and Frame, J. (eds) *Grassland and Society*, Wageningen Pers, The Netherlands, pp. 343–360.

Wortmann, C.S. and Kaizzi, C.K. (1998). Nutrient balances and expected effects of alternative practices in farming systems of Uganda. *Agriculture, Ecosystems and Environment* 71, 115–130.

Index

Acacia spp. 186
Accuracy 77, 78
Acid
    deposition 184–186
    load 187
Acidification 183–186, 187, 297, 302
Aerosols 185
Afforestation 186
Agricultural production 101, 247–262
Agricultural systems 57
Agroecology 296
Agronomic efficiency 12, 16, 17
Alternative agricultural system *see*
    Farming systems
Animal 204, 209, 210, 215, 218, 223, 306
    feeds 216, 221, 229, 240, 300, 310
    production 59, 63–69, 216, 272, 278, 300
    see also Cattle; Fish; Livestock; Sheep
Anthropogenic nitrogen
    enrichment 237–238, 240, 242, 243,
Anthropogenic nutrient enrichment 304
Anthropogenic pollution 229, 235,
Apparent nitrogen recovery (ANR) 259
Application efficiency 11, 13, 103
Aquaculture 230–231, 239

Ashes 50, 52
Availability index 10

Bacteria
    cyanobacteria 144
    heterotrophic 106
    photosynthetic 106
Barley 251–252, 260
Bat guano 192
Biases 77, 307
    data manipulation biases 82
    due to fraud 82
    measurement biases 87
    personal biases 80, 81
    sampling biases 81
Bio-availability 7, 10
Biodiversity 178, 188, 248, 261
Biodynamic agriculture *see* Farming
    systems
Bio-industry 53
Biomass 59, 168, 209, 251
    earthworm 255
Bioturbation activity 146
Breeding of trees, improved 178
Budget approach 298
Buffering capacity 42, 184
Burning *see* Fire

Calcium (C) removed per harvest 175
  see also Nitrogen; Nutrients;
    Phosphorus; Potassium; Sulphate
Carbon
  balance 134, 135, 297, 298
  see also Soil organic matter
Carbon dioxide (CO_2) 146, 297
Carnivore 6
Cash flows 211, 227
Cassava 15
Cation balance 184–186
Cattle 63, 69
  beef production system 265–266, 273–275, 304
CEC 45
Chemical occlusion 14, 306
Chemical precipitation 34
Climate 47, 50, 178, 183, 204, 208, 212, 251–253, 297, 307
  tropics 107
  weather conditions 46
Cocoa 127
Conventional farming system *see* Farming systems
Cotton 127, 201, 204
Crop failure 204, 206, 208
Crop growth 38, 46
Crop improvement 107
Crop production 65, 69, 158, 216
Crop residues 65, 192, 195, 197, 198, 201, 204, 205, 208, 216, 225, 248, 278, 308
Crop yield 251, 261, 300
Currency devaluation 126, 128

Dairy production system 265, 267–273, 304
Data availability 58
Data gaps 297
Decomposition 5, 6, 144
  fungi 183
Denitrification 47, 48, 86, 105, 197–198, 210, 269–270, 272, 273, 274, 275, 277, 298, 300
Deposition 216, 267, 269, 304
  atmospheric 47, 49, 78, 84, 91, 175, 187, 238, 259, 280
  dry 6, 31, 180, 181, 281

  wet 105, 181, 194, 196, 199, 278
  see also Acid
Desorption 34
Dimension rules 17, 19
Double cropping 165
Draining 124
Dung *see* Manure
Dust 42

Ecological agriculture *see* Organic agriculture
Ecological performances 247–248, 261
Economic analysis 216
Economic benefits 207
  from labour 169
  and sustainability 169
Economic environment 50
Economic performance 211, 220–227
Economics, competitive 240
Ecosystem soil 168
Edible fish product 230, 239, 241
Edible food product 230, 236
Energy 187, 260, 261, 304
Environmental cost 296, 310
Environmental quality 167–168, 261
Erosion 5, 30, 31–34, 45–52, 107, 122, 123, 144, 145, 159, 168, 180, 187, 194, 195, 198, 206, 209, 216, 227, 271, 273, 281, 298, 304, 307
  control 30, 33, 126, 306, 308
  water 216
Eucalyptus grandis 173, 182, 186
Errors 77, 307
  measurement error 83, 87, 89
  relative error 84
  sampling error 83
Eutrophication 76, 240, 242, 271, 302
  lakes 168
Excreta production 146
Export of harvest 306
External input 158

Fallow 191, 298
Fallow containing systems 206, 284
Farm gate balance 78, 79

Farm household 122, 132, 135
Farming systems 214–215, 247–262, 265–287, 305
Feedback mechanism 297
Fertilizers 4, 49, 52, 59, 61, 65, 91, 93, 124–130, 143, 259, 261, 265, 267, 273, 285, 303, 306, 307, 310
   application 36, 46, 70
   balanced use 173
   chemical 15, 33, 62, 102, 104, 122, 123, 125, 126, 127, 144, 147, 157, 159, 161–163 170, 172, 195, 199–200, 205, 207, 209, 216, 221, 227, 240, 243, 247, 286, 299, 308
      balanced use 168
      production increase due to 172
      solubility 206
   consumption 162
   efficiency 172, 173, 278
   forest 178
   inorganic see Fertilizers, chemical
   mineral see Fertilizers, chemical
   nitrogen 2, 21, 30, 42, 60, 65, 161, 209, 230, 269, 284
      unbalanced use 164
      production increase due to 172
   organic 105, 107, 194, 196, 199–200, 205, 216, 227, 250, 299, 303
   phosphorus 21, 32, 192, 271, 278, 280, 303
   potassium 271
   production 304
   rates 281, 309
   rock phosphate 42
   strategy 249
   subsidies 127, 128, 131–135
   sulphate (free) 165, 280
   unbalanced use 165–167, 170, 173
   uneven use and distribution 171
   urea 144
   use efficiency 157
Field heterogeneity 17
Fire 5, 14, 50, 52, 105, 184, 192, 197, 202, 205, 208, 297
Fish 145, 146, 229–243, 303
   carnivorous 229, 239
   herbivorous 239
Fish catch 231–234, 242
Fish farming 142, 229, 234–236, 239
Fish feeding efficiency 241–242
Fish for feed production 234
Fish recycling 232
Fluctuations, seasonal or annual 35
Food balance 133
Food production 128
Food secure 299
Forage 84
Forest plantation 178, 186, 187
   see also Sustainability
Fruit trees 188
Fungi see Decomposition
Fungicides 251

Garbage 199, 216, 218, 310
   recycling 240, 305, 308
Gaseous losses 14, 30, 31, 48, 49, 195, 197, 201, 216, 221, 227, 287, 306
   see also Denitrification; Fire; Volatilization
Grain 72, 73, 161, 173
Grass clover 251–252, 260
Grassland 178–180, 183–184, 186, 188, 282
   productivity 266
Grazing system 227
Growing season 31, 32, 34, 45

Hard science approach 296
Harvested product 15, 195, 197, 201, 202, 206, 303, 306
Harvest efficiency 12, 103, 107, 187
Herbivore 6
High input 303, 304
Human faeces 216
Human fish consumption 229, 232–242
Human nutrient intake requirement 218
Human nutrition 229, 230, 240

Illegal practices 238

Immobilization 34, 47, 87, 181–183, 287, 306
Income 215
Industrial origin 186, 305
Input cost 158
Integrated agriculture aquaculture (IAA) systems 142, 303
Integrated nutrient management (INM) *see* Nutrient, management, integrated
Integrated plant nutrition system (IPNS) 299
Intensification 297, 300
Inter-cropping 107, 187
Irrigation 108, 124, 195, 196, 216, 278, 280, 283
　　causing nitrogen losses 164
　　water 144

Land use shift 171
Land use zones (LUZ) 211–214, 218–227
Land utilization intensity 193–194
Lawes and Gilbert experiments 2
Leaching 5, 30, 31, 47–53, 60, 83, 105, 106, 108, 145, 148, 216, 221, 225–227, 266, 306
　　of fertilizers 206
　　of nitrogen 164, 168, 201, 230, 269, 273, 274, 275, 277, 300
　　of nutrients 14, 195, 197, 225, 287, 298, 307
　　of phosphorus 278, 279, 284
　　of sulphate 280, 281, 287
Leaf banding system 186
Leakages 85
Legumes 173, 269, 284, 285
Leguminous crops 106, 187, 196, 199–200, 204, 205, 206, 208, 308
Leguminous trees 298
Less developed country (LDC) 121
Levelling 124
Limiting factor 204
Litter 181–183, 297
Litterfall 5
Livestock *see* Animal
Livestock systems 107, 208, 303, 305

Low education level 172
Low external input agriculture (LEIA) 49

Maize 15, 127, 165, 166
Management practices 172
Manure 32, 53, 58, 60, 107, 128, 192, 204, 205, 208, 212, 215, 247, 259, 261
　　animal 29, 60–65, 69, 93, 94, 122, 126, 196, 207, 216, 273, 306, 308
　　grain 197
　　green 106, 122, 197
　　organic 33, 160–163, 216, 225
　　production 64, 105
　　recycling 126
Market orientation 227, 298
Meadow 63
Method of Dabin 43
Microbial activity 42
Milk production 63
Millet 127
Mineralization 34, 35, 47, 59, 91, 197, 260, 269, 287, 297, 304
　　rates 47, 108
Mixed cropping system 192, 198, 265–266, 282–286, 303–304, 308
Mobility of phosphorus 197
Models
　　bio-economic models 130
　　ECOPATH 146, 147
　　modelling 10
　　NUTMON 211, 216
Mulching 105
Multidisciplinary system 169–170
Multipurpose crop 197
Mycorrhizas 5

Nitrogen (N) 20, 57, 59, 64, 81, 106
Nitrogen
　　balances 61, 109, 134, 201, 211, 218–226, 231, 236, 251–258, 268–270, 276–278, 282–285, 303

biological fixation 49, 60, 65, 78, 92, 144
cost 239–242
cycle 274
deficiencies 269
effect on ecosystem productivity 185
efficiency 63, 69, 163, 274, 277, 285
excretion 63, 64
fertilizer *see* Fertilizers, nitrogen
fish catch 229
fixation 6, 23, 104, 144, 195, 265, 269, 273, 275, 283, 286, 308, 310
  biological 216, 259, 277, 303
  (non-)symbiotic 196, 267, 278
fixing soil organisms 91, 184, 187
flows 232
harvest 283
immobilization 81
inorganic 229
internal flow 299
losses 30, 31, 63, 81, 89, 90, 92, 146, 164, 237–237
microbial 45
mineralized 45, 62
mining 211
nitrate 10
non-symbiotic fixation 106
not immediately available 33
plant-available 62
recovery 247, 259
recycling 277
soil organic 30, 33
storage in soil 89, 90
supply 59, 60, 299, 303
symbiotic fixation 31, 33, 106
see also Nutrients
Non-renewable input 287
Nutrients 295
  accumulation 297, 300–303
  available 27, 32–38
  accounting system MINAS 76, 91–94
  accumulation in plants 42, 86, 87
  acquisition 10
  additions 103
  availability 31–34, 306

balance/ imbalance 27, 31, 34–36, 43, 45, 50, 58–63, 69–71, 123, 125, 130–132, 147, 161–163, 183, 187, 201–208, 216, 227, 285, 286, 287, 296–298, 303, 308–310
  see also Nutrients, in- and outflow
budget 27–30, 47–50, 57–59, 75–79, 83, 84, 89, 90, 146, 266, 295–300, 304, 306, 307
  of plantation 178–183
consumption 160
content 253–254, 266
cycle 5, 57, 58, 256, 266, 275, 287, 295–296, 305, 307
deficiency 38, 296
depletion 31, 42, 45, 87, 109–111, 129, 130, 161–162, 165–167, 175, 202, 208–209, 211, 212, 265, 297–300, 303
emission 38, 39, 47, 101, 113
equilibrium 183, 286, 305
extraction *see* Production; Harvested product
flow 31–34, 47, 75, 76, 78, 79, 89, 103, 107, 108, 130, 146, 199, 211, 212, 215, 267, 295, 297, 303, 304
in- and outflow 183, 193, 195–198, 216–218 273, 306
  see also Nutrients, balance/inbalance
input 261
internal flow 216, 272–273
losses 6, 38, 49, 58, 70, 76, 85, 92, 93, 108, 109, 126, 129, 187, 300
management 101, 102, 110, 126, 191, 208, 298, 303, 308, 310
  improved 170, 308–310
  integrated (INM) 212, 296, 308–311
  sustainable 163–164
  technologies 308–310
mining 225, 296, 298
monitoring 218
not immediately available (NIA) 27, 32–34, 45

Nutrients *continued*
    pool 35, 37, 41, 43, 75, 78, 89, 178, 215
    recovered in animal product 230
    removal 178–180
    retentivity of ecosystem 185
    replacement 183
    residence time 38
    solution 52, 85, 86
    sources
        inorganic 158–163
        organic 158–163
    stock 34, 35, 39, 76, 105, 130, 193, 198–199, 202, 206, 215, 296, 304
        in cattle 94
    supply 178
    surplus 300–302
    transfer 193, 310
    transformation process 78
    use efficiency (NUE) 11, 12, 28, 49, 50, 69–71, 76, 89, 103, 107, 163, 241, 297, 303, 305, 308, 310
    wasting 53
    see also Calcium; Nitrogen; Phosphorus; Potassium; Secondary and micronutrient deficiencies; Sulphate

Oil palm 15
Optimum yield 234, 242–243
Optimum treatment (OPT) 166
Organic agriculture 303
    see also Farming systems
Organic matter 169
    recycling 298
    see also Soil, organic matter

Pasture 218
Pastoral system 265–287
Pesticides 123, 125, 129, 261
pH 43, 45
    of soil 144, 175, 185–186, 198
    of water 146
Phosphorus (P)
    accumulation 300
    balances 109, 209, 218–226, 251–258, 270, 271, 278–280, 284
    fluxes 261
    fractionation analyses 43, 44
    in- and outflow 175–178
    labile 43
    phosphate 10
    surpluses 302
    see also Nutrients
Photosynthesis 59
Pinus patula 179, 182, 186
Plantation 178–183, 186, 303
Policy 262, 300, 305, 308, 310
Pollution 76, 106, 164, 271, 272, 311
    control 187–188
    (ground)water 168, 171, 261
Pool 37, 38, 42, 45, 46
Potato 251–252, 260
Potassium (K) 45
    balances 218–226, 251–258, 271–272, 280
    depletion 157
    exchangeable 45
    recycling 272
    removal by animal products 272
    removed by harvest 175
    supply 251, 285, 299
    utilization 272, 303
    see also Nutrients
Power generation 185, 187
Precipitation 60, 86
Precision 77, 78
Prices of inputs/outputs 211
Production 7
    see also Harvested product
Productivity 187
Profit margin 187
Pyrodenitrification *see* Fire

Regeneration 191
Rehabilitation 188
Relative crop density 198, 205, 206
Relative dry matter yield 166
Relief 47, 50
Replenishment efficiency 12, 13
Resource use efficiency 134
Rhizosphere 14

Rice 2, 141–152, 159–160, 196,
    fields 144, 146
    rice–fish integration 146
Risk perception 308
Rooting depth 106
Root zone, buffering of nutrients 5
Rotation
    crop 128, 247, 250, 251, 258, 282, 285
    forest 178, 186, 187
Rothamsted Experimental Station 2
Rubber 15
Runoff 31, 47, 48, 83, 84, 122, 145, 148, 271, 278, 279, 281

Secondary and micronutrient deficiencies 171
Sedimentation 31, 33, 48, 195, 196, 201, 216, 218
Self-sufficiency 191
Sheep production system 265–266, 275–282, 304
Shifting cultivation 194, 205, 206, 208
Silvicultural practice 178
Slash and burn techniques 6
Slurry 84
Small-scale operation 172
Socially acceptable 287
Societal cost 261
Socio-economic 296
Soft scientific approach 296
Soil
    chemical characteristics 255–256
    content 106, 169
    depletion 180
    fertility 38, 39, 49, 109–113, 126, 144, 157, 187, 193–194, 207, 253–257, 262, 296, 297, 303, 310–311
        improvement 127, 128
        maintenance 101, 159
        management 161, 167–169, 204–206, 208, 298
        regeneration 204, 282, 299
        research 2, 103
    formation 5
    internal flows (SIF) 36
    microbial characteristics 256
    minerals 29
    mining 39
    moisture
        content 36
        regime 108
    organic matter 29, 60, 61, 126, 282, 284–285
    physical parameters 248, 255
    pool 183
    quality 128
    repair 39
    resource 168–169
    soil fertility level (SFL) 39, 45
    solution 29–35, 38, 41, 42, 46, 47
    stock 35
    texture 45, 142
    tillage 47
Sorghum 127
Sorption 34, 35
Spatial extrapolation 17
    scaling rules 18
    stratification 18
Spatial interactions 19
Spatial scales 296–297, 303, 304–306, 311
Spatial variability 16, 17
Species composition 185
Structural adjustment programme (SAP) 122
Subsidies on fertilizer 301, 309
Subsistence farming 303
Sulphate
    budget 280–282
    deficiency 165
    effect on ecosystem productivity 185
    see also Nutrients
Surface balance 78, 79
Sustainable land use 122
Sustainable production 168, 296
Sustainable soil fertility 31
Sustainable system 304
Sustainability 4, 28, 31, 41, 43, 50, 169, 209, 287
    forest plantation 187
Straw/ stubble
    removal 163
    return 169
Sugarbeet 102

Sugarcane 15, 188
Systematic approach 165, 305
System boundaries 58

Target inputs (TI) 27, 39
Target soil fertility (TSF) 27, 39, 45, 49, 53, 103, 296, 306
Target yield (TY) 27, 39, 41, 49
Temporal scales 296–297, 304–306, 311
Terracing 107, 124
Tillage 107, 108
Timber harvest 183
Topography 142
Trade liberalization 296
Traditional forestry 187
Traditional practices 212, 305

Unfertilized control *see* Farming systems
Uptake efficiency 11, 13, 103
Utilization efficiency *see* Nutrient, use efficiency

Vegetable 171, 197
Volatilization 30, 47, 50, 60, 63, 83, 84, 145, 164, 179, 197–198, 269, 272, 273, 274, 275, 281, 300, 307
Volcanism 304
Von Liebig theory 2

War 204
Waste *see* Garbage
Water availability 142
Water erosion 60
Water nitrate content 167–168
Water percolation 144
Water supply 59
Water turbidity 146
Weather conditions 32, 46
Weathering 34, 35, 180, 304
Weed growth 144
Wheat 165, 251–252, 260
Wind erosion 14, 60